Modern Data Engineering with Apache Spark

A Hands-On Guide for Building Mission-Critical Streaming Applications

Scott Haines

Apress®

Modern Data Engineering with Apache Spark: A Hands-On Guide for Building Mission-Critical Streaming Applications

Scott Haines
San Jose, CA, USA

ISBN-13 (pbk): 978-1-4842-7451-4   ISBN-13 (electronic): 978-1-4842-7452-1
https://doi.org/10.1007/978-1-4842-7452-1

Managing Director, Apress Media LLC: Welmoed Spahr
Acquisitions Editor: Jonathan Gennick
Development Editor: Laura Berendson
Coordinating Editor: Jill Balzano
Copyeditor: Kezia Endsley

Cover image designed by Freepik (www.freepik.com)

Distributed to the book trade worldwide by Springer Science+Business Media LLC, 1 New York Plaza, Suite 4600, New York, NY 10004. Phone 1-800-SPRINGER, fax (201) 348-4505, e-mail orders-ny@springer-sbm. com, or visit www.springeronline.com. Apress Media, LLC is a California LLC and the sole member (owner) is Springer Science + Business Media Finance Inc (SSBM Finance Inc). SSBM Finance Inc is a **Delaware** corporation.

For information on translations, please e-mail booktranslations@springernature.com; for reprint, paperback, or audio rights, please e-mail bookpermissions@springernature.com.

Apress titles may be purchased in bulk for academic, corporate, or promotional use. eBook versions and licenses are also available for most titles. For more information, reference our Print and eBook Bulk Sales web page at http://www.apress.com/bulk-sales.

Any source code or other supplementary material referenced by the author in this book is available to readers on GitHub.

Printed on acid-free paper

To my wife Lacey, thanks for putting up with me during the long nights and weekends, and for being my rock during the ups and downs of the writing process. Thanks for always supporting my wild ideas.

Table of Contents

About the Author

 **Scott Haines** is a seasoned software engineer with over 20 years of experience. He has worn many hats during his career, across the entire software stack, from front to back. He has worked for a wide variety of companies, from startups to global corporations, and across various industries, from video and telecommunications, to news, sports, and gaming, as well as data, insights, and analytics. He has held positions at notable companies, including Hitachi Data Systems, Convo Communications, Yahoo!, Twilio, and joined Nike in early 2022. Scott most recently a senior principal software engineer at Twilio, where he split his time between architecture and applied data systems and platform engineering. Scott has enjoyed working on distributed systems, real-time communications platforms, and enterprise-scale data platforms for over a decade and was foundational in helping to drive Apache Spark adoption for stream processing at Twilio. He is an active member of the Apache Spark community, a Databricks Beacon, and speaks regularly at conferences like the Data+AI Summit, Open Data Science Conference, and others.

In his free time, Scott enjoys reading, learning, writing, teaching, and mentoring. Follow him on Twitter (`@newfront`) to get tips and tricks for working with data systems at scale.

About the Technical Reviewer

Ed Elliott is a data engineer who has been working in IT for 20 years, the last 15 years focused on data. He uses Apache Spark at work and has been contributing to the Microsoft .NET for Apache Spark open source project since it was released in 2019. Ed has been blogging and writing since 2014 on his own blog as well as for SQL Server Central and Redgate. He has spoken at a number of events, such as SQLBits, SQL Saturday, and the GroupBy conference.

Acknowledgments

There are so many people who have helped me in my career and enabled me to learn and thrive. There are also many people who helped shape the contents of this book, implicitly and explicitly. First off, I want to thank the folks who helped build the Voice Insights product and the Insights Platform over the years at Twilio (Marek Radonsky, Rohit Gupta, Nivedita Muska, Harshit Rastogi, Amogh Rao, Bhargav Kumar, Maruthi Kuchi, Kenny Law, Monica Ravichandran, Pavan Kotikalapudi, Jason Dielman, and Vijay Jagannathan). Thanks for taking an early chance with Apache Spark, and for adopting Protobuf (even though it was counterculture at the time). Thanks to Oliver Pedersen and the Yahoo! Games team (Henrik, Arash, Alex, David, Dan, Sal, and Yuan) for introducing me to C# and Google Protocol Buffers, and for letting me choose my own adventure. Without your support, I wouldn't have learned to love streaming systems and wouldn't have ever talked with the Boson/Tachyon team (thanks Satish and Shushant for your patience), where I discovered Apache Storm and learned to love and trust Redis.

A special thanks to Nivedita and Angela (at Twilio) for helping to workshop the hands-on material in this book, and for making suggestions and telling me what was confusing. Thanks to the Apache Spark community, for continuing to make Spark such a joy to use and to Databricks for helping to keep the spirit and community alive. A special thanks to Holden Karau for all you've done to help make Spark approachable and testable, and to Denny Lee, Jules Damji, and Karen Bajza-Terlouw, for helping me feel like part of the Spark community and for the many conversations we've had over the years. I want to also give a special thank you to Ed Elliot for putting up with endless chapters and still providing a great technical review. Your comments and suggestions have helped bring things together. Thanks as well to the folks at Apress (Jill Balzano and Jonathan Gennick) for the advice and help while I was stuck overthinking things on my first complete book. Thanks for everything.

Introduction

I set out to write this book during the uncertainty of the COVID-19 pandemic. It was at a time where we, as a people, spent more and more time indoors (if we could) distracting ourselves from the outside world in the form of increased digital connectivity and entertainment. In the place of social interactions in real-life, we went online. It became routine to meet up with friends and family for calls over Zoom and we focused on the people (and things) that mattered most to us. We shared our lives and worlds the only way we could: digitally. Fast forward to today, more than a year later, and while the pandemic is still here, there is hope now. I am hopeful.

This is definitely not the way I imagined I would start my first book, with a retrospective, but as we look toward the future, it is hard not to see the lasting impact the last two years (2020-2022) has had on the very way humanity interacts with the world. As more experiences shift to virtual and more and more collaboration takes place online, we need to take a fundamentally different approach to how we architect our next-generation platforms, frameworks, and technologies. The magnitude shifts in system scale and processing power require evolutionary thinking, revolutionary data platforms, and data-centric ecosystems to emerge to power this digital future. Hyper connectivity and intelligent automation will power real-time, consistent data systems at an incredible scale and today is just the beginning.

This book is not about the future, but rather it teaches you how to build reliable data systems today, harnessing the incredible power of Apache Spark to orchestrate consistent scalable streaming applications built from the fundamental building blocks of the modern data ecosystem, by you, the modern data engineer. Using a hands-on approach, you learn to leverage many of the best practices and problem-solving skills necessary to make it in the exciting field of data engineering.

About the Book

This book is an introduction to building consistent, "mission-critical" streaming applications using Apache Spark. You will not immediately start writing streaming applications on page 1, but rather you will work hands-on, solving small problems using Spark and a wide array of tools to help you along the way. Each chapter introduces a critical foundation, a new tool in your data engineering toolbox, and as the book progresses, you will gain exposure to many of the common data systems and services that work well with Apache Spark. By the end of the book, you will have written and deployed a fully tested Spark Structured Streaming application on Kubernetes. You will have an entire containerized local data platform at your disposal, to take the ideas and implementations covered in this book with you to your next project. For some readers, what you learn throughout this book may be a refresher, and for others the lessons learned next may feel like a glimpse into the future of data. Either way, the journey of discovery will be worth it.

Who Should Read this Book?

You should read this book if you are a software engineer looking to transition your career toward data engineering, or if you are a data scientist or analyst, looking to get ramped up in using Apache Spark. This book teaches you the core APIs and common components of Spark, with a focus on using Spark SQL and writing reusable, compiled applications. This book also acts as a gentle introduction to writing Scala to build reliable Apache Spark applications. This book can be used as a resource by data platform engineers or software architects who are looking to understand how to build reliable streaming systems using open-source software, and as a primer for how to build composable distributed data systems. Finally, this book is a hands-on journey, and can be experienced by anyone with an interest in data systems, as long as you have a computer and are curious about learning new things.

Hands-On Material and Setup

This book contains hands-on exercises from Chapter 3 to Chapter 15. To follow along and get the most from the book, you need to download the book materials from GitHub. Once you have downloaded the materials, the chapters will be much more fun and feel more interactive.

GitHub Location for the Book

`https://github.com/newfront/spark-moderndataengineering`

Caveats

The contents of this book are written with macOS and Linux systems in mind. Although you can run a good amount of the book material using Docker on Windows, the examples and runtime scripts are written for Bash execution, and have not been fully tested on Windows.

PART I

The Fundamentals of Data Engineering with Spark

## CHAPTER 1

Introduction to Modern Data Engineering

Given you now have this book in hand, I can make two predictions about you. My first prediction is that you are currently working as a data engineer and want to understand how to hone your craft and take full advantage of the Apache Spark ecosystem to do so. However, if my first prediction fell flat then I believe you are interested in what it takes to become a data engineer and would like a friendly hands-on guide to get you there.

Regardless of the accuracy of my predictions, our journey begins in the same place—on this very page with the same goal in mind. The goal is that, by the end of this book, you'll have grown in new and hopefully unexpected ways. It is my honest hope that the additional knowledge, tricks, tips, and hands-on material will have a profound impact on your career as a data engineer.

This first chapter is simply a history and an introduction to data engineering. You'll discover the origins of the specialization, the foundational technologies, and evolutions that helped shape and steer the modern data revolution. Together, we'll identify the common tools and emerging technologies that will be part of your foundational data engineering toolbox. There is of course a special emphasis on technologies that work well inside the Apache Spark ecosystem. However, along the way, you'll discover the common roles, responsibilities, and challenges required of the modern data engineer and you'll see how these essential skills knit together to paint a picture of the modern data engineering ecosystem.

Note Over time I've learned that the best way to remember anything is through stories. Let's start from the beginning and rewind (remember VCRs!) to where data engineering began. This will establish a foundation from which we can fast-forward to cover the current (and ever-evolving) umbrella we call *data engineering*.

© Scott Haines 2022
S. Haines, *Modern Data Engineering with Apache Spark*, https://doi.org/10.1007/978-1-4842-7452-1_1

The Emergence of Data Engineering

I started my career in software almost 20 years ago. I remember reading books and articles written by engineers reminiscing about when they started their careers and how the very landscape of software had transformed and evolved during their careers for the better. This was often due to evolving best practices and design patterns, as well as emerging technologies and programming languages that simplified common problems across the board.

My history lesson begins with how things were and then moves to catalog how things have changed over time. It reveals the technological waypoints, novel innovations, and seemingly simple improvements that make our jobs today more enjoyable, while also providing new challenges and thrills that we all tend to seek in our careers. So, without further ado, let's rewind our clocks and go back to a critical point in time—the dawn of the data engineer.

Before the Cloud

I was both fortunate and unfortunate to start my career in what I like to call the awkward *pre-cloud* phase. I feel fortunate because I can appreciate the modern conveniences introduced with good version control, simplified and reliable cloud orchestration, remote debuggers (debuggers in general), log aggregators, rich unit and integration test frameworks, the general availability of the cloud, and so much more. With that said, let's look at where things were pre-cloud.

Before the general availability of *cloud computing*, the world looked a bit different, and the pace of things was a bit slower. In fact, that would be an understatement—things were many orders of magnitude slower. Rewinding back to early 2000s, we had servers and spinning disk hard drives (HDDs) and decent Internet connectivity between data centers (or at least within racks at the data center). However, the pace of most projects felt glacial from the initial idea and proof-of-concept, through development, testing, manual quality assurance (QA) sign-offs, until that special point in the future when you finally got your service fully deployed. Today, what used to take multiple months, if not years, can be done in a matter of weeks and sometimes days.

Back in the early 2000s, the server-side stack looked similar running the Linux or Windows operating systems, but the biggest difference came with disk space, compute (CPU), and memory (RAM); these were contented resources. I remember waiting for six months to complete the process of purchasing new hardware. These things always started with proposals, specifications, purchase orders, shipping, waiting, getting the servers installed (in the data/colocation center), IP address allocations, and finally the result was network addressable computers. However, people made the most of the simple tools in their toolboxes, piecing things together by using the general availability of websites, text-based messaging, text-based email, and fancy early HTML-based email. They also used other forms of connected communications, from text to audio, and even low-resolution video.

It was also not uncommon at the time for companies to host their services in a single location, on servers they owned and maintained, using network hardware they also owned, while paying mainly for reliable power, secured connections to the Internet, and a safe facility that had staff around to restart servers, or do minor diagnostics and triage.

This may feel like a far cry from what we're used to today, but it was not uncommon back then for large teams to work on large yearly software releases, instead of the sometimes-daily continuous releases we see today. This time was a mere steppingstone, or waypoint, on the journey toward the platforms we run today.

Automation as a Catalyst

Around this time (late 90s-early 2000s), there was a second option for hosting websites and running databases (rather than doing it yourself), and that was to use hosting providers. Companies like *Rackspace*, *GoDaddy*, and *Media Temple* reigned supreme and helped paved the way for the foundational cloud computing powerhouses of today. They essentially did this by simplifying the steps necessary to take a service online, which reduced the complexity of running systems and services for their customers.

These providers offered a slew of packaged solutions, making it simple to allocate server space on dedicated virtual machines, assign static IP addresses, register for domain names, and easily apply and modify the DNS settings to point a domain name to the services running in these hosted environments.

These are all things that were much harder to do when running your own hardware, and these providers enabled smaller companies without a dedicated IT staff to get online and have a web presence. These companies provided essential services, which sped up the time it took from ideation to having a web experience running online. This would also become the blueprint for cloud-hosting providers as companies looked to reduce the complexity of running their systems further, given most still had to operate and maintain full developer operations teams (DevOps).

As more and more companies were taking their businesses online, there was a rapid rise in the quality and demand for better design and overall aesthetic, and a push toward more feature-rich web experiences. These richer experiences helped to mold more cutting-edge offerings, as we will see in the next section.

The Cloud Age

With the rise in better online experiences, there was also a demand for more personalization and customization of content tailored for the users of a particular website or *web application*. The demand for better, smarter experiences also meant a need for more compute power, more uptime and better reliability, and what would also end up being massive amounts of data to track the key metrics of these experiences.

Note   For those of you wondering, web applications were all the rage in the mid-2000s. For simplicity, they can be compared to mobile applications or standard desktop software. There was a time when Adobe Flash and HTML5 existed in harmony and enabled unified experiences to be experienced on *most web browsers*.

The Public Cloud

The company that changed the game and helped start the cloud era was Amazon. They envisioned and built one of the most widely known and successful cloud infrastructure-as-a-service offerings, known today as Amazon Web Services (AWS). With the public debut of AWS in 2006, the modern cloud was born. This was the propellant that fueled the rapid rise of data-first and data-centric ways of thinking. Now anyone and any company could build on top of Amazon's foundation.

Netflix is probably the best known for taking a risk and going all in on their AWS bet. As you probably know, their bet paid off, but at the time no one could possibly have fathomed what Netflix would be able to achieve today. By shifting focus away from managing infrastructure, they could focus instead on building a compelling product that today has achieved fully AI driven personalization, a unique viewing experience available from almost any device, all without the viewers having to do more than just watch content. What they built today would have felt like it was science fiction not so long ago.

At the time, Netflix was in its infancy. It was shipping DVDs to their customers in the mail and had just launched an initial streaming service using Adobe Flash and Microsoft's Silverlight (relics of the web media players of yesteryear) for a small selection of their films. Netflix however was working on rich analytics systems capable of collecting, analyzing, and harnessing the power of their customer's usage data. In order to analyze and understand the behavior of their customers and generate personalized recommendations, Netflix leaned on some cutting-edge data stacks and also played an integral role in the success of Cassandra, which was one of the main players in the NoSQL revolution.

Note   We'll review NoSQL databases later in this chapter, but it is important to state now that data comes in all shapes and sizes. It is very important for data engineers to be knowledgeable about many alternative database technologies, and to understand the pros, cons, and risks associated with the different offers.

The thing that Amazon, Netflix, and other successful companies have in common is that none of their business ventures would have been successful without teams of dedicated data engineers, data analysts, and data scientists working together to solve complex problems at massive scale.

In the next section, we look at how data engineering played a critical role, if sometimes behind the scenes, in the rise of these massively successful companies. You'll learn how the roles, responsibilities, and even titles changed over time.

The Origins of the Data Engineer

Strictly speaking, data engineers have been around for some time. They just went by different names and were associated with different supporting pillars of the data umbrella. As a specialization, data engineering emerged from the roots of traditional *software engineering*, *data analytics*, and *machine learning* due to an ever-increasing demand for data and the evolving data needs by companies and organizations over time. The need to bridge these disciplines led to a new engineering specialization focusing on data quality, availability, accessibility, governance, maintainability, and observability at scale.

Back in the day, access to data was a more complex process, and one that required very specialized skills, especially since databases, at the time, were typically single points of failure. Given the dependence and importance of data to most organizations, data access hurdles were lined up, leading to the data finish line. Anyone requesting data, say for a quarterly report, had to go through a good deal of effort to get some answers to their data questions (queries). This was done to defend the scarce and fragile data resources and to protect these data investments, as compute and storage space were much more expensive. I'm sure you've heard horror stories or maybe you've even been the one who took out a database with a single bad query. While this is harder to do today, it wasn't so hard just a decade ago.

Data engineering today is essentially all about controlling the flow of data through services and between different databases, data streams, data lakes, and warehouses. This flow of data through these data networks supports different access and processing styles that can be applied to the captured data, ranging from historic data analysis, insight generation, model training for machine learning and intelligence, as well as fast access to data for metric serving, operational insights and monitoring, and for real-time model serving and intelligence.

To be successful and meet the current data demands, data engineers must understand how to reliably capture, process, and scale the torrents of data that are generated across the entire data platform, and within the full data ecosystem at large. This requires a solid understanding of software engineering best practices, distributed systems, a concrete understanding of SQL (including data modeling, querying, and query optimization), as well as knowledge pertaining to analytical processing and querying, and even the basics of statistics and machine learning.

While data engineering overlaps many pillars of the data umbrella, the need to hire specialists across the board remains. The key difference I've seen over time is that the specialists work to automate processes and remove redundancies. They're tuned into and understand how to solve complex problems generically within their domain of expertise, and as a result these optimizations reduce the need for as many specialists. Seemingly, the one thing that has remained constant over the years has been these cycles of evolution and innovation that continuously improved on the prior technology, software, best practices, and solutions. With respect to data, this means that every year it is easier to store, access, process, and learn from our data in a safe and secure way.

Do you remember the last time you talked to a database administrator (DBA)? Or interacted directly with a data analyst? How about the last time you needed to pull or schedule a report? Did you use tools that automated some of these processes?

Having a firm grasp of the myriad technologies, operating systems, cloud vendors, services, frameworks, CPUs, GPUs, or insert the container technology of the week, can be a Herculean task. All the while, you're trying to work, study, and learn everything, or even simply trying to keep up with regular, on-going changes. So, for simplicity, I am going to let you cheat, and we'll go through a curated set of core technologies and concepts that are integral to the data ecosystem at large. They will, of course, be viewed through the lens of what is important to know for use later in the book, as well as what works especially well with Apache Spark.

The Many Flavors of Databases

The database has been around for almost as long as computers have had storage devices. We can thank Edgar F. Codd for pioneering the relational database (while at IBM), and for providing us with a means to describe our data world through connections and their relationships. While there are many types of databases out in the wild these days, I believe there are essentially four flavors of database you will commonly encounter.

OLTP and the OLAP Database

I am willing to bet that the *online transactional processor* (OLTP) style database is what most people first associate with the notion of a database. This style of database is used for bank account transactions, for booking a seat on an airline, and so on. It is especially well suited to deal with small, isolated (atomic) transactions within the system. These transactions are inserts, updates, and deletions, to data. Atomic transactions aim to achieve the smallest isolated change possible, such as decrementing or incrementing a number, such as the number of seats left on a flight, or the account balance in someone's account.

OLTP databases are traditionally architected and optimized for handling transactions. This means there is a single centralized process running, which achieves atomicity, consistency, and durability, using *write-ahead logging (WAL)*. The WAL enables the database to accurately track changes to each tabular row of data so that a consistent state can be guaranteed. To achieve fast, durable transactions, this style of database is non-distributed and runs in a single location. By removing any network IO (distributed communication/transfer), each transaction can run incredibly fast.

The only problem with running as a non-distributed database is that if you want to increase the size of the database, with respect to hard drive space or working memory (RAM), then the only option is to go find a bigger box and migrate to the new server. Given the OLTP database is a vertically scaling database, it can get expensive to scale to meet increasing demands.

The Trouble with Transactions

The *online analytical processor* (OLAP) style database is built from the same core technology as the *OLTP database,* the key difference being the way in which the database is optimized. There is a *cost to transactions,* which means that the database needs to lock table rows that are being written to during a transaction. This process is good for consistency, as it ensures that concurrent modifications, or possible out-of-order writes, don't screw up the integrity of the data.

However, analytical queries are traditionally done against an append-only style database, which means that *data can never be mutated,* so there is no need for transactions. When you remove transactions from the equation, you also remove the need for row-level locking, since each row of data in the database will always be in its immutable, or final, form. This is especially important when dealing with analytical queries, which we look at next.

Analytical Queries

Analytical queries got their start under the umbrella of business intelligence (BI). These queries typically involve multiple phases of processing to provide data in the form that can be used to answer analytical queries. We'll explore *roll ups, pivots,* and *windowing* in Chapter 12, which are common analytical and aggregations computations necessary for analytical processing.

For example, say you were asked to write a query to determine the top sales numbers per day of the week over the past year, broken down across key geographic regions, and you also needed to provide the top five categories of goods sold for each day. This query may require the database to process all rows (data) across all target tables to produce multiple datasets, defined as views or temporary tables, that would need to be stored as intermediary results, from each logical phase of the query execution, in memory for fast results.

A large, analytical query can be an operational burden that consumes a lot of the database resources (CPU/RAM). These queries can also be hard to maintain (given their complexity), and difficult to scale up due to the law of large numbers. The law of large numbers in a nutshell states that as the number of rows and scale of the data (number of columns) being queried increases (which is normal in the world of Big Data), the response time of the query will increase proportionally to the number of rows queried (linearly) in the best case and (exponentially) in worst case. While being slow, these queries would have a chance to eventually complete; however with the overhead of row-level locking, due to transactions, a large analytical query may not even be possible.

While relational databases capable of running these *OLTP/OLAP* queries like *MySQL*, *PostgreSQL*, *DB2*, *SQL Server*, and *Oracle* are popular for asking questions across large, related/connected data, there is another style of database that handles independent/denormalized data that's accessed through keys rather than queries across indexed tables. This next database flavor that's important to know about is called the *NoSQL* database.

No Schema. No Problem. The NoSQL Database

I remember going to a meetup at Carnegie Mellon in California (with the Silicon Valley Cloud Computing Meetup Group) and the folks presenting were all excitedly geeking out on the pros/cons of this new style of database. It was optimized for incredible speed (fast random access) and horizontal scalability (enabling the database to be distributed over many machines storing previously unseen amounts of data with a hive of machines). The last bit of icing on the cake was that these databases removed the need for explicit schemas, or table definitions, prior to writing your first record.

For me, this was day one of the NoSQL revolution. This notion that data didn't only have to be relational, or even adhere to a concrete schema, spoke to me. It still speaks to me but not in the same way as it did at first. No schema also means that the guard rails are removed, enabling data to be written in non-consistent ways. Without a semi-defensive system in place to validate and maintain some kind of data consistency, it can be difficult to build the foundation required for running reliable data systems at scale. This means you can easily have large sets of corrupt data sitting in this hyper-fast, scalable, NoSQL database. However, the NoSQL databases also enabled new access patterns, including the ability for data readers to essentially discover the structure of the data on-the-fly using schema inference (which we'll see in Chapter 3) and other schemas on read techniques (which Apache Spark supports as well).

The NoSQL database is a distributed database. This means that a cluster of one or more machines encapsulates the database. While this enables the database to scale horizontally, this also means that some tradeoffs need to be made regarding how to handle common problems that arise due to the distributed network. These tradeoffs are encapsulated by *CAP Theorem*, which declares that any distributed system can deliver on only two of the following three traits at any given point in time: **c**onsistency, **a**vailability, and **p**artition tolerance.

This means that the database needs to sacrifice something since we can't have our cake and eat it too. It is common to see a database described as being Cp (consistent and partition tolerant), or Ap (available and partition tolerant). You won't see a system described as being Ca (consistent and available) since distributed systems are subject to network partitioning. Network partitioning is like a road closure with no alternative routes and is also referred to as split brain since a subset of the distributed system is cut off and can't communicate with the rest of the system. In order for systems to be described as Cp, it means that sometimes the data in the database isn't available, and the alternative is for systems to be Ap, meaning they may not be completely up to date (consistent), but they are still available and can serve data (which may be stale).

The most well-known databases from that meetup still in use today are *MongoDB, Redis,* and *Cassandra.* You'll be using Redis with Spark later in the book (Chapter 10). The exercise there uses the best of NoSQL combined with the best of the ridged structures that make for reliable systems at scale (aka schemas).

Along the line of combining the best of two worlds, we come to the NewSQL database. This is a hybrid of the traditional relational database (OLTP/OLAP) with the benefits of the fault-tolerance and elasticity of the NoSQL database.

The NewSQL Database

The *NewSQL* database is a hybrid of the best of both SQL worlds, marrying the consistency, reliability, and durability of OLTP transactions, with the fault-tolerance and horizontal scalability of NoSQL, with a minimal (and usually configurable) tradeoff between availability and consistency (Cp|Ap).

These databases were rearchitected from the ground up to run well inside the cloud, where network in-availability can take nodes offline, out of view via network partitioning, and where servers can go offline with little notice, making replication of data across nodes a first-class necessity of these systems at scale. These databases replicate data across nodes in order to ensure consistency and availability in the case of node failures or network partitioning.

CockroachDB and Google Spanner are two of the heavy hitters in the NewSQL market, and while we won't be working directly with any NewSQL services in this book, it was worth mentioning these technologies. You can connect and interoperate with these services using Apache Spark by adding a library dependency.

Thinking about Tradeoffs

Working with data almost universally means you will be working with databases at one point or another, and knowing what your options are, and why to choose one kind of technology over another, is always a big win in terms of a successful project. I tend to go with the golden oldies and select a technology that is firmly established in the market. Also, since it tends to work out in your best interest, it is wise to use a technology that your company supports. That way, there is a community of people to engage regarding any growing pains or for support along the way.

Cloud Storage

Amazon's Simple Storage Service (S3) is probably the most widely known cloud storage solution in use today. Microsoft's Azure Blob Storage and Google GCP's Cloud Storage are similar solutions available from these other cloud providers. Regardless of the cloud provider you use, the technology aims to achieve the same results.

At the heart of the general cloud storage paradigm is the promise of a simple to use, inexpensive data storage solution that provides high availability and reliability, alongside almost unbounded elastic scalability, meeting the ever-increasing data storage needs of any sized company.

Prior to scalable cloud storage offerings, companies were forced to run, maintain, scale, and operate large *Hadoop* clusters to achieve the same sort of elastic scalability using the *Hadoop Distributed File System* (*HDFS*). Today, armed with almost unlimited elastic data storage, companies have found more and more reasons to store more data than ever before. There are pros and cons to just storing everything, and there are even more challenges related to working with blindly stored data, especially since data formats evolve regularly over time.

Data Warehouses and the Data Lake

You likely will have already come across or at least heard of the concepts of the data warehouse and the data lake. While similar in name, the conventions and utilities provided by these two data storages and serving technologies are on two separate sides of the data platform. On one side, you have the data lake, which is a distributed elastic file system, and on the other side, you have an analytics database that solves the issues of siloed data. However, they can still play well off each other in practice.

The Data Warehouse

The data warehouse was born from the common problem of siloed data. Data silos literally just mean that the data is distributed across multiple, logical SQL tables that reside in different databases or data stores, making it difficult to access the data necessary to answer queries from the context of a single database system.

The ETL Job

Given that data was siloed across many individual databases, a common technique emerged alongside the data warehouse, and that was the *extract-transform-load* (ETL) job. Figure 1-1 shows at a high-level the sequence of steps required to move data out of a siloed database and into the common data warehouse.

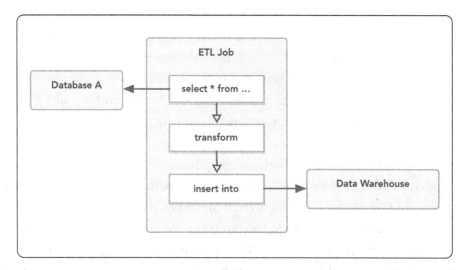

Figure 1-1. *The extract-transform-load (ETL) Job*

ETL jobs are usually configured to run as batch jobs. These batch jobs can be scheduled to run once, which is common in the case of doing a one-time data import, or they can also be scheduled to run on a recurring, time-based schedule. This means that each job (the work of moving data from one database and loading it into the data warehouse) is isolated, and for a job to be successful. it is expected to run to completion.

The ETL job shown in Figure 1-1 starts by connecting to a remote database and running a query that *extracts*, or reads, rows of data as specified by the given query. It is worth mentioning that these queries are not bound to the simple task of moving data from one place to another; this step can also include more complex queries that join data across many tables to produce wide tables that can be used in the data warehouse for analytical querying.

The next step, which is the *transform* step, prepares the data for insertion into the data warehouse. It is common at this step to apply a specific schema, or format, to the data to match common naming conventions (taxonomy), and to improve joinability across different tables in the data warehouse. This step is optional, since data doesn't always need to be transformed when being replicated into the data warehouse.

The final step in the job is the *load*, or write, process. It concludes the process of moving data from a siloed source into a table in the main data warehouse.

The data warehouse acts as the much-needed central database (typically a relational, non-transactional database), where *data analysts* or different personas would go to run their business intelligence queries and generate insights or explore the data looking for interesting trends and patterns.

While the data warehouse was this centralized, relational database, the data lake emerged to become the centralized staging ground for data. Companies could offload the scalability and operations work necessary to manage these large, distributed file systems to the cloud vendors, at a cost much lower than owning and operating the servers, disks, and staff necessary to maintain these data infrastructures. The data lake rose as a unified location for storing literally everything. Let's look at that now.

The Data Lake

The *data lake* emerged to solve a common problem associated with the siloed storage of raw, source-of-truth, data. Source-of-truth data just means data that has not been cleaned up, or joined with any other data. It exists exactly how it was initially emitted, or provided by, a data source, which we'll cover more in Figure 1-2. Similar in spirit to the data warehouse, which emerged to alleviate the issue of data silos and fragmentation within an organization, the data lake stepped in to provide a solution to the problem of fragmented and siloed "raw" data. It is a centralized staging location for *any kind of data*, from raw binary data (images, audio, video files) called *unstructured data*, to *semi-structured data* (CSV, JSON data), and even *fully structured data* (data that has explicit types and guarantees).

At its core, the data lake is simply an *elastic, distributed file system*. This is important because elasticity enables the data lake to scale (stretch) horizontally to meet the size and storage requirements of your current data demands today, while simultaneously being able to flex and increase the underlying total storage allocation, in order to handle the storage demands of the future *with zero downtime*. Using commodity cloud storage (Amazon S3, etc.), or by running Apache Hadoop's HDFS, you can provide the foundation for running your own data lake. Ultimately, files (also referred to as objects within the cloud stores) are stored across multiple servers for reliability (fault-tolerance), and to reduce contention across many concurrent readers and writers. You'll read more about the uses of the data lake when you read about MapReduce and get into using Apache Spark in Chapter 2.

Data Formats   As you know, data comes in all shapes and sizes, and "raw" data spans many styles and structures, from semi-structured text-based data formats like log data, CSV files, and newline separated JSON, to fully unstructured binary data, including image and video data, sensor data emitted from the Internet of things (IOT), and even fully structured data protocols like Apache Avro and Google's Protocol Buffers, which you learn to use in Chapter 9.

Figure 1-2 shows a high-level overview of the data lake. Starting on the left side, you'll see many external data sources, which are the initial upstream data sources. They represent the different providers of the source-of-truth data. While each upstream data source is responsible for providing its own datasets and data formats, all the disparate data sources can take advantage of the same underlying "data lake" to provide a unified, or central, location for downstream systems to get their data. Downstream systems can be thought of for now as any number of systems or services that ultimately rely on the data generated by one or more (if joined) of the upstream (source-of-truth) data sources.

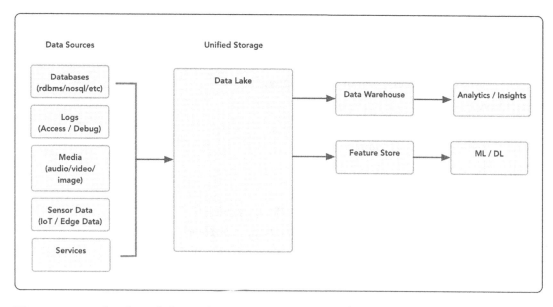

Figure 1-2. *The data lake architecture. Data flows (in from the left) into the data lake, through any number of automatic or manual processes and can be reliably used to generate new sources of data to meet the demands of the company, product, or feature. Once data lands in the data lake, it can hang out for as long as it needs to before being used, or eventually deleted*

The data lake itself is nothing more than a reliable elastic file system capable of scaling horizontally to meet the storage needs of many datasets. It is common to organize the data stored in the Data Lake using directories. Just like with a file system, or a package in Java, these directories exist for contextual purposes and to create namespaces so that tables don't collide or become corrupted accidently. In Listing 1-1, you can see how distributed SQL tables are defined using hierarchical trees. These tables are examples of some of the tables you'll create in the chapters to come.

Listing 1-1. File System Layout of Three Distributed SQL Tables (customerratings, customers, and coffee_orders) Inside a Data Lake

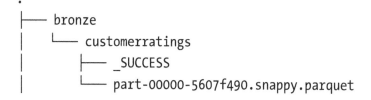

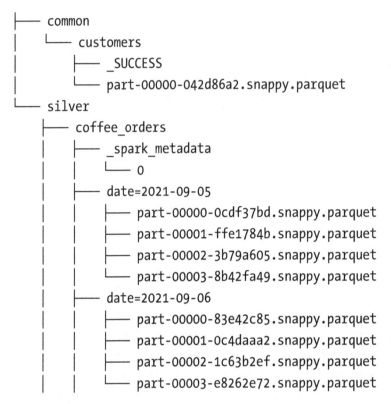

```
├── common
│   └── customers
│       ├── _SUCCESS
│       └── part-00000-042d86a2.snappy.parquet
└── silver
    ├── coffee_orders
    │   ├── _spark_metadata
    │   │   └── 0
    │   ├── date=2021-09-05
    │   │   ├── part-00000-0cdf37bd.snappy.parquet
    │   │   ├── part-00001-ffe1784b.snappy.parquet
    │   │   ├── part-00002-3b79a605.snappy.parquet
    │   │   └── part-00003-8b42fa49.snappy.parquet
    │   ├── date=2021-09-06
    │   │   ├── part-00000-83e42c85.snappy.parquet
    │   │   ├── part-00001-0c4daaa2.snappy.parquet
    │   │   ├── part-00002-1c63b2ef.snappy.parquet
    │   │   └── part-00003-e8262e72.snappy.parquet
```

As shown in Listing 1-1, the data lake on its own is simply a way of organizing source-of-truth data. To process (or query) the data stored in the data lake, you must first read the data (represented by one or more files in the distributed table layout) into a system capable of querying the data directly (such as Apache Spark SQL, Amazon Athena, and others), with the caveat being that the underlying storage format must be directly queryable (such as Apache Avro or Apache Parquet). In Listing 1-1, the distributed tables shown are stored in the Apache Parquet columnar format using Snappy compression. You'll learn more about Apache Parquet as the chapters unfold.

If the data is unfortunately in a format that can't be queried easily, there is still hope. These are many technologies supported by myriad programming languages that carry out the process of initially reading (opening), parsing (translating), and transforming the raw source-of-truth data into a format capable of being loaded into a physical database, or more commonly, into a managed cloud-based data warehouse.

Note   You will use Apache Spark to process simple workloads in Chapter 3, involving reading, transforming, and querying data stored on the file system. By Chapter 6, you use managed tables, which allow Apache Spark SQL to handle the operations required to manage the distributed table's metadata. This, for example, contains the distributed database locations, table names, ownership and permissions, and the metadata annotating the tables and databases to assist in the process of data discovery.

Aside from simply loading the data from the data lake into external data systems, the process of reading and transforming the raw source-of-truth datasets can also be used to produce more specialized datasets, which can be saved back into the data lake as children of the upstream datasets, or dependent datasets of the parent(s). These specialized data subsets (views) are like child nodes, with common upstream parent(s) acting like the relationship between a universal dataset (raw), and the many forking child subset(s) and their decedents.

A dependent dataset can be as simple as a subset of one parent dataset, or can be more complicated, as in the case where data from many datasets are joined to create more complex children. Adding to the complexity is also the fact that each child dataset also becomes a source-of-truth data for any new downstream forking from it, and each child dataset can also be the output of joining one or more additional datasets (nodes) from alternative "data family trees." This is where the concept of data lineage comes from, and for the mental model it is easier to think of a tree structure rather than what ends up often becoming a larger graph-like structured of inter-dependent datasets. Just think of this all as a family tree, but with a larger number of potential parents, and of course, data rather than people!

Considering that all possible datasets can be derived (or rederived) from the myriad source-of-truth datasets, things can quickly go from simple to complex. It is for this reason that distributed SQL tables became a popular way of working the large amounts of data. Large data quickly became "Big Data" which was eclipsed again with the notion of "Unbounded Data," but the metadata and storage mechanics have remained mostly static, using cloud data lakes for unbounded storage, reinforced with metadata stored in the popular Apache Hive Metastore.

As you will see later in the book, Apache Spark SQL makes it simple to scale out these common SQL workloads, connecting to data stored in one, or many, locations, in many different processing and access modes, from near real-time continuous streams to realer-time structured micro-batches, to simple based batch processing, on-demand, and even ad-hoc. This process often begins with the first step of moving raw data out of the data lake and into a more usable data store using traditional ETL (extract-transform-load) jobs, as well as ELT (extract-load-transform), which you learn about later. You'll see how simple it can be to run your workloads directly against this source-of-truth data with *Apache SparkSQL*.

Tip   Having source-of-truth data stored in a central place can be one of your more critical data assets. Not only can you use the data to solve many problems and answer many questions, but you can also use this data if you need to rebuild tables or recover from problematic issues (such as a bad deployment or other downstream corruption) within your data infrastructure. You learn more about how to set yourself up for data success as you venture into the second half of the book.

Data lakes and data warehouses have their roles to play in the fabric of the modern data ecosystem. They are like the data yin and yang. Using these different data capabilities together encapsulates a holistic data continuum and caters to the many needs of the data platform.

The Data Pipeline Architecture

One of the key responsibilities of data engineers is very simple in theory—they are responsible for the infrastructure and management of *data pipelines*. The data pipeline is just a fancy name encapsulating the reliable movement and transformation of data, at scale, between sources and sinks. This idea of *sources* and *sinks* is a mental model for thinking about systems that move and transform data, and you can also think about this in terms of *upstreams* and *downstreams*.

If you zoom out and look at things at a high enough level, data is essentially passing through a directed lineage of jobs and through function transformations in a journey across a large network graph. The *source* is a reference to an upstream source of data, for example, a database. The *sink* is a reference that can be mapped physically to a real-life sink. This is an outlet, or egress point, where a flow of data is collected.

Tip   Data flows are commonly represented by water, since this is something people can easily understand. Consider how water functions in nature. Drops of water collect and coalesce to create puddles and pools of water. Depending on the rate of collection and the container in which the water is being collected, small pools of water can easily become streams. These streams flow and join to create rivers, spread out across deltas, into lakes, and even oceans. At a micro level, each molecule of water is comprised of a structure, and can be represented by a data point, and from here the analogy starts to become overkill. However, as a mental model, it just works.

I like to think about how each job in the lineage of a data pipeline works in a similar fashion as one would describe an API. An API has an agreed upon contract that specifies what data must be included in an API request, and the API owner makes strong guarantees about how to handle and respond to good and bad requests, as well as how it handles errors and how these errors can propagate to back to the callee (client) when something goes wrong. An API that doesn't act in the way it is documented will make internal or external customers very angry, and in the same way, data that was "supposed" to look a certain way or have specific fields, but doesn't, will anger anyone working with the data downstream to a pipeline job owned by you or your team.

The Data Pipeline

A data pipeline can be defined in a few separate ways. In the simplest form, a pipeline can be defined as a job or function that reads from a single upstream data source and moves (writes) a dataset in its original, or transformed form, to another location (sink). On the extreme side, a data pipeline can also be used to describe a complex network of distributed data transformations over time, which we will look at when we cover workflow orchestration in the next section.

If you consider the case of the ETL jobs from earlier in the chapter, they were introduced as playing an essential role in the reliable movement of data between a source database table and a destination database sink. A modified example is presented in Figure 1-3. It showcases how the ETL job plays a foundational role in defining the data pipeline.

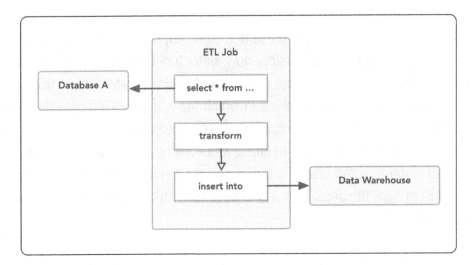

Figure 1-3. *A simple ETL pipeline job*

Things start to get more exciting as these simple jobs are joined across a series of pipeline steps and multiple phases of execution. These multistep jobs have become so common that workflow orchestrators like Airflow have become commonplace with the data platform.

Workflow Orchestration

Workflow orchestration is an essential way of controlling the reliable movement of data across a large data network, as defined by a series of inter-dependent jobs and data pipelines. Figure 1-4 shows a multi-step distributed network of directed data transformation jobs. Jobs can be triggered within the workflow manager at a specific time, such as in the case of scheduled cron jobs, and they can also be triggered by the completion of an upstream job. Using upstream job completion as a trigger to kick off the next job (or jobs) in a series of data transformation stages is typically defined as a directed acyclic graph (DAG), or network of upstream and downstream jobs. Chapter 8 is dedicated to workflow and covers setting up and running Apache Airflow.

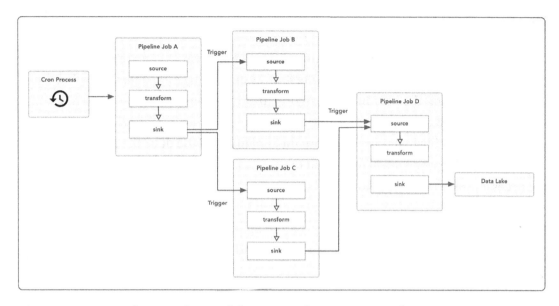

Figure 1-4. *A multi-step directed data transformation pipeline*

Figure 1-4 is a high-level view of a directed acyclic graph (DAG) of pipeline jobs with a common upstream initial trigger. If you only have a few jobs to run and most of those jobs can run independently (without the knowledge of the successful completion of any upstream job), then workflow orchestration may be overkill and can be delegated to simpler cron mechanisms.

The benefit of operating data pipelines, either as simple transformations or as a series of inter-dependent pipeline jobs defined through DAGs, comes in the form of a single source-of-truth for the location (source), format or schema, and expected of behavior of the data being emitted at each stage (sink) of execution within the data pipeline. Having sources-of-truth data is not just beneficial, it is critical for operating large networks of data dependencies. Just as workflow management and orchestration services steps in to help reliably run distributed data transformation jobs and pipelines, data catalogs step in to provide a mechanism for discovering sources or data, the upstream and downstream dependencies of those data sources, as well as the schemas, versions, owners, and even the data lineage and access policies of the data.

The Data Catalog

Data catalogs provide essential views into the data available within an organization and can be extended to encapsulate ownership, as defined by metadata describing the producer of the data (the person, team, or organization). These are sometimes referred to as the *data stewards*. The data catalog can also act to define a public *data contract* that can be used to established rules to govern the timeliness of the data, the availability of each field (such as when it will be default or null), as well as rules around the governance and access to each field, and more.

We explore *data catalogs* in Chapter 6 with an introduction to the Spark SQL Catalog. You learn to use the Apache Hive Metastore and Spark SQL in Chapter 5 to create and manage SQL tables that can be easily discovered and utilized directly in Apache Spark.

Data Lineage

This high-level view of how the data flows through the myriad data pipelines within an organization, as defined by the logical data sources and sinks, creates the means to publish the data lineage represented as a series of data transformations across the data pipeline.

Knowing and understanding how data flows through these data pipelines is one of the harder things to manage holistically and taking these steps early on can help with the common problems that arise when building a reliable data platform. Without the right common tools in place, managing data can become a nightmare as things become more and more complex.

We look at data catalogs and data lineage and introduce the notion of data governance in more detail later in this book. Next, we look at stream processing, which enables the pipeline architecture to move in a more fluid fashion that acts as a push-based system based on data availability rather than the pull-based systems common to batch-based systems.

Stream Processing

Stream processing evolved from the simple mechanism of messages passing between operating system (OS) processes across buses. They were more popularly known simply as *pub/sub systems*, as system processes would publish or subscribe to messages in a form of cross-application communication within a single machine.

Interprocess Communication

This simple, albeit useful, construct allowed for magnitude gains in parallel processing by allowing processes (applications and kernel tasks) within a single computer OS to take part in a conversation. The beauty of this was that processes could participate in conversations in either synchronous or asynchronous fashion and distribute work among many applications without the necessary overhead of locking and synchronization.

Network Queues

This approach to parallel processing via pub/sub message buses on a single system allowed engineers to expand and evolve these systems in a distributed-first approach. This led to the advent of the network *message queue.*

The basic message queue allowed for one or many channels, or named queues, to be created in a distributed *first-in, first-out (FIFO)* style queue that ran as a service on top of a network addressable location (e.g., `ip-address:port`). This enabled network-based applications and services to communicate asynchronously, which meant that the producers of the messages didn't need to know about the subscribers (consumers) directly. Instead this was offloaded to the network queue, which simplified distributed systems greatly.

From Distributed Queues to Repayable Message Queues

Apache Kafka is a household name within the tech community and is as common a component in the data platform as boutique seltzer and ping pong are to any startup. Kafka started at LinkedIn, and was later gifted to the Apache Foundation, with the goal was to create a distributed, fault-tolerant platform for handling low-latency data feeds, or *streams of data.*

Kafka was built with a fresh approach to the distributed network queue. However, it had a different approach to handling partial failures from consumers. As you may imagine, if we have a distributed queue and we take a message from that queue, then we could assume that the queue would purge that message and life would go on.

However, in the face of failure, no matter where you point the blame, there would be data loss associated with a message that has been purged from the queue if the process that took the message was lost due to a fault in the application code. It is common for systems reading messages to come across bad data, and without the appropriate defense mechanisms in place to handle corrupt data, the process could terminate unexpectedly. Without a means of recovering. Now as you can imagine, this was not good for business.

Fault-Tolerance and Reliability

To protect consumers from the inevitable, Kafka was architected to ensure that downstream failures would not immediately lead to data loss. The solution was to treat the messages in a similar fashion to how databases handle reliability. That is, through durable writes. Like magic, the distributed write-ahead log (WAL) emerged. We go over the key terminology, architectural components, and Kafka nomenclature now. If you have experience with Kafka, you can skip ahead to how Spark comes into play as the final piece of the modern data puzzle.

Kafka's Distributed Architecture

Kafka stores data emitted by producers in what are called topics. A Kafka topic is further broken down into one or more logical partitions, which enable each topic to be scaled to handle variable read and write throughputs. Each partition is guaranteed to store the data it receives (known as a *Kafka record*) in time based sequential order. See Figure 1-5.

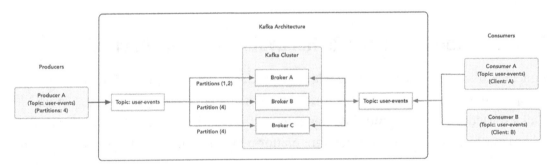

Figure 1-5. *The Kafka architecture*

Given that each partition maintains this synchronous stream of records, Kafka is commonly used to store event streams from contiguous timeseries events (such as users placing items into their shopping carts or transactions being made across a credit card). These timeseries event streams enables the system to analyze data that would otherwise be hard to group efficiently.

Since you can guarantee the ordering of the events within each partition, and events being written into each partition are distributed based on each record's key, this kind of distribution helps to solve common problems associated with timeseries analysis. The problem of out-of-order data, stored across many locations and across different data stores, feels like the data silo conversation from earlier in the chapter.

Instead of first running a job to collect all the data by key, across all partitions or across different data stores, an application can work across individual partitions, or slices, of the Kafka topic without having to shuffle data around first.

This can greatly reduce the stress on a system processing credit card fraud or handling recommendations based on a user's real-time activity. With the right selection of keys for each record, you can guarantee how the data lands within each topic.

Kafka Records

Records are simple rows of data, written out to disk in a durable fashion, like with the write-ahead log (WAL). Each record is identified by a key, which helps Kafka pass the data to the correct partition for the topic being written to. The record carries a main message, or payload, which is called the message value. In addition to the key/value pair, there is an internal timestamp (UNIX epoch milliseconds) stamped when the message is received and written into the topics log. This ensures that the insert order is maintained without issues with time (clock drift, time zone differences, etc.). Lastly, in more recent versions of Kafka, each record has an optional map of headers that can be used to annotate the data further. It can apply metadata about the producer, the payload schema, and just about anything you can dream up.

Brokers

Kafka manages producers and consumers of data in a similar fashion to that of a common pub/sub system. Data stored in each partition of a topic is managed by an active broker within the Kafka cluster.

These brokers manage the replicas of each record written into the topic and are kept in sync so that if a broker is lost due to the server going offline, or from network partitioning, that there is zero data loss. The next available in-sync replica quickly takes over the serving of the data, and Kafka will find another available broker to assign as the next backup to prevent a data loss scenario.

Why Stream Processing Matters

Kafka is the most widely used stream-processing framework and given the flexible, low-latency, highly available nature of the framework it isn't difficult to see why. Kafka however isn't the only player in the field. Apache Pulsar, which evolved from an internal project at Yahoo!, is another open-source alternative to Kafka. While we won't be using Apache Pulsar in this book, it is always good to know there are alternatives out there. Chapter 11 introduces using Apache Kafka with Spark.

Summary

In this first chapter I set out to talk to you about what steps had to be taken and what technology had to advance for us to be able to be successful in the exciting field of data engineering right now.

You saw how the rise in cloud first-solutions enables us to focus on what we really want to work on and prioritize. Having operated Hadoop clusters, and having used Amazon S3, I have to say that not having to worry about disk and host failures, data replication, backups, or recovery and disaster mitigation has allowed me to focus on using the data and create more useful tools that work *with the data*, versus working on the operations of what the data runs on.

We looked at how databases evolved over time and discussed some of the basic concepts and paradigms with respect to the traditional OLTP/OLAP style databases, the NoSQL database, and lastly the NewSQL databases. One thing that is interesting here is that there was this monumental move away from SQL during the days of NoSQL, and while document stores (fully denormalized datasets) have immense value to organizations, being able to express how we work with data through SQL-style queries and operations has returned to dominate the market.

Given the myriad styles of database available, and the fragmentation within these data siloes, we looked at the rise of the ETL job and the data pipeline architecture. This notion of being able to move data efficiently from one source to another was a catalyst that brought on the data warehouse, but also which spawned the invention of the unified data lake. As companies were working with more and more data, and as pipelines increased in size and complexity, workflow managers like Apache Airflow came into existence. These technologies were like the missing pieces in the data orchestra and the workflow manager became the conductor assisting in bringing harmony to the full data platform.

We wrapped up the chapter by talking about the evolution of message queues and brokers. We talked about stream processing and how this distributed game of telephone established a basis for the exciting step from batch-based processing to more real-time, stream-based processing with the birth of Kafka.

Over the course of the rest of the book, we'll dive deeply into the Apache Spark ecosystem, taking a hands-on, step-by-step, approach to solving real-world data engineering problems. You'll progress through the many layers and technologies supporting the data engineering umbrella as you solve foundational problems and build your own fully containerized, working data platform along the way.

## CHAPTER 2

Getting Started with Apache Spark

Apache Spark is the data engineer's Swiss Army knife. As a unified framework, it provides essential libraries to effectively connect and establish a common data narrative for engineers to work together cross-discipline. From ingestion and validation of raw data to data cleansing, transformation, and aggregation, as well as analytical exploration of trends and generation of insights, Spark connects the dots between the various constituents in any successful data operation. It also supports consistent (serializable) pipelines for feature engineering and robust machine learning.

In this chapter, you learn how Apache Spark acts as the centralizing core component within the data platform. Much in the same way that the data warehouses and data lakes solved the central problem of common access to data, Spark solves the problem of both centralizing and unifying how data can be processed, at scale, across the entire data platform. This simplifies the number of tools you need to use, so you can focus on less and work more efficiently. As a side-effect, Spark presents a wonderful opportunity to streamline many systems and processes in your organization.

The goal of this chapter is simple; together we will build a solid foundational understanding of the Spark ecosystem and firmly establish the role that Spark plays for the modern data engineer. We begin with the foundational concept of the Spark programming model, explore the core fundamentals and components within the Spark Application Architecture, and see how Spark elegantly handles separation of concerns via delegation.

This chapter is also a primer including the necessary steps to get up and running with Spark locally. You will complete your first hands-on exercise with Spark. The exercise is useful to build on your understanding of how to approach some basic problems in the realm of business intelligence, using the spark-shell (or the *Spark*

S. Haines, *Modern Data Engineering with Apache Spark*, https://doi.org/10.1007/978-1-4842-7452-1_2

playground, as I like to call it). By the end of this chapter, you'll have run your first Spark application, and for those more seasoned with Spark, this exercise may provide a different perspective on how you work with Spark.

The Apache Spark Architecture

Apache Spark evolved from the *MapReduce* programming model. To understand how this evolution occurred, let's begin by looking at what MapReduce is. MapReduce was made infamous by Jeffrey Dean and Sanjay Ghemawat (Google) in their seminal 2004 paper, "MapReduce: Simplified Data Processing on Large Clusters." This programming model led to the success of Google's search engine, and the paper helped pave the way for open-source MapReduce frameworks like Hadoop to come into existence. This enabled companies other than Google to begin working efficiently on large cluster-compute data problems, aka *Big Data*, in standardized ways.

The MapReduce Paradigm

MapReduce, at a very high level, shares some parallels to the ETL job example from Chapter 1. Each job begins with a specific set of input data (files stored on a distributed file system), and a workflow is written using the MapReduce APIs in two (or more) distinct parts.

A typical workflow begins with some transformational logic (mapping phase). This process essentially extracts information from the distributed file system, as defined by the encoding of the file(s) being read and is responsible for the conversion of the input data into collections of key/value map data. *You can say the input data is parsed and mapped to specific known types.* Next, the output of the mapping phase is then handed off to the reducers, which processes the resulting sorted collections of data from the mapping phase, as organized by common grouping keys.

Essentially the whole process effectively maps a large amount of data (larger than could reliably fit into memory), using a network of interconnected servers (cluster), to sort and group data into distributed sets of data. These datasets are then further reduced to accomplish a given outcome, such as counting the number of inbound references to a given website, which is what helped Google generate their PageRank algorithm. The output of the whole process is then written again to reliable storage for reuse elsewhere.

Rather than reading directly from a database, like with the traditional ETL job, MapReduce jobs required access to more data than could fit into any database in existence at the time (2004). To reliably process these gigantic amounts of data, a distributed execution model was born. It could process these huge datasets using data stored on reliable, distributed file systems and could automatically track complex distributed workflows.

The MapReduce programming model is introduced in Figure 2-1. By look at MapReduce conceptually, it will be easier for you to see how Spark evolved from this programming model.

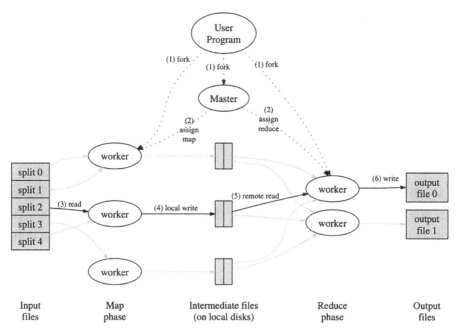

Figure 2-1. *The MapReduce programming model as conceptualized by Jeff Dean and Sanjay Ghemawat. Data moves from left to right through an acyclic, distributed workflow*

Mappers

The MapReduce programming model is built around *jobs*, which are essentially encoded routines that read and process data in a horizontally distributed way. Each job can process large amounts of data (sourced from the file system) by taking a divide-and-conquer approach that begins with a finite set of input files. These input files (stored in the distributed file system) are then distributed within the MapReduce cluster for processing as a series of partitions named *splits* (shown on left side of Figure 2-1).

All the required splits are then converted into Map Phase tasks by the MapReduce engine through assignment to *task*. Each mapping task then runs in complete isolation, enabling each task to work toward completion in a loosely coupled way.

Mental Model   The MapReduce execution pattern can be thought of as a distributed *divide-and-conquer* approach, as to handle increasingly large datasets, more machines can be added to share the work horizontally. Each task can be conceptually treated as a black box that takes its assigned input and writes the results of its work to a specific output location.

Durable and Safe Acyclic Execution

After each isolated task completes, the resulting output is written back to the distributed file system for durability and to acknowledge the completed task.

Note   Durability, consistency, and reliability tend to go hand in hand in distributed systems. When talking about the durability of data systems, you can almost always distill this line of questioning down to fault-tolerance and consistent data behavior.

From the MapReduce side, durably writing all intermediate files (seen in Figure 2-1) to disk establishes a concrete barrier between data being processed at each phase of job execution, much like the series of jobs defined in the example multi-step data pipeline (Figure 1-4 in Chapter 1). Lastly, durably writing the intermediate files also ensures that large, complex jobs can handle disk failure or network partitioning, without having to restart the whole job again. By default, data stored between stages of execution has implicit availability guarantees, thanks to the replication and fault-tolerance of the backing distributed file system (HDFS).

As you can image, the MapReduce jobs running to update the Google's search index (in 2004 and now as well) would need to run to completion, with consistent guarantees, across all phases of execution, to ensure that the result of each job is deterministic and trustworthy.

Reducers

We conclude this quick tour of MapReduce with the *reducers* (reduce phase tasks) seen on the right side of Figure 2-1. Each reducer task reads the output from the mapper task (you can consider this the upstream data preparation phase), which were written as intermediate key/value maps, and for each key, the map values are grouped, sorted, and then fed as input into a reducer task.

The reducer task receives as input a `key` and an `Iterator<value>` and processes the collection of data to achieve a desired result. Lastly, the output of each task is then written back to disk, and when all work has been processed, the job is considered a success.

MapReduce was a success because the programming model enabled large cluster-compute jobs to run without the engineers needing to understand the low-level complexities of writing large distributed systems and applications. I am not saying it was simple to build intuitive MapReduce programs; there was always some friction when it came to easily adopting the mental model. I guess you can say there was a need for more flexibility or creativity in the APIs. This is where Spark comes into the picture.

Apache Spark takes the best of the MapReduce paradigm while also enabling engineers to intuitively control how data is accessed, processed, and cached within the context of each job or series of jobs. Next, we explore how Spark set out to enable parallel computations that could use the standard distributed (disk based) file system, but that could also store (cache) intermediary data in memory as well. This greatly speeds up distributed data processing, analytics, and iterative algorithms as we know them today.

From Data Isolation to Distributed Datasets

Apache Spark evolved from the seeds of MapReduce, introducing a revolutionary new way for working with distributed data. It took the best of the MapReduce paradigm and introduced an entirely new programming model by enabling parallel computations and iteration across shared, distributed, sets of data.

In 2010, the seminal paper, "Spark: Cluster Computing with Working Sets," introduced the Spark programming model to the world and focused on improvements to common problems experienced when working with MapReduce. The novel idea was to reuse working sets of data across multiple parallel operations (in memory), removing the need to continuously go back to disk (and incur additional IO/file system open costs) between each phase of execution and processing. So how does it work?

The Spark Programming Model

To understand the Spark programming model, you need to adopt the good parts of MapReduce. The core design principle for MapReduce is that of isolated data access within the working memory of loosely coupled tasks. Due to the fact that MapReduce jobs divide and conquer work using an acyclic data flow model, similar to the flow of common data processes tasks encapsulated by a data pipeline, this enables MapReduce jobs to focus on the decomposition of large data problems across smaller functional black box stages. However, it also means that to process the same dataset across multiple jobs, each job has to repeat work, loading the same data into working memory. This is efficient but can also cause other side-effects like increased pressure on the MapReduce cluster's finite resources.

Did You Never Learn to Share?

To understand the drawbacks commonly associated with MapReduce, consider the following. Say you are tasked with writing a routine that needs to count all unique users who visited your company's web store, on a particular day. You also calculate the average number of products added to each customer's cart, while finding the top ten most common products added across all carts for the day. How would you go about answering these questions? Questions like this are common for analysists working in business intelligence (BI). Figure 2-2 shows a solution using MapReduce.

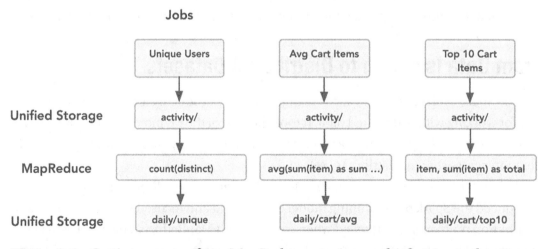

Figure 2-2. *Getting answers from MapReduce requires multiple trips to the origin dataset*

To solve each of the three problems presented in Figure 2-2, you need to create three distinct jobs:

- The first job in Figure 2-2, *Unique Users*, reads the data from the records stored in the `activity` directory of the distributed file system. For simplicity, let's assume that this directory exists in a daily directory like `/YYYYmmdd/activity`. This job loads the daily user activity event data and transforms it into a map, which can be used in the reducer stage, simply as a means of fetching the unique users for the day by key. These IDs would then be summed to calculate the daily unique users.

- The second job from Figure 2-2, *Avg Cart Items*, reads the data stored in the `activity` directory as well. It maps the *user id* to the *item id* for each split, then these maps are sorted and grouped to feed into the reducers. The reducers then sum the cart items, for each unique user, and output this data as a map of the user to the sum of their daily cart items. This data would be written to another location where an additional MapReduce job could read the data and generate the average cart items for all users. It finishes by writing these results to the `daily/cart/avg` directory.

- The last job from Figure 2-2, *Top 10 Cart Items*, again reads the origin data from the `activity` directory, maps the data as *item id* to total counts of that item per cart, and writes to intermediate reliable storage. It's sorted and grouped by item ID and fed into the reducers so they can compute the total count of each item. These running daily total counts are then written into reliable storage, and yet another job picks up from here and can select the top ten from this dataset.

The solution to common iterative problems like this are commonplace nowadays within the realm of analytics, and within other more complex data domains like Machine Learning.

Note   If the total number of steps required to solve the problems in Figure 2-2 using MapReduce felt like too many steps, you wouldn't be wrong. In retrospect, there is a good deal of unnecessary complexity and many moving parts, and this is what Apache Spark set out to change.

Apache Spark changed the way data is loaded, transformed, and processed across parallel operations within the same application context (such as with our BI queries in Figure 2-2). It did so by introducing a flexible programming model that enabled engineers to write code locally that could be distributed when run within the cluster, that was incredibly fault-tolerant, and that could be written using a functional programming style. At the heart of the Spark programming model is the *Resilient Distributed Dataset*, or *RDD*.

The Resilient Distributed Data Model

At a high level, you can think of the *RDD* as a *read-only, immutable collection of data* partitioned across a set of network-connected servers that are bound to a Spark application, called *executors*. The RDD object itself encodes the lineage of transformations required to achieve a desired outcome (such as the queries in Figure 2-2), beginning with the input data, as represented by a distributed data source and the source metadata, as well as each transformation along the journey from input to output. The input data can be files physically located on a machine, within a distributed file system, or partitioned across topics, such as with Apache Kafka.

Note   If you are new to the notion of partitions, that is okay. Partitions within files are simply range-based pointers that allow a file to be divided and read by many parallel processes. If you have a 1GB file and you split it across ten equal-sized partitions, then each partition would be 100MB. Partitionable files typically mean a file format that makes it easy to split the file. Such as a new line separated JSON file where each line is a record. You may not be able to partition the file exactly, but you can treat this similar to the concept of pagination for search results. Each partition encapsulates a range of data, say from line 0 to line 400 for partition 1, and so on.

However, the RDD goes beyond the MapReduce paradigm and achieves better performance (velocity) and fault-tolerance by embedding, or recording, the lineage of all distributed transformations recorded across a series of explicit steps within the runtime of a Spark application. This approach means that Spark can handle failures elegantly during the runtime of a job since there is a lightweight recording or transcript of everything of note that has occurred within a job. Think of the RDD as the blockchain ledger. This distributed lineage of transformations (transactions) across all partitions of an RDD can be used to synchronize access to shared variables and parallel access to data stored in cache, but it can also be used to recreate lost partitions by traversing the lineage of a job in reverse from a point of failure. The RDD can be reconstructed at any point of a Spark job, for any of its partitions, by reading the chain of transformations, back up to the origin data source again, if need be, or to any other cached waypoint along the way. This process enables Spark to fill in and recover missing data due to hardware failure, without having to restart the job or needing to reload all the data all over again. This saves both time and money!

Additionally, the elements represented by the RDD are not required to be physical objects like with the MapReduce splits and intermediate file. Rather the RDD acts like a graph of transformational pointers, so in the case of a partial failure, the RDD representing a specific phase of processing can be recomputed efficiently. The RDD itself stores no physical data, simply metadata, making it a means to coordinate data processing across a distributed cluster of network connected computers.

Lastly, while the RDD isn't specifically a shared data model, Spark enables programs to cache RDDs at specific points within the data lineage. This enables parallel processing, or multiple passes, over the data from that point on, without going back to the origin data source and needlessly retransforming data or doing costly joins more than once. We cover caching and checkpointing later in the book, so for now being aware of the RDD model is enough to get you thinking about how Spark operates.

Note   The RDD caching paradigm took me a while to fully grasp. Many programs don't need to access a dataset more than once, so storing the contents of the data in memory does not make sense for these single-pass applications. Spark caches lazily, so in order to cache the data, Spark needs to do a first pass over the data. Therefore, caching makes sense only if you want to access the data more than once.

We revisit the example BI use case from Figure 2-2 again at the end of this chapter. It is my hope that you will see how simple it is to use Spark to solve the same problem more efficiently, and in fewer steps. Before we get to the exercise, let's go over the Spark architecture, so we have a reference to circle back to. We will also go over the steps necessary to install Spark in order to run the sample code in the exercise that follows. This will be exciting.

The Spark Application Architecture

Writing distributed software can be a difficult undertaking given the complexities inherent to designing even simple applications. Why is this so hard? I'm glad you asked! Distributed computing requires a program to be written by taking specific actions to ensure that complex portions of the program can be run efficiently across a network of connected servers, or nodes, in a parallel fashion. All the while, it must coordinate both cross-node communication and marshal (serialization/deserialization, or simply put conversion to and from a binary representation) data between different phases of the program's runtime.

I realize I am hand-waving over the other complexities, such as how you handle network partitioning, which causes nasty blind spots within the cluster, or how you handle node failure, and how this gets more complicated due to partial error scenarios. The nice thing about writing applications with Spark is that you don't have to spend a lot of mental cycles strategizing how to write distributed applications. You simply write your application, and Spark will take it from there (for the most part).

Given the complexities of writing fully distributed applications, the approach to writing Spark applications simply abstracted away the complexities inherent to distributed applications so that engineers can focus on solving problems rather than running and managing complex distributed applications. Essentially, the Spark framework enables application authors to write programs using Spark that, for all intents and purposes, look no different than non-distributed software. However, when they run, the programs operate in a fully distributed, highly parallelizable fashion, with the heavy lifting being implicitly done behind the scenes.

Spark programs, more commonly just called applications, are made possible due to the following three main components:

- The driver program
- The cluster manager
- The program executors

The driver program is simply the Apache Spark application. Each application is supervised by the driver program, and the stages of execution (the work) is divided and distributed across the program executors using simple RPC communication for each stage of execution, typically some kind of *transformation*, along the journey to the application's desired outcome, which is referred to as an *action*. Behind the scenes, the cluster manager is hard at work simply keeping tabs on the state of the cluster, checking in on the running applications, and watching the available compute capacity remaining in the cluster. We will deep dive further into Spark Standalone and Spark on Kubernetes in the last half of the book.

Figure 2-3 shows these three main components, and how the Spark application (driver program) interacts with the other distributed components (cluster manager and executors) at runtime.

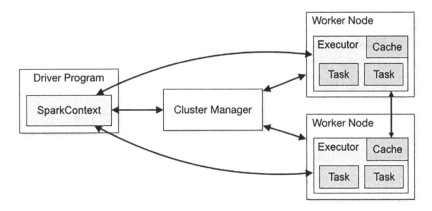

Figure 2-3. *The core components of a Spark application and its runtime*

The Role of the Driver Program

The driver program is the heart and brain of any Spark application. Earlier when I said that distribute software is complicated, well the driver abstracts away most of these complexities and selflessly works on your applications' behalf to coordinate the execution of your app in a highly fault-tolerant way.

Figure 2-3 shows the driver program on the left side. The driver simply acts as a delegate for the application runtime. As a representative of your Spark application, the driver program wears many hats and manages the complexities of running your application with the help of the *SparkContext*. At a high level, the SparkContext controls the following aspects of the driver program:

- **Resource management**: The Spark driver requests and releases cluster resources by interfacing with the cluster manager (see Figure 2-3). This enables your program logic to be distributed across a sub-cluster of assigned compute nodes, called executors. Given that the driver also handles the distribution of work across these nodes, it can also coordinate with the cluster manager in the case of lost executor nodes (if these machines go offline or become unreachable), or in the case of task failure. Work can be rescheduled again on any of the available executors to ensure that partial failures don't result in full application failure (within reason).

- **Application state management**: The SparkContext keeps track of the active and completed jobs. This way, the application state machine can control the execution of the driver program while also handling failures within the application.

- **Configuration manager**: The SparkContext is also responsible for synchronizing the initial application configuration within an object, named the *SparkConf.* This is an initial snapshot of the configuration of a Spark application, and it can be modified at runtime in a separate object named the *RuntimeConfig* (which we will work with in later chapters as well).

- **Job scheduler**: The SparkContext controls the scheduling work across the cluster as a series of *jobs*, *stages,* and *tasks*. We go into much more detail in Chapter 3 as you start to write your first Spark applications.

We could spend the rest of this chapter highlighting the myriad responsibilities of the *Spark driver program* and the *SparkContext,* but it is easier to peel back the layers as we move through more practical explorations in the chapters to come. We will press pause for now and become acquainted with the Spark cluster manager.

The Role of the Cluster Manager

The cluster manager (middle of the diagram in Figure 2-3) has a central role to play, just like the Spark driver. If you consider each abstraction as supporting a pillar of the Spark architecture, then the cluster manager is the *cluster coordinator* and the delegate in charge of managing and maintaining the state of the cluster, as well as the executors assigned to each active Spark application.

For a Spark application to be started inside a cluster, the driver program must first request resources from the cluster manager. This is a deliberate *separation of concerns* that divides the responsibilities of management of the cluster itself from the control of the application runtime environment.

Bring Your Own Cluster

To achieve an even higher level of granular control for the actual Spark cluster itself, Spark clusters can run across a few popular distributed cluster compute frameworks, in a pluggable style, which enables organizations to reuse technologies that are already well established across the industry. Or they can choose to run the cluster on Spark's native cluster orchestrator, called Spark Standalone.

The compute frameworks that Spark currently supports are:

- **Spark Standalone**: The native spark cluster manager. This operating mode requires specific resources to be carved out to coordinate the cluster state, including a *Spark Master,* which acts as the *cluster manager*, and an optional *standby* master for high availability (HA) in the case of the *active* master going offline. This mode requires cluster compute nodes, which are known as *worker* nodes and these run one or more *executor* JVMs. We look at Spark Standalone in more detail in Chapter 14.

- **Mesos**: Mesos is a popular, generic, distributed computation framework co-started by Matei Zaharia, who is also one of the core creators of Spark. We don't look at how to deploy applications using Mesos in the book but there are plenty of resources online for those interested.

- **Kubernetes**: Kubernetes is a household name that has become incredibly popular over the past few years as a way of managing infrastructure-as-code. Originally built inside of Google and then open-sourced, this is the newest addition to the clusters that Spark can run on top of. We look at Spark on Kubernetes in Chapter 15 as we look at deployment options.

- **Yarn**: *Hadoop* is still very popular in many organizations for running MapReduce style jobs. The Yarn manager enables Spark Applications to run executor processes at-the-source or co-located as close to the data stored within HDFS for a specific task as possible. We don't look at how to deploy applications using Yarn in the book, but there are plenty of resources online for those interested.

Last but certainly not least, we come to the final component of the Spark application architecture from Figure 2-3, and this is the *executor nodes*.

The Role of the Spark Executors

The executors, as seen on the right side of Figure 2-3, have a simple, yet important responsibility, and that is to act as the compute delegate for tasks assigned by the Spark driver program. Simply speaking, most of your work will be run across the executors and not the driver program itself.

Each executor instance operates in a disposable (transient) fashion. This means workloads can easily be rescheduled when a node in the cluster goes offline, or if an exception is encountered in the runtime of a Spark application, requiring an application to stop. Because executors are loosely coupled, you can ignore much of what happens behind the curtains, knowing that Spark applications are resilient and that executors themselves are easily replaced.

As a side note, because Spark is so resilient that many companies opt to run lower-priority workloads on lower-cost cloud compute resources like Amazon EC2 Spot instances (which are inexpensive and highly unreliable) but will still get the job done at a 90% cost reduction. The reason for this resiliency bubbles back to the RDD, which you were introduced to earlier in the chapter, and to recap it enables the fast recovery of data to ensure that a program can continue as soon as a new executor instance is assigned by the driver program. Remember that the RDD stores a distributed recording of all steps taken to transform data, for all partitions, and it acts as the catalyst to power fast and reliable recovery during failures.

Depending on the cluster manager in use for the Spark cluster, the executors will either be either running within a *worker* process, as is the case for Spark Standalone, or are controlled by the resource manager, when backed by Yarn, Mesos, or Kubernetes. As mentioned earlier, we look at deployment strategies for Spark applications later in the book (Chapter 13), and we cover Spark Standalone and Kubernetes deployments specifically at that time.

Tip   Executors can be run on low-cost or high-cost instances. Running your executors on expensive hardware with good SSDs capable of high IOPS, many CPUs to distribute the tasks, and a large amount of working RAM can help jobs complete faster. But depending on the way your application is configured, bigger doesn't always mean better. For some subsets of problems over provisioning, a Spark application will actually hurt the overall performance.

My hope is that, by covering the core components of the Spark application architecture (the driver, the cluster manager, and the program executors), and by taking a look at Spark's component-based delegation, you are starting to build a cursory understanding of how Spark works from a high level. As we continue this journey together, we will continue to revisit these central components as well as widen our view and understanding of how to use Spark as a central framework for handling the common problems that come up for the modern data engineer.

To finish this tour of the features and components of Spark, it is important to talk about the Spark ecosystem and the APIs that extend the core capabilities of Spark.

The Modular Spark Ecosystem

The core of Spark is written in Scala and Java and is architected to run on the Java Virtual Machine (JVM), but this doesn't stop you from using Python, R, or even SQL to run Spark operations. Each of these alternative ways of writing Spark applications is made possible through the use of a command interface or gateway. This means that Spark interpreters can be written to further extend the languages supported by Spark in the future and add idiomatic functional support for these new languages. This also means that you can choose how you want to interface with Spark when writing data engineering pipelines and jobs. As you will see next, you can also choose how to layer additional capabilities on top of your Spark jobs by mixing in additional modules.

The Core Spark Modules

The core capabilities of Spark are distributed across a few core modules. These libraries enhance the capabilities of Spark, thus enabling different ways of working with data within your Spark applications. From *SparkSQL*, which enables you to query your data like you would with a traditional OLTP or OLAP database, to *Spark Structured*

Streaming, which enables you to work in either micro-batches or across a true stream of data in continuous mode, to *Spark MLib* (for training and embedding machine learning models), and the querying of graphical relationship through *GraphX* and the community driven effort behind *GraphFrames*. Each of these libraries enhances Spark's capabilities and enables you to handle complex use cases, all within the same unified platform. You get a sneak peek at SparkSQL later in this chapter.

From RDDs to DataFrames and Datasets

We discussed *Resilient Distributed Datasets* (*RDDs*) earlier in the chapter, and although RDDs were critical to the early success of Spark, you will mainly be working with a higher-level abstraction over the RDD, called the *DataFrame*, and its strongly typed sibling the *dataset*. We explore the DataFrame in more detail during the exercise at the end of this chapter and see how it also plays nice with SparkSQL to simplify how we ask questions from our data. We will spend a good amount of time as well with the datasets later on in the book. Now if you are ready, let's get up and running with Spark.

Getting Up and Running with Spark

We've covered a lot of ground, and history, in our short time together and in only a few dozen pages. I think it is safe to say we are all ready to get started with Spark. The goal of this next section is simple. You'll see how to install Spark locally and then go through an introductory exercise to wrap up the chapter.

Note   The following installation instructions use Homebrew for macOS. To use Homebrew for Linux or the Windows subsystem for Linux, refer to `https://docs.brew.sh/Homebrew-on-Linux`.

Installing Spark

Installing Spark is a breeze. We'll go through the steps for getting up and running on Spark 3 now. If you want to go through the book using Spark 2.x, you'll have to stay on Java 8, but it is worth mentioning that the source code for the book is written and tested for Spark 3.

The only requirements necessary for running it are as follows:

- The Java Development Kit (JDK)

- Scala 2.12 (the default for Spark 3)

If you have Java and Scala installed, you can skip ahead to downloading Spark.

Downloading Java JDK

The Java JDK comes in two flavors, the *official Oracle Java JDK* and *OpenJDK*. Spark 3 works with both Java 8 and Java 11, and the flavor is up to you. For this book, we will be running the open-sourced OpenJDK 11. You can download the official JDK by going to Oracle's website at `https://www.oracle.com/java/technologies/javase-jdk11-downloads.html` and selecting the build for your environment. Or if you are running on a Mac laptop with Homebrew, you can simply run the following command from a terminal window:

```
> brew install openjdk@11
```

You should now have Java installed on your laptop. Now let's make sure you have Scala installed as well.

Downloading Scala

If you are going through these exercises on a Mac laptop, you can use Homebrew and run the following command to get this requirement out of the way:

```
> brew install scala@2.12
```

If you are on planning to work in a different environment (Windows/Linux), refer to the download instructions at `https://www.scala-lang.org/download/` to get Scala installed.

Okay. Now that you have the prerequisites out of the way, it is time to download and install Spark.

Downloading Spark

You are ready to download the Spark 3 release. Figure 2-4 shows an example of the download page from the official Spark website.

Download Apache Spark™

1. Choose a Spark release: 3.0.1 (Sep 02 2020) ⌄
2. Choose a package type: Pre-built for Apache Hadoop 3.2 and later ⌄
3. Download Spark: spark-3.0.1-bin-hadoop3.2.tgz

Figure 2-4. *Download Spark from the stable release packages at* `https://spark.apache.org/downloads.html`

The steps are simple:

1. Select the Spark 3.x release.

2. Choose the package type that is built for Apache Hadoop 3.2 and later.

3. Click the Download Spark link.

This will download the compressed Spark release to your laptop. Okay, so far so good. Now you just need to decompress the package and move Spark into a convenient home location. Open your favorite terminal application (I'm on a Mac laptop using the default Terminal app) and execute the following commands;

```
1.    mkdir ~/sources
2.    cd ~/sources
3.    mv ~/Downloads/spark-3.0.1-bin-hadoop3.2.tgz ~/sources
4.    tar -xvzf spark-3.0.1-bin-hadoop3.2.tgz
5.    mv spark-3.0.1-bin-hadoop3.2 spark-3.0.1
6.    rm spark-3.0.1-bin-hadoop3.2.tgz
```

If you followed these steps correctly, you should now have Spark installed in the ~/sources/spark-3.0.1/ directory. Let's test that everything and spin up our local Apache Spark.

> **Note**   The version of Spark available when you read this book will be subject to the release date of new versions. The book is written for Apache Spark 3.x. The versions used throughout the book might differ slightly, so just keep that in mind as you follow the exercises.

Taking Spark for a Test Ride

Now that Spark is installed, you should run a simple test to make sure that things are working correctly. Open a new terminal window and execute the commands in Listing 2-1.

Listing 2-1. Exporting the Required JAVA_HOME and SPARK_HOME Environment Variables and Launching the Spark Shell to Test the Installation Process

```
export JAVA_HOME=/usr/local/opt/openjdk@11/libexec/openjdk.jdk/
Contents/Home/

export SPARK_HOME=~/sources/spark-3.0.1/

$SPARK_HOME/bin/spark-shell
```

> **Note**   The paths on your laptop may be different from those on my setup. Refer to the installation instructions if the Spark shell does not come up.

Given that everything worked out as planned, you should now see some debug information in the terminal window, including an ASCII version of the Spark logo, the runtime Java and Scala version information, and the release version of Spark running. Here is an example from my machine:

```
Spark context Web UI available at http://192.168.1.18:4041
Spark context available as 'sc' (master = local[*], app id =
local-1609119924108).
```

```
Spark session available as 'spark'.
Welcome to

      ____              __
     / __/__  ___ _____/ /__
    _\ \/ _ \/ _ `/ __/  '_/
   /__/ .__/\_,_/_/ /_/\_\   version 3.0.1
      /_/

Using Scala version 2.12.10 (OpenJDK 64-Bit Server VM, Java 11.0.9)
Type in expressions to have them evaluated.
Type :help for more information.

scala>
```

The Spark Shell

Think of the *Spark Shell* as a utility for testing ideas quickly, or simply as a Spark playground. I say this because you don't need to write a single line of code or compile anything before getting started with a new idea. The magic of this shell program comes from the way it handles the dynamic loading of classes. For each new line of input, the shell saves its prior state, reads and reacts to the input, dynamically compiles and links new classes and functions, and saves its new state if no errors were thrown. If you are curious to see exactly how the REPL works, you can look at the SOURCE CODE FOR THE Spark REPL. However, it is easier to get an idea by building a mental model while doing something first, and then let your curiosity take you the rest of the way when you are ready. For now, let's start with some of the basics to get warmed up with the first of the hands-on exercises of the book.

Exercise 2-1: Revisiting the Business Intelligence Use Case

Do you remember the hypothetical business intelligence problem that was presented in Figure 2-2? To recap, we needed to create a series of jobs to read event data pertaining to user activity on a fictional ecommerce site, with the goal of coming up with the following three common daily reports:

- Find the daily active users (or daily unique users).

- Calculate the daily average number of items across all user carts.

- Generate the top ten most added items across all user carts.

Defining the Problem

Solving most data problems begins with the *data* and a specification of what we want to accomplish. In this case, we have a rough idea of what we need to accomplish, as noted in the introduction to this exercise, and for now we can get started. But in a more formal setting, you would also want to think about the following:

- What sources of data do you have access to?

- Who owns the data? Do you know who produces the data and controls the format (or schema) of the dataset?

- Can the data be read natively into Spark?

This is just a sample of some of the questions teams need to ask before starting to work with data. Luckily for us, we can just plow ahead and solve the problem since we are going to create the data we work with.

Solving the Problem

If you are not running the `spark-shell`, go back to the steps in Listing 2-1 to start up the shell. With the `spark-shell` spun up, we'll create a few rows of data to work with. You can copy and paste the code (if reading on your laptop) by pasting directly into the `spark-shell`, or you can simply write each line into the `spark-shell`.

Tip   To enable paste mode, just type `:paste` and press Enter. You should see `// Entering paste mode (ctrl-D to finish)`. To exit paste mode and interpret the code block from Listing 2-2, just press Ctrl+D. You should see `// Exiting paste mode, now interpreting.`

Listing 2-2. Importing Spark Helpers, Defining the Activity Case Class, and Generating Some Data

```
import spark.implicits._
import org.apache.spark.sql.functions._

case class Activity(
  userId: String,
  cartId: String,
  itemId: String)

val activities = Seq(
  Activity("u1", "c1", "i1"),
  Activity("u1", "c1", "i2"),
  Activity("u2", "c2", "i1"),
  Activity("u3", "c3", "i3"),
  Activity("u4", "c4", "i3"))

val df = spark.createDataFrame(activities)
            .toDF("user_id", "cart_id", "item_id")

// output the tabular rows
df.show()
```

The code in Listing 2-2 sets the stage for our first micro-application. The example begins by importing some of Spark's implicit helper functions from the "magic spark object". The spark-shell on startup creates what is known as the *SparkSession* and assigns it to the variable spark. Next, we import the functions necessary for using SparkSQL and define our data model using the Activity case class.

Things get more exciting as we generate some Activity data, and finish by using Spark to interpret our activities and create an initial DataFrame. There is a lot of magic going on here, but for now just know that spark.implicits._ enables us to convert and encode our Scala objects into the format needed by Spark for the DataFrame APIs. We'll cover DataFrames in much more detail in the chapters to come.

The output of the code block from Listing 2-2 begins when you call `df.show()`. At this point, Spark will present you with a formatted table view of your activities' DataFrame.

```
+-------+-------+-------+
|user_id|cart_id|item_id|
+-------+-------+-------+
|     u1|     c1|     i1|
|     u1|     c1|     i2|
|     u2|     c2|     i1|
|     u3|     c3|     i3|
|     u4|     c4|     i3|
+-------+-------+-------+
```

Armed with some data, let's go ahead and walk through a solution to the three questions proposed at the beginning of this exercise.

Tip   If you are new to Scala, then the case class will need some introduction. In a nutshell, case classes create templated classes extending the Scala product type, and automatically generate an immutable constructor, a companion object (helper template), field level accessors, as well as the copy, equals, hashCode, and toString methods. Case classes save you time when writing new classes and the immutable objects represented by case classes are important for functional programming as well as distributed systems. See the Online Scala Book for full details.

Problem 1: Find the Daily Active Users for a Given Day

Now that we have our DataFrame as defined in Listing 2-1, getting the unique users is as simple as counting the distinct `user_ids` within the DataFrame. The code snippet from Listing 2-3 will count the unique users.

Listing 2-3. Getting the Distinct Users

```scala
scala> df.select("user_id").distinct().count()
res1: Long = 4
```

You should see the number 4. You will notice that you immediately get an answer to this problem.

Problem 2: Calculate the Daily Average Number of Items Across All User Carts

This next problem is a bit more complicated, as it requires the data to be grouped and aggregated. But as you will see, it can be expressed intuitively as well. Paste the code from Listing 2-4 into the spark-shell (remember to use :paste, then copy, then interpret with Ctrl+D).

Listing 2-4. Calculating the Daily Average Number of Items Across All Carts

```
val avg_cart_items = df.select("cart_id","item_id")
  .groupBy("cart_id")
  .agg(count(col("item_id")) as "total")
  .agg(avg(col("total")) as "avg_cart_items")
```

When you run the code from Listing 2-4 you will notice that you don't get an answer as you did in Listing 2-3. This has to do with the way that Spark executes, and it is an explicit design choice, called *lazy execution*. Spark sets up a plan of attack but doesn't immediately execute anything unless there is an *action* (in Listing 2-3, count() was the action). We cover query plans later in the book, but for a sneak peek, use the explain() method on avg_cart_items, as shown in Listing 2-5.

Listing 2-5. Using the Explain Function to See What Spark Is Doing Behind the Scenes

```
scala> avg_cart_items.explain()
== Physical Plan ==
*(3) HashAggregate(keys=[], functions=[avg(total#87L)])
+- Exchange SinglePartition, true, [id=#109]
   +- *(2) HashAggregate(keys=[], functions=[partial_avg(total#87L)])
      +- *(2) HashAggregate(keys=[cart_id#57], functions=[count(item_id#58)])
         +- Exchange hashpartitioning(cart_id#57, 200), true, [id=#104]
            +- *(1) HashAggregate(keys=[cart_id#57], functions=[partial_
               count(item_id#58)])
               +- *(1) LocalTableScan [cart_id#57, item_id#58]
```

Note   The output from running `explain()` will probably look like gibberish at this point. That is expected if you are new to Spark or the DataFrame API. Remember earlier in the chapter when we talked about how the Spark RDD stores the data lineage for all transformations leading back to the original data source? Well, the DataFrame operates on top of the RDD, and this plan is an example of that transformational data lineage in a stepwise way. To read the output, begin at the end of the explain output *(1) LocalTableScan and walk backwards up to *(3) HashAggregate. You'll see that Spark automatically analyzes a query to create an execution plan to achieve an optimal outcome. In this case, it is a simple (1) local table scan of the activity data from the DataFrame, (1) select the fields we need, groupBy and aggregate (2) the total item's per cart, and finally (3) generate the average of the aggregated cart totals. What a trip!

Now to get the average of the total cart items, we need to add an action. The easiest thing we can do here is to call `show()`, which is an action on the DataFrame. This will execute the statement from Listing 2-4 and is shown in Listing 2-6.

Listing 2-6. Using the show() Action to See the Results of the Computation

```
scala> avg_cart_items.show()
+--------------+
|avg_cart_items|
+--------------+
|          1.25|
+--------------+
```

We will cover SparkSQL, DataFrames, and aggregations more in the chapters to come, consider this a sneak peek for what's to come. Now, let's solve the final problem for the exercise.

Problem 3: Generate the Top Ten Most Added Items Across All User Carts

This problem is like Problem 2 in that it requires us to group and sort results across the DataFrame. Listing 2-7 shows the full solution.

Listing 2-7. Grouping, Aggregating, Sorting, and Limiting the Result Set in Order to See the Top N Items in Users' Carts

```
scala> :paste
// Entering paste mode (ctrl-D to finish)

df.select("item_id")
  .groupBy("item_id")
  .agg(count(col("item_id")) as "total")
  .sort(desc("total"))
  .limit(10)
  .show()

// Exiting paste mode, now interpreting.

+-------+-----+
|item_id|total|
+-------+-----+
|     i3|    2|
|     i1|    2|
|     i2|    1|
+-------+-----+
```

Given there are only three items in the carts (unless you added more) then we only get three results to our query. However, in the real-world, your Spark job(s) may have millions of rows of data to process, yet Spark makes it incredibly easy to scale from five rows of data to millions.

Now that we are finished with this exercise, we can safely stop the spark-shell. To do so, all you need to do is press Ctrl+C to close the session, or press :q to quit without any error codes. This will tell the driver application to stop running the spark-shell.

Exercise 2-1: Summary

This exercise gave you some hands-on experience with Spark using the spark-shell. While the code in these exercises is simple, my hope is that it got your gears turning and excited for what is to come.

Summary

This chapter covered the history of Apache Spark, which came into existence from the knowledge and learnings of MapReduce. You learned about the Spark programming model and how it enables greater flexibility than MapReduce, while achieving fault-tolerance and reliability. It has the bonus of data caching and reuse through the sharing of in-memory datasets between processes running within Spark applications.

You learned how Spark is architected in a truly modular fashion, enabling a true separation of concerns between cluster management, the runtime for Spark applications, and even the pluggable libraries and APIs required by the Spark applications.

It covered how Spark was built to run on the JVM, but how different programming languages such as Python and R can control the actions of the Spark driver using command-processing gateways. This means that team members with different skill sets or language affinities can all cooperate and write programs that run on top of the same underlying Spark cluster and ecosystem.

We concluded the chapter with your first exercise, where you learned how simple and expressive Spark makes it to answer common business intelligence problems. In fact, solving the distinct user counting problem (Listing 2-3) can be further simplified using Spark SQL and temporary inline views, which is shown in Listing 2-8.

Listing 2-8. Creating a Temporary View to Find the Total Count of Distinct Users

```
df.createOrReplaceTempView("activities")

spark.sql("""
select count(distinct(user_id)) as unique_users
from activities
""").show
```

As mentioned, Spark enables many ways of accessing and working with data. In fact, Spark SQL can be run using the Spark Thrift Service, allowing you to connect to Spark using Tableau or Apache Superset for data analysis and visualization.

The next chapter focuses on the foundational capabilities enabled by the core Spark APIs that assist you when working across a wide variety of external data sources, and natively across a wide variety of common files formats, protocols, and compression types. You'll learn how the core Spark APIs can be used to solve many problems that come your way as a data engineer, through reinforcing exercises and examples that help you gain familiarity with the core Spark APIs. If you are ready, let's go.

CHAPTER 3

Working with Data

The last chapter introduced you to the Spark architecture and programming model. We took a quick tour of the core Spark components and APIs and finished up with an exercise that introduced you to the `spark-shell` and the DataFrame API. You also saw your first glimpse of the Spark SQL API, which empowers you to express complex analytical queries quickly and easily in a structured way. It also that cleanly abstracts away the underlying complexities when composing difficult SQL expressions.

In this chapter, we continue right where we left off and build upon the first two foundational chapters. We explore the process of reading, transforming, and writing structured data through the DataFrame API and take another look at using Spark SQL. Throughout this journey, you'll be working hands-on across introductory exercises focused on using Spark's core data sources capabilities.

As a data engineer, your job hinges on the ability to easily wrangle data, and I am a firm believer this is one of the most important core skills required for day-to-day excellence. You will be expected to think quick on your feet and figure out new ways to ingest, parse, and validate new sources of data that may not always be in the most optimal or easily ingestible format. But nonetheless it will be your mission to transform the data into clean, well structured, and above all reliable data. Luckily for us, Apache Spark supports many of the more commonly used data formats in use today and as you will soon find out, working across data sources and formats doesn't need to be a difficult task at all.

Before you begin, you will be learning about and installing two additional pieces of infrastructure within the development environment. The first is the *Docker* runtime, and the second is a notebook environment called *Apache Zeppelin*. These tools will become instrumental in your journey as a data engineer and that process beings now!

© Scott Haines 2022
S. Haines, *Modern Data Engineering with Apache Spark*, https://doi.org/10.1007/978-1-4842-7452-1_3

Docker

For those of you who have yet to work inside a *containerized* ecosystem, Docker is one of the more widely used container platforms on the market today. Docker has also been instrumental in helping to establish the *open-container initiative (OCI),* which has defined the standard for how container formats and runtimes should work and interoperate. So what exactly are containers?

Containers

Let's start by understanding why containers are so important. If you've worked or have a background with traditional DevOps or have simply tried to bootstrap your local environment across a few laptops, or a small handful of servers, then you know the struggle associated with streamlining the steps required to set up and deploy each physical environment. Even using package managers (such as apt-get, yum, or homebrew), the necessary steps to get each environment up and running efficiently can quickly become unruly. The biggest common pain points for this process revolve around consistency and runtime guarantees. Often, what works locally for you may not operate with the same level of finesse (or at all) for other engineers. Lastly, and most important, even when things operate for you and everyone else locally, things can differ significantly when they're finally deployed.

It is for these reasons that containers have become a must-have for mission critical applications. With containers, you can focus on writing software, knowing that what runs for you will also run for everyone else, without having to worry about conflicting runtime packages, or unmet system requirements, when sharing the same underlying containers and file system layers.

Docker, and containers in general, solve the common problem of deploying software by making it easy to run in a lightweight, standalone, executable environment. Your software is packaged alongside everything necessary to run reliably, regardless of the underlying operating system and hardware configurations. This is achieved through virtualization on Macs or PCs to create a Linux environment. Within the Linux OS, control groups (*cgroups*) enable these lightweight, isolated environments to be cordoned off (jailed) with respect to the shared host operating system. CGroups are outside the scope of this book, but if you are curious to understand how containers work, just take a gander online. It is fascinating.

Lastly, containers enable a unified development experience across Linux, Mac, and Windows, as well as across different CPUs like the common Intel x64 (64-bit), x86 (32-bit), and the ARM processors on the market today. Everything will simply run without the need to compile and maintain multiple targeted builds of your software for the various release environments.

Installing Docker is as easy as downloading the runtime, or you can take the path of least resistance and install the *Docker Desktop,* which is covered in the next section.

Docker Desktop

Follow the download instructions from `https://www.docker.com/products/docker-desktop` and in no time at all you will have all the components necessary to run Docker on your computer. Docker Desktop will automatically keep your system up-to-date and notify you when new releases or security patches are available for installation. If you are running on Linux directly, you can install the Docker engine.

Note   At the time of writing, Docker Desktop is currently in beta for anyone using macOS on the M1 chipset. To install the M1 preview, you have to use this link `https://docs.docker.com/docker-for-mac/apple-m1/`.

Configuring Docker

The Docker runtime is used to coordinate and run containers using a shared OS. This means your system needs to share resources with the Docker runtime and with the host machine (e.g., your laptop, desktop, or server). To effectively run the Docker environment, you allocate a slice (subset) of your CPU cores and RAM memory for Docker upfront.

I recommend allocating the following minimum set of resources for Docker to effectively run. The more CPU cores and memory, the better.

- CPUs: 4

- Memory: 8GB

- Swap: 2GB

- Disk Image Size: 100GB

Note   As a rule of thumb, it is beneficial to leave at least 1GB of RAM and at least one CPU available in order to run the main OS on your host machine.

To reconfigure your Docker preferences, follow these steps:

1.   Open Docker Desktop.

2.   Click the Config icon (Preferences) and click the Resources menu item (see Figure 3-1).

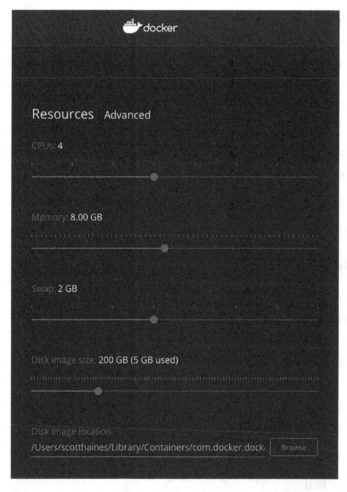

Figure 3-1.   *Docker Desktop: Preferences ➤ Resources view from the UI*

3. Update the configurations.

4. Click Apply & Restart.

Now that you have the Docker runtime installed, you can move onto Apache Zeppelin. You use Zeppelin to run the chapter's exercises.

Apache Zeppelin

Apache Zeppelin is an open-source *notebook* environment. If you are not familiar with what a notebook environment is, that is okay, as you will be by the end of this chapter. You may have heard of *Jupyter*, which is the standard experimentation environment for data scientists and machine learning engineers alike. Like Jupyter, Zeppelin is a web-based interactive development environment, but it ships with native support for Apache Spark, as well as many other connectors, which are called interpreters.

Interpreters

Within Zeppelin, an *interpreter* is a common interface that allows the Zeppelin runtime to control local libraries and remote services using a common *remote procedure call (RPC)* framework. RPC enables a software client to make external calls to a backend RPC service, via a lightweight messaging protocol, which enables the RPC client to run code on the server in an abstract way that feels no different than if the client were running the same routine locally. This abstraction makes it easy to create efficient interfaces for interoperating with many additional services without having to change the way the client-side calls out to various connected services. In the case of Zeppelin, you can think of the Zeppelin runtime like a router to many backend services, but with a common user interface. We use the Markdown, SQL, and Spark interpreters in the exercises to come.

Notebooks

The magic of the notebook environment is rooted in its ability to dynamically compile and run code (on-demand). A *notebook* is a collection of *notes*. Each note is broken down across a series of one or more *paragraphs*. Each paragraph contains individual snippets of code, documentation, graphs, etc. Within each notebook, code can be interpreted and run either as a procedural cascade, where each paragraph is run in sequential order, also called *playing the notebook*, or by running individual paragraphs in isolation, in an async fashion. Zeppelin works in a similar fashion to the `spark-shell`, in that it reads and compiles your code on-the-fly, but it also enables you to go beyond the terminal. For instance you can:

1. Store the progress of your session within a notebook, so you can quickly fire your environment back up and pick up where you left off. Your notebooks auto-save so you will never be caught off guard if your computer turns off unexpectedly.

2. Visualize the data stored in your DataFrames through the embedded SparkSQL interpreter. This can be a useful tool since the human brain isn't tuned to read text-based output of data and make sense of it. Graphing your data can be critical to the success of a project since you can uncover missing data and other characteristics of your data that may have been missed otherwise.

3. Interact with multiple users in the same notebook in real-time. *This is available only if you are running hosted Zeppelin or open an `ngrok` tunnel to your localhost.

4. Easily import and export notebooks to share common solutions to problems without requiring anyone to compile any code.

In the next section, we spin up Zeppelin and explore how to make the most of this valuable tool.

Book Source Code   If you have yet to download the source code for the book, then I suggest doing that now. It is as easy as popping over to `https://github.com/newfront/spark-moderndataengineering` and downloading the contents there. You will be using the starting point located at `/ch-03/start` for the exercises that follow.

Preparing Your Zeppelin Environment

To follow along with the exercises, you need to have Spark installed (and working) locally. This process was covered in Chapter 2.

Assuming everything is installed, and you were able to test the `spark-shell` in the previous chapter exercises, then one thing you may remember is that you manually exported environment variables to the terminal session before starting the `spark-shell`. It is a better practice to export these variables into your `.bashrc` or `.zshrc`.

Listing 3-1 shows a reference of the environment variables required to run Spark. The same process can be done using a `.bashrc` or `.bash_profile`. Adding these parameters manually each time you want to run Spark will quickly become annoying, so instead you can copy them into your Bash environment profile. Just make sure the paths are correct for your local setup. The reason for testing using transient environment variables (like the ones shown in Chapter 2) is to ensure that the paths are defined correctly and that things work before exporting them.

Listing 3-1. An Example from a .zshrc Profile

```
export JAVA_HOME="/Library/Java/JavaVirtualMachines/zulu-11.jdk/
Contents/Home/"
export SPARK_HOME=~/install/spark-3.1.0
```

Now that you have the environment set up, make sure you call `source ~/.zshrc` before continuing to refresh your Bash session (from the terminal). Now you're ready to party!

Running Apache Zeppelin with Docker

Execute the following two simple steps to start the Zeppelin process. This process will install and then start up Zeppelin. If this is the first time installing the Zeppelin Docker image, it will take a few minutes depending on your Internet connection. Once this Docker image is cached locally, starting and stopping the Zeppelin environment will take no time at all.

1.   `cd ch-03/start/docker`
2.   `./run.sh start`

At this point you will have Apache Zeppelin running on your laptop.

Peeking at the `ch-03/start/docker/run.sh` script gives you more complete details regarding how the environment is set up. The `start` function is shown in Listing 3-2.

Listing 3-2. The start Function in the run.sh Script Ensures Your Local Environment Is Set Up As Expected, then Delegates the Actual Running of the Zeppelin Container to the Docker Compose Process

```
function start() {
  sparkExists
  createNetwork
  docker compose -f ${DOCKER_COMPOSE_FILE} up -d --remove-orphans zeppelin
  echo "Zeppelin will be running on http://127.0.0.1:8080"
}
```

We'll continue with Zeppelin after a brief pause to talk about the Docker runtime.

Note   The following section dives further into the processes at work when you call `run.sh start`. If you are familiar with Docker and containers, you can skip ahead.

Docker Network

The `run.sh` script creates a network within Docker called `mde`. This bridged network allows containers running within the Docker runtime to communicate with one another only if they also share access to the same Docker network, and the network is also protected from the external host network. This means that unless further action is done (using port forwarding to the host machine), you have essentially created a private network that is protected in the Docker runtime.

```
docker network create -d bridge mde
```

Given that Docker runs a self-contained Linux executable, that means that the containers running inside of Docker are isolated by both their container and within the Docker runtime. This is good since Docker takes a secure, share-nothing style to the container runtime, but if you want to establish connections from one container to many different containers—say for example to run a database, distributed file system, or streaming data source like Kafka or Redis Streams—you wouldn't be able to do so without opening a shared network between the containers.

As we work through the chapters ahead, we will be using the Docker network to run everything we need for various streaming pipelines and to handle orchestration of different batch jobs.

Note If you want to use the hostname associated with your container from your localhost network, you can modify your /etc/hosts file to include a line at the bottom of the record for Zeppelin. For example: 127.0.0.1 zeppelin will enable you to view the Zeppelin UI at http://zeppelin:8080.

Docker Compose

The process for composing the Zeppelin environment is fairly simple. In this first use case, the heavy lifting is mostly done via ch-03/start/docker/docker-compose-all. yaml. The Docker Compose configuration is presented in Listing 3-3. The configuration shown here is a simple example of a single service named zeppelin.

Listing 3-3. The Docker Compose File docker-compose-all.yaml Creates a Consistent Configuration to Reliably Run One or More Containers

```
version: '3'
services:
  zeppelin:
    image: apache/zeppelin:0.9.0
    container_name: zeppelin
    volumes:
      - ${PWD}/notebook:/notebook
      - ${PWD}/logs:/logs
      - ${PWD}/data:/learn
      - ${SPARK_HOME}:/spark
    environment:
      - SPARK_HOME=/spark
      - ZEPPELIN_LOG_DIR=/logs
      - ZEPPELIN_NOTEBOOK_DIR=/notebook
      - ZEPPELIN_ADDR=0.0.0.0
```

```
    - ZEPPELIN_SPARK_MAXRESULT=10000
    - ZEPPELIN_INTERPRETER_OUTPUT_LIMIT=204800
  healthcheck:
    interval: 5s
    retries: 10
  ports:
    - 8080:8080
    - 4040:4040
    - 4041:4041
  networks:
    - mde
  hostname: "zeppelin"

networks:
  mde:
    external:
      name: mde
```

The Docker Compose configuration is a declarative resource interpreted by the Docker Engine to reliably run consistent services. Looking at the services block in Listing 3-3, you'll see a series of nested configurations that allow the Docker runtime to assemble the service. For example, the zeppelin service instructs Docker to use or pull the apache/zeppelin:0.9.0 image (container) and associate the resulting bound service to the container named zeppelin. The container_name property can be used as an alias to a specific name. By default, the docker runtime will create a uniquely named container.

Note   0.9.0 is the image tag and the release version of Zeppelin at the time of writing this book.

Volumes

The volumes block is used to mount external volumes into the runtime of the Docker container. If you look at the configuration of the zeppelin service, you will see we are forward-mounting specific directories from our host file system into the zeppelin container. These declarations enables us to automatically import the introductory

Zeppelin notes using relative paths ${PWD}/notebook:notebook. You'll also see that a data directory is included in ${PWD}/data:/learn, as well as the output of our Zeppelin logs via ${PWD}/logs:logs. While all of the resources beginning with ${PWD} are located in our local file system, the resulting path is the absolute location of the mounted volumes within the container (Zeppelin, in this case).

Tip   Containers are immutable. This is an important tidbit because all files created in the Docker container are isolated (or trapped) in the container by design. If you don't remember to export changes made at runtime, then by design if you stop and remove a container, all changes made after starting the container will be lost, including any new notes, changes to any configuration, or anything else done at runtime. Using volume mounts, you can share fully working experiments, without modifying the containers themselves. This concept is at the heart of the container ecosystem, and you'll be learning many techniques and best practices (and some interesting hacks) in the chapters to come.

Environment

Environment variables are a simple way to modify the behavior of Docker containers at runtime. The environment block enables us to override some of the Zeppelin defaults to create (or plug) the base container to achieve our desired outcome. This technique is important since the alternative means of customizing containers results in creating a new Docker image. In the docker-compose-all.yaml we use the host system environment variables to reference and mount the local Spark installation path using $SPARK_HOME. This trick provides a local container-based mount $SPARK_HOME=/spark, which can be seen under the volumes block. The environment variables in the Docker container are then updated to point to your mounted Apache Spark, which lets you bring your own Spark in the Zeppelin runtime. Neat.

Ports

Each Docker container runs in its own isolated environment. In order to allow access to the outside world, we must purposefully expose any ports we want to make available explicitly. Using the ports block, we enable port-forwarding from the internal docker network to our host system. This allows the Docker runtime to forward external requests to specific ports to be forwarded to the internal containers. This step is necessary to expose the Zeppelin UI, and to expose the ports for the Spark UI too.

You'll be using docker-compose for most exercises in the book. Visit this section again if you want to go back over the basics of the Docker network or view the structured of the docker-compose YAML. Now let's get back to Zeppelin.

Using Apache Zeppelin

Open the Zeppelin home page (using your favorite web browser) on localhost port 8080 (http://127.0.0.1:8080). This is the result of executing the ch-03/start/docker/run. sh start command from earlier. Confirm that the home page loads, and then we can run our first note together. The service home page is shown in Figure 3-2.

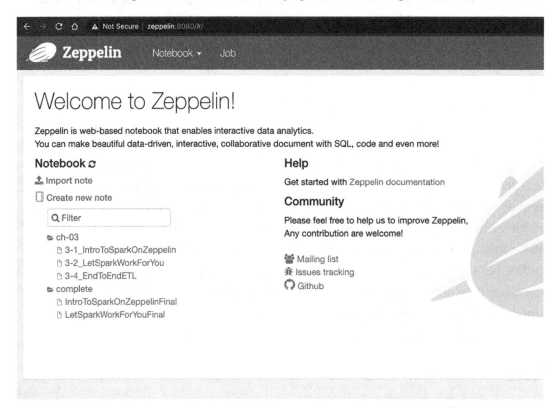

Figure 3-2. *Loading the Zeppelin homepage will confirm that everything has started up correctly*

With the Zeppelin environment running (you should see a similar screen as that of Figure 3-2), you will see that the home page is separated with a section labeled Notebook on the left. Within this section, you will see that some notes are already loaded for use in the hands-on exercises. The local notebook directory is volume mounted to your host system, so you can add new notes and modify existing ones, without worrying if the changes will persist.

From the Zeppelin home page, click the notebook under ch-03 called 3-1_ IntroToSparkOnZeppelin. This will open the initial starting note. You'll notice the note is broken into individual text blocks. These individual sections of the note are called paragraphs. Each paragraph can be run individually, or you can choose to run all paragraphs in a synchronous cascade using the play icon located at the top of the notebook. Figure 3-3 shows how to run all paragraphs.

Figure 3-3. *Use the Play icon at the top of the note to run all paragraphs*

Running all paragraphs is a simple way to ensure that the contents of the note have been executed, and that the Spark interpreter is working as expected. This initial step may take a little time, as there are some background processes at work that are readying the notebook environment.

Tip   I prefer to walk through the notebook in the same way that I would scroll through an article online, and that is using a top-down approach. This just means that I like to manually run each paragraph in sequence. This exploration tactic can help you understand how the notebook is architected, what the intentions of the author are, and to also double-check that there is no destructive code that made it into the note. If you jumped the gun and already clicked Run All Paragraphs, that is completely fine. As a rule of thumb, though, it is always better to look before you leap, especially when it comes to running a note the first time.

Binding Interpreters

Apache Zeppelin uses interpreters to enable each paragraph within the note to interact with different data sources using a single unified UI. Behind the scenes the complexity of keeping connections alive is delegated to the Zeppelin service itself. It is common for each note to connect to a smaller subset of services, but as you get used to using Zeppelin, you may find it more useful to interweave many data sources to create holistic dashboards or to debug and investigate data from many locations.

The way that Zeppelin denotes which interpreter is associated with each paragraph is with the initial modulus line, for example %{interpreter} is seen on each paragraph. We use a mix of Markdown %md and Scala-based Spark %spark in the introductory note you currently have open. We'll be focused on using the %spark interpreter for all new paragraphs for the remainder of this chapter. Figure 3-4 is used as a reference for interpreter binding.

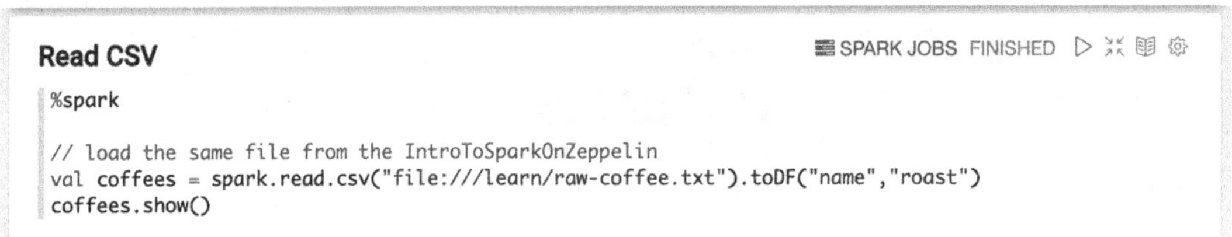

```
Read CSV                                    ≣ SPARK JOBS  FINISHED  ▷ ⨯ 📖 ⚙

%spark

// load the same file from the IntroToSparkOnZeppelin
val coffees = spark.read.csv("file:///learn/raw-coffee.txt").toDF("name","roast")
coffees.show()
```

Figure 3-4. *An example of the Spark interpreter in Zeppelin*

You can go ahead and click the Play icon on the Read CSV paragraph within the note now if you haven't clicked Run All yet.

As a quick recap, we just learned that a note contains one or more paragraphs, and that each paragraph is bound to an individual interpreter. For example, Spark/Scala %spark or Markdown %md. Essentially, the separation of concerns handled by the Zeppelin notebook environment enables a more consistent experience than simply using the Spark shell directly, as you experienced in the last chapter.

Let's officially begin the chapter exercises with the goal in mind to learn how Apache Zeppelin can be used to speed up the process of Apache Spark exploration and experimentation.

Exercise 3-1: Reading Plain Text Files and Transforming DataFrames

As I alluded to at the beginning of this chapter, the work you do as a data engineer is not always based upon well-defined, well formatted, or well documented data. In some cases, like this first exercise, you'll be exploring unknown data, and there will be some custom parsing required to format the dataset into something that can be efficiently processed and analyzed with Spark. So, you'll put yourself into the shoes of a new data engineer working at a boutique coffee company named CoffeeCo. It is day one and you've been asked to find a simple way to process and clean up plain text files for reliable use to downstream data systems.

Converting Plain Text Files into DataFrames

From the intro Zeppelin note, scroll down to the paragraph titled Read and Analyze. The contents of the paragraph plain text file go into a DataFrame. The paragraph code will be prepopulated for you and shown for reference in Listing 3-4. The DataFrame schema is printed to standard out to give you an idea of how Apache Spark interprets plain-text files.

Listing 3-4. The Code Snippet Generates a DataFrame by Instructing the SparkSession to Read a Plain Text File

```
%spark
val df = spark.read.text("file:///learn/raw-coffee.txt")
df.printSchema
```

The simple snippet from Listing 3-4 uses the `DataFrameReader` class. The `DataFrameReader` class enables you to load various datasets into Spark, as DataFrames, using a simple interface. We'll use this paragraph (Zeppelin's term for the runnable section in the notebook) as the starting point for the exercise. From here on out you'll be adding new paragraphs, writing code, and running the notebook as you proceed.

Go ahead and run the paragraph by clicking the *Play* button located to the top-right of the paragraph. You'll see the following output after the Spark engine is initialized.

```
root
 |-- value: string (nullable = true)
```

The output is the result of calling `df.printSchema`. Spark associates all data sources with a schema, which defines the typed structured of the data. In this case, we are reading a text-based file, so the resulting DataFrame will have a single typed column called value that is represented by a StringType. We cover schemas, which are encapsulated by the Spark StructType and underlying StructFields, in more detail toward the end of the chapter.

You may notice that the output schema is a bit ambiguous. Mainly you may be asking yourself if the whole file is represented by this single value, or what. To discover what the schema represents, let's create a new paragraph and call the show() action on this simple DataFrame.

Peeking at the Contents of a DataFrame

Hover underneath the Read and Analyze paragraph with your mouse, and you will see a prompt show up to Add Paragraph. Click the prompt and you'll see a add new paragraph show up. This new paragraph will automatically fill in the interpreter based on the prior interpreter, which means that the first line will be %spark. Now simply call show() on the DataFrame (df) like the reference shown in Listing 3-5.

Listing 3-5. Calling show() on the DataFrame Will Read Up To the First 20 Rows and Output It in a Simple Console Formatted Table

```
%spark
df.show()
```

Now run the code in Listing 3-5. You'll see that Spark has taken a basic first pass over the unknown text file, splitting each line of the text file based on the newline (\n) separator. You'll also notice the output is formatted to represent the tabular structure of the DataFrame (rows and columns), which in this case, is just a single column (named value) across a total of five rows.

```
+--------------+
|         value|
+--------------+
|   folgers, 10|
|     yuban, 10|
|nespresso, 10,|
|     ritual, 4,|
|four barrel, 5|
+--------------+
```

This output gives us better insight into the data itself. We can observe that the value Column represents a name and a numeric value as a comma-separated pair. It is common as a data engineer to write custom parsers when you are normalizing and cleaning data. We'll walk through how to parse each row of the DataFrame to transform it into a pair of named columns representing a coffee brand (e.g., Folgers, Yuban) and a corresponding roast level (or simply how dark the beans are).

DataFrame Transformation with Pattern Matching

Using what you know of the underlying data format (obtained with Listing 3-5), you can build a custom transformer to format the DataFrame. In a nutshell, this transformation will convert each raw row of text (from the value column) into a DataFrame representing a coffee brand (String) and a roast level (Int).

Create a new paragraph just below the paragraph from Listing 3-5. Then copy the contents of Listing 3-6 into the paragraph and run it.

Listing 3-6. Creating a New DataFrame Using a Series of Transformation

```
%spark
val converted = df.map { r =>
  r.getString(0).split(",") match {
    case Array(name:String,roast:String) =>
      (name,roast.trim.toInt)
  }
}.toDF("name","roast")

converted.printSchema
```

Listing 3-6 is an example of using Scala to create a functional transformation (map) married with a technique called pattern matching, which results in our custom parser.

In a nutshell, the transformation code in Listing 3-6 does the following:

1. By calling the map function on the DataFrame, you can generate an inline functional transformation (using Scala) that will be applied to every row (represented by the *r* =>) of the DataFrame.

2. Given that a DataFrame is a collection of rows, (represented by the Row class), and since we know that each row is composed of one String column, we can explicitly call the r.getString(0) accessor method. This will provide a String that we can call split on to manually subdivide the String.

3. Next we use pattern matching to deconstruct, or unapply, the two-item String array to the name and roast variables. This technique allows us to use named variables that can be dropped into a tuple. In order to convert the value of roast level to an integer, we first remove any unintended padding (space) on the String using the *trim* method in order to ensure the conversion to int (roast. trim.toInt) will succeed. It is worth mentioning that this method is shown for simplicity and is not fully defensive. This example will break for any row that has more than two commas. The complete notebook (in the chapter material) shows how to add defensive fallback using the catch-all pattern matching case _ =>.

4. Lastly, using the toDF("name","roast") method will convert the Scala tuple back into a DataFrame. Because we are converting from a tuple, the default column names of the DataFrame would be _1, _2, which is not really human readable.

After running this new paragraph (locally via Zeppelin), you'll see that the resulting DataFrame (called converted) has the desired schema.

```
root
 |-- name: string (nullable = true)
 |-- roast: integer (nullable = false)
```

Create one additional paragraph and call converted.show() to see the effects of the custom transformation.

```
%spark
converted.show()
```

```
+-----------+-----+
|       name|roast|
+-----------+-----+
|    folgers|   10|
|      yuban|   10|
|   nespresso|  10|
|     ritual|    4|
|four barrel|    5|
+-----------+-----+
```

While this first exercise isn't necessarily rocket science, it should have gotten you excited for learning to use Apache Zeppelin to rapidly test ideas, and even to learn the ins and outs of Apache Spark.

Exercise 3-1: Summary

This exercise introduced the basics of running Apache Zeppelin on Docker. You learned to read and parse raw text files using Spark and along the way learned that the DataFrame is essentially comprised of rows of data with strongly typed columns. Next, we dive a little deeper into the DataFrame, uncover how schemas work, and learn how to make Spark work for us so our data engineering lives become easier, and ultimately so we can work more efficiently.

Working with Structured Data

Continuing along in the spirit of exploration, we are going to look at how to use Spark to do the heavy lifting for us when it comes to content parsing and transformation. This will essentially remove the need for complicated custom parsers and instead you can learn to lean more on Spark as it reads and applies structure to your data.

Exercise 3-2: DataFrames and Semi-Structured Data

Head back to the Zeppelin home page. Open the notebook titled 3-2_ LetSparkWorkForYou, which is the starting point for Exercise 3-2. You learn to use the CSV DataFrameReader, as a follow-up to Exercise 3-1, where you created a simple typed

CSV parser. While you could have used the CSV reader in Exercise 3-1, you wouldn't have learned the valuable skills that came with manually parsing and transforming the data. Moving on!

The contents of the paragraph shown in Listing 3-7 create a DataFrame using the CSV reader. Now you can load the coffee data without needing to manually split the columns.

Listing 3-7. Augmenting Listing 3-4 by Replacing the Text-Based Reader with the More Powerful CSV Reader

```
%spark

val coffees = spark.read.csv("file:///learn/raw-coffee.txt").
toDF("name","roast")
coffees.show()
```

The output after running the paragraph from Listing 3-7 shows us that the coffee data appears to be correctly interpreted (parsed) on the first pass.

```
+-----------+-----+
|       name|roast|
+-----------+-----+
|    folgers|   10|
|      yuban|   10|
|  nespresso|   10|
|     ritual|    4|
|four barrel|    5|
+-----------+-----+
```

Although it seems like everything went well, that isn't really the case. Let's see why by adding a new paragraph to test the schema (`printSchema`) of the `coffees` DataFrame. Use the code snippet from Listing 3-8 as your guide.

Listing 3-8. A Simple Zeppelin Paragraph That Tests the Schema Generated Automatically by Spark via the CSV Reader

```
%spark
coffees.printSchema
```

Well, it turns out the backing schema of the `coffees` DataFrame is only simple strings.

```
root
 |-- name: string (nullable = true)
 |-- roast: string (nullable = true)
```

While Spark saves us the effort of manually splitting each line of CSV data into strings and is more than capable of doing some heavy lifting on our behalf, we are still missing one important piece of the puzzle, and that is how to coerce Spark into providing the correct schema for us.

Schema Inference

By design, Spark ships with a balanced configuration. This means you can get started quickly and the engine will make decisions based on the collective common needs of the Spark community. This also means that many options are turned off by default, to optimize the general performance of the engine. This is the case with schema inference.

If you think about what Spark needs to do when attempting to infer the schema for a given input dataset, then it isn't much different than how your brain works when looking over and scanning the same file. You instinctually look for patterns and can easily say, just by looking, that a file is string data or numeric data, but Spark isn't so efficient. Spark must open the file, read, parse, and test the data residing in each column to generate a consensus, or come to an agreement, regarding the different types to bind to each columnar row.

For a small file like the one we are using in this exercise, that is not a problem. It is cost effective to test everything, as there are only a few rows of data. However, if you wanted to infer the schema of a large file that also happens to contain a very large number of columns per row, then you could easily find yourself running out of memory.

Using Inferred Schemas

Add a new paragraph to the notebook ("3-2_LetSparkWorkForYou"), underneath the paragraph you just added in Listing 3-8. This new paragraph, shown in Listing 3-9, will be used to augment how Spark interprets and infers the schema when loading the contents of the plain text.

Listing 3-9. Setting the inferSchema Option on the DataFrameReader

```
%spark
val coffeeAndSchema = spark.read
  .option("inferSchema", true)
  .csv("file:///learn/raw-coffee.txt")
  .toDF("name","roast")

coffeeAndSchema.printSchema
```

By adding the optional configuration, you can now unleash the power of the Spark engine to infer (test and generate) schemas automatically. The resulting DataFrame (coffeeAndSchema) learned on its own that each row contained a String column and a Double column. The inference process will select the best possible match when reading semi-structured data like CSV or JSON.

```
root
  |-- name: string (nullable = true)
  |-- roast: double (nullable = true)
```

Using this technique enables Spark to learn from a smaller subset of your data, say the first 10,000 rows of a large file. In practice, data can change upstream in unexpected ways, so often it is in your best interest to use an explicit schema (declarative) when working with semi-structured data.

Using Declared Schemas

You can use schema inference to generate and then export a concrete schema for later reuse. When working with critical datasets, using strict schemas enables you to ignore (skip) corrupt data, or to fail fast and kick back an exception, when encountering data that doesn't conform or parse correctly. Lastly, using specific schemas allows Spark to perform more optimally since schema inference requires all columns across a subset (sample) of the dataset to be tested to come to a consensus for each columnar type. I'll introduce you to a fun technique I like to call "steal the schema," which you can use to generate and export strict schemas.

Steal the Schema Pattern

Create a new paragraph and use Listing 3-10 as your guide. When you are finished, go ahead and run it.

Listing 3-10. Export an Inferred Schema Using the DDL Method on a DataFrame

```
%spark
import org.apache.spark.sql.types._

// steal the schema
val coffeeSchema: StructType = coffeeAndSchema.schema
val coffeeSchemaDDL: String = coffeeSchema.toDDL
// `name` STRING,`roast` DOUBLE

// create an explicit schema (from the stolen schema)
val coffeeDDLStruct: StructType = StructType.fromDDL(coffeeSchemaDDL)

// read the coffee csv using the stolen schema directly
val coffees = spark.read
  .option("inferSchema", false)
  .schema(coffeeDDLStruct)
  .csv("file:///learn/raw-coffee.txt")
```

The steal the schema pattern from Listing 3-10 shows how you can extract an inferred schema as a DDL string. When applying this technique, you would export the value of `coffeeSchemaDDL` directly to the file system, or simply copy and paste it manually into a configuration file. The important part is that you can load the DDL string directly into the `DataFrameReader`, which enables you to by-pass schema inference and ultimately add more explicit strict rules for handling data in your Spark applications.

```
spark.read
  .schema("`name` STRING,`roast` DOUBLE")
  .csv("file:///learn/raw-coffee.txt")
```

This pattern can save time and enable your applications to enforce a strict schema without having to rely on inference when running in production. While this process isn't necessary in the case where you begin with a well-known schema, it is worth understanding what is possible and how the `StructType` can easily convert between DDLs and internal Spark schemas. You can also use the Spark DSL to create data definitions manually.

Building a Data Definition

Using the steal the schema pattern (Listing 3-10) resulted in an inferred schema that was slightly different from the resulting schema in Exercise 3-1. The difference being the

81

inferred data type associated with the roast column. We wanted an integer and Spark interpreted it as a *double,* aka the Spark DoubleType. We can correct the schema by manually construct the StructType.

Add another paragraph to the Zeppelin notebook and copy the code snippet from Listing 3-11 into the paragraph; click run. The added benefit of this manual legwork provides documentation to each field in the schema.

Listing 3-11. Construct a coffeeSchema StructType to Use Directly When Reading the Coffee Data as a DataFrame

```
%spark
val coffeeSchema = StructType(
  Seq(
    StructField("name", StringType,
      metadata = new MetadataBuilder()
        .putString("comment", "Coffee Brand Name")
        .build()),
    StructField("roast", DoubleType,
      metadata = new MetadataBuilder()
        .putString("comment", "Coffee Roast Level (1-10)")
        .build())))

val coffees = spark.read
  .option("inferSchema", "false")
  .schema(coffeeSchema)
  .csv("file:///learn/raw-coffee.txt")
```

Defining the StructType manually opens additional options available directly on the StructField object, including declaring the DataType of each named column as well as the all-important metadata option. Let's look behind the curtain and get to know the StructType, StructField, DataType, and metadata a little more.

All About the StructType

As we saw in Listing 3-11, when we compose a schema definition, we are really generating an instance of the StructType class. The StructType is a metadata object storing field-level information for each column represented by a DataFrame. This is analogous to a table definition in SQL. The StructType itself is just a container object composed as a sequence of StructFields.

StructField

The StructField is a simple object associated with a COLUÇMN in a DataFrame. Each StructField has a unique field name, a DataType, a Boolean declaring if the column is nullable, and the optional, albeit important, metadata.

Spark Data Types

Spark supports many types of data that can be represented within a DataFrame. Each type can be converted automatically between the Java JVM and native memory. We dive deeper into the Catalyst engine in Chapter 6, but we can use this section as an introduction to the types available out of the box. Table 3-1 can be used as a quick reference during future exercises. While this table isn't a comprehensive list of all available data types, it is more than enough to get started.

Table 3-1. *Apache Spark Can Natively Convert Between Scala Types and Native Encoded DataTypes*

DataType	Value Type in Scala	Spark API
ByteType	Byte	ByteType
ShortType	Short	ShortType
IntegerType	Int	IntegerType
LongType	Long	LongType
FloatType	Float	FloatType
DoubleType	Double	DoubleType
DecimalType	java.math.BigDecimal	DecimalType
StringType	String	StringType
BinaryType	Array[Byte]	BinaryType
BooleanType	Boolean	BooleanType
TimestampType	java.sql.Timestamp	TimestampType
DateType	java.sql.Date	DateType

You may be familiar with some (or all) of the Scala types from Table 3-1. We've even used a few of the native `DataTypes`. Do you remember which ones? By the end of the book, you will have worked with most of the native types in this table, albeit implicitly, in addition to even more complex types as we get into the later chapters of the book. For now, just keep this table in the back of your mind, as we take a brief moment to discover how to work with the Metadata class.

Adding Metadata to Your Structured Schemas

The `Metadata` class is a simple wrapper over `Map[String, Any]` and enables you to add descriptive metadata, at the field level, for your schemas. The `comment` field is a reserved metadata key name that is automatically carried along to your table definitions. Ultimately, the use of metadata enables you to document your datasets and fix the discovery issues that arise with poor data documentation. Using a simple built-in construct, you can help pay things forward for other engineers, systems, and analysts.

Exercise 3-2: Summary

At this point, you are probably beginning to understand, and maybe even getting excited about, what Spark is capable of. I find it is helpful to attack a problem from the inside out, and in the exercises so far, we have done just that. You learned how to read and transform plain text into a clean and well-defined DataFrame in Exercise 1, and you discovered how to take advantage of schema inference to skip the effort of writing custom, field-level parsers. Then you learned that there are some additional benefits for manually constructing a schema (`StructType`) providing field-level metadata and declarative `DataTypes`. We've covered a lot already, but we still have further to go, so if you are ready, we are going to get our feet wet with a little Spark SQL.

Using Interpreted Spark SQL

Who here likes SQL? One of the ways in which Spark makes our lives easier as data engineers is by enabling us to process data as if we were simply querying a table in a traditional SQL database. How does this work, you ask? Well, by simply taking advantage of the structure of the data known to it. This is a roundabout way of saying that Spark leans on the `StructType` to materialize views for use in your applications. Let's see how with a practical example.

Exercise 3-3: A Quick Introduction to SparkSQL

For this exercise, you can continue using the Zeppelin note from Exercise 3-2 ("3-2_LetSparkWorkForYou"). When we left off (see Listing 3-11), you had just created a manual schema and rebuilt the `coffees` DataFrame.

Creating SQL Views

Add a new paragraph to the note and enter the contents of Listing 3-12.

Listing 3-12. Create a SQL View Using a DataFrame with createOrReplaceTempView

```
%spark
coffees.createOrReplaceTempView("coffees")
spark.sql("desc coffees").show(truncate=false)
```

In one line (Listing 3-12), you can easily generate a temporary, in-memory view using the `coffees` DataFrame. This will register a temporary table in the context of your `SparkSession` and will exist for the duration of the session, or until you explicitly drop it. Line 2 of Listing 3-12 introduces the SQL interpreter, which is available through the `SparkSession`.

```
spark.sql("desc coffees").show(truncate=false)
```

The output you'll see after clicking run on the paragraph (Listing 3-12) will show the table metadata. Do you notice anything interesting?

```
+--------+---------+-------------------------+
|col_name|data_type|comment                  |
+--------+---------+-------------------------+
|name    |string   |Coffee Brand Name        |
|roast   |double   |Coffee Roast Level (1-10)|
+--------+---------+-------------------------+
```

As mentioned in Exercise 3-2, the metadata has a reserved key named `comment`, the value of which will be applied to any table you create. One of the main benefits of defining the metadata for your schemas is to pay things forward to other engineers by documenting your data. This simple example shows why field-level documentation can be beneficial. Now anyone describing the `coffees` table can immediately understand the intention of each column. That is worth its weight in data gold.

With the temporary view (coffees) available within your active Zeppelin session, it is time to learn how to use the %sql interpreter.

Using the Spark SQL Zeppelin Interpreter

Add another new paragraph to your note and then add the simple SQL command from Listing 3-13.

Listing 3-13. Using the Spark SQL Interpreter in Apache Zeppelin

```
%sql
select * from coffees
```

Click run and you will see the temporary view from Listing 3-12 rendered directly in the Zeppelin UI. The rendered output is presented in Figure 3-5 for reference. Visualizing your datasets can be a helpful tool when exploring new datasets, and even if you are just viewing the output as a basic table, you can lean on the SQL expression itself to modify the output or do basic sorting using the Zeppelin UI.

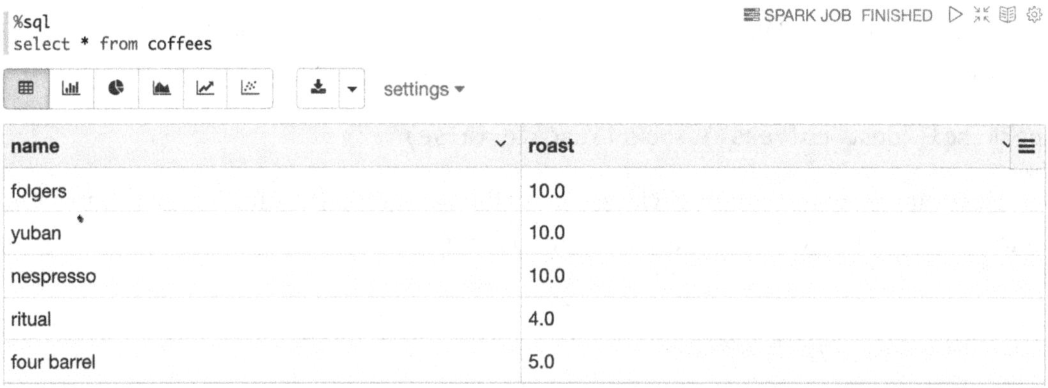

```
%sql
select * from coffees
```

SPARK JOB FINISHED

name	roast
folgers	10.0
yuban	10.0
nespresso	10.0
ritual	4.0
four barrel	5.0

Figure 3-5. *The nicely formatted output for the coffees table via the Zeppelin UI*

Behind the scenes, Zeppelin uses the active SparkSession to pass your SQL command through the SQL interpreter. The SQL entry point on the SparkSession is an instance of the SQLContext.

Let's look at a more interesting example. Let's use the Spark SQL interpreter to compute the average roast level for the coffees table.

Computing Averages

The task of computing an average is straightforward with Spark SQL (and SQL). You need to simply call the `avg` expression on a column.

Create a new paragraph in the note and enter the Scala code snippet from Listing 3-14.

Listing 3-14. Computing the Average of a Column Using Spark SQL

```
%spark
spark.sql("select avg(roast) as avg_roast from coffees").show
```

Run the code in Listing 3-14 and you'll discover the average roast is `7.8`. This is a simple analytical query. What other reasons can you see for utilizing Spark SQL?

Depending on your background with SQL and where you are on your Apache Spark journey, you can use the SQL interpreter while you are gaining familiarity writing Spark applications. At the end of the day, the SQL parser is powerful enough that you won't really see any major difference. In most cases you also won't hinder the overall performance of your application. It is also worth mentioning that the return type of calling `spark.sql("select * from coffees")` is a DataFrame, so you can mix modes within your applications where you see fit.

Exercise 3-3: Summary

You learned to generate temporary SQL views, which enabled you to query DataFrames using SQL. While we only scratched the surface here, the main point of the exercise was to show you different ways of working with data with Spark. As a recap, a DataFrame has a schema, and the DDL of the schema is used when generating a table or view within the context of a SparkSession. While you will be working with richer and more complex sets of data in the chapters to come, you can continue to explore this first dataset.

I suggest taking some time to try some additional commands with the `coffee` data that you loaded. Here are some suggestions for continued explorations:

- Find the `min` and `max` roast values in the table. Hint: `min(roast) as min`

- Try to `sort` the table using the `ORDER BY` clause. Hint: `order by roast desc`

- Try sorting the data by coffee name

When you are finished, there is one final exercise left, and that is to construct a simple end-to-end ETL application.

Your First Spark ETL

You learned to read, transform, and view data in this chapter. To construct an official ETL, you will need to write the results of the transformation to a reliable location.

Open the complete ETL example (from the Zeppelin home page, under "3-4_EndToEndETL").

Exercise 3-4: An End-to-End Spark ETL

Once the note is loaded, you'll see we defined the coffeeSchema (copied from Exercise 3-2), and that there is a main paragraph containing the code referenced in Listing 3-15.

Listing 3-15. The Full ETL Job

```
%spark

spark
  .read
  .option("inferSchema", "false")
  .schema(coffeeSchema)
  .csv("file:///learn/raw-coffee.txt")
  .write
  .format("parquet")
  .mode("overwrite")
  .save("file:///learn/coffee.parquet")
```

The first half of the ETL job should be familiar from Exercise 3-2. The section in bold contains the write flow, which introduces the DataFrameWriter.

Writing Structured Data

Once we have data loaded into a DataFrame our data story can begin. Each project, feature, or use case drives the steps and desired data outcomes required by your Spark applications. In the case of the ETL job from Listing 3-15, there isn't much going on aside from writing clean, structured data in a format that will make it easier for downstream processing. In the case of this example, the final output of the job writes parquet data to the file system, in a simple process that converts the semi-structured CSV data into fully structured parquet data.

Parquet Data

Parquet data is a fully structured columnar data format. This means there is no need to infer any types since the schema itself is encoded within the file. This means you can simply read a Parquet file, and Spark will automatically apply the schema on the resulting DataFrame.

Reading Parquet Data

Working with the results of the ETL job from Listing 3-15 is as easy as referencing the file location. This process is shown in Listing 3-16.

Listing 3-16. Spark Natively Supports Reading Parquet Using the DataFrameReader

```
%spark
spark.read
  .parquet("file:///learn/coffee.parquet")
  .createOrReplaceTempView("coffee")
```

Because you went through the effort of cleaning, transforming, and writing the final transformation as fully structured Parquet (`coffee.parquet`), you have reduced the effort of anyone needing to work with the same dataset in the future. While the exercises this chapter have been simple, there is one more interesting side effect that is worth pointing out. The Parquet file you generated in the ETL inherited all the metadata of the `StructType`, including the field-level comments. You can add another paragraph to the note and simply describe the coffee table (based on the replaced temp view from Listing 3-16).

```
%spark
spark.sql("desc coffee").show(false)
+--------+---------+------------------------+
|col_name|data_type|comment                 |
+--------+---------+------------------------+
|name    |string   |Coffee Brand Name       |
|roast   |double   |Coffee Roast Level (1-10)|
+--------+---------+------------------------+
```

The fact that your field-level descriptions are encoded and preserved even after writing your data out as Parquet is an effective delivery device. It ensures that anyone working with this data (presumably written into your data lake) can easily understand what each field represents.

Exercise 3-4: Summary

This last exercise was intended to reinforce what you've learned over the course of this chapter. Ultimately, with a little up-front effort it is possible to not only effectively extract, transform, and load data, but it also reduces the effort of anyone working downstream to your data pipeline.

Summary

This chapter covered a lot of ground. It introduced you to practical examples for using Apache Spark as a data engineer and hopefully introduced you to a new style of working, which is procedurally with data using Notebooks. I have found the techniques presented in this chapter of incredible use when teaching as well as in my actual day job. Together, Zeppelin and Docker can save you a lot of time and money, by enabling you to quickly get gut checks and test ideas. The added benefit of being able to share fully working environments and applications, which require a few simple steps to get up and running, can be invaluable, especially when working remotely (like many of us have done during the pandemic).

This chapter introduced some useful patterns for working with data within the Apache Spark ecosystem. The intent here was simply to start showing you how to do something, and then present ways in which you can improve on the earlier solutions. We could have started off reading the Parquet data from the end of the chapter, but we would have missed out on learning to read plain text and transforming the DataFrame by hand. Thinking about how to solve problems while learning about the core Spark classes and ways of working with data can help you build a solid foundation for the years to come.

The next chapter looks at data transformations. It is a continuation of the foundations established in these first three chapters, focused on the core functions and operations available for structured datasets in Spark.

## CHAPTER 4

Transforming Data with Spark SQL and the DataFrame API

The previous chapter introduced you to using Docker and Apache Zeppelin to power your Spark explorations. You learned to transform loosely structured data into reliable, self-documenting, and most importantly, highly structured data through the application of explicit schemas. You wrote your first end-to-end ETL job, which enabled you to encode this journey from raw data to structured data in a reliable way. However, the process we looked at is just the beginning and can be looked at as the first step of many in a data transformation pipeline. The reason we began by looking at raw data transformations is simple—there is a high probability that the data you'll be ingesting into your data pipelines starts at the data lake.

This chapter continues to focus on the process of data transformation. You'll begin with an exercise that explores manipulating data using simple *selection* and *projection* as a means of filtering and reshaping data. Afterward, we'll explore using *joins* to compose new views of data from multiple sources, without the need for writing complicated code. Lastly, we'll look at intuitive ways to problem-solve using nested selects (inner and outer selections), expression columns, and select expressions. Throughout the chapter, we will look at how to express these transformations using both Spark SQL and the DataFrame APIs.

Data Transformations

As with any tool, the more you use it, the more proficient you become with it. This is sometimes referred to as *honing your skills*. We all hone, or sharpen, our skills each and

© Scott Haines 2022
S. Haines, *Modern Data Engineering with Apache Spark*, https://doi.org/10.1007/978-1-4842-7452-1_4

every time we write code in our favorite programming languages, learn and memorize common usage patterns across frameworks and APIs, and solve a new problem. In order to react to a problem that comes your way, you must know which tools are available to you and understand how to take advantage of similar problems you have solved in the past in order to tackle these problems head on.

We are all aware that there is no one-size-fits-all solution to most problems in life and this is also true when it comes to transforming data. However, many effective techniques with data problem-solving begin by reducing the size and complexity of the data. This can happen across many different steps in a data pipeline or even in the logical stages of a single application. It commonly relies on the basic techniques of selection and projection to get the job done.

Basic Data Transformations

In the last chapter, we worked on some introductory tasks on the simple coffee dataset (coffee brands and roast level) for our job with CoffeeCo. In the spirit of continuity, we will be continuing to work on problems related to this job at the coffee company (I guess that means we're still employed!).

Exercise Content   You can find the materials for this chapter under `ch-04/` of the book's GitHub `https://github.com/newfront/spark-moderndataengineering/tree/main/ch-04`. Just start Zeppelin using the run script and head on over to `http://localhost:8080`.

```
cd ch-04/docker && ./run.sh start
```

Exercise 4-1: Selections and Projections

Begin by opening the note titled "4_1_DataSelectionAndProjection." You will see a single paragraph titled Data Generation at the top of the screen. Go ahead and run this paragraph to get Spark initialized. The remainder of this exercise will follow along from here.

> **Note**   This exercise can be followed by adding new paragraphs and running each step inside of Zeppelin, just like we did in Chapter 3. If you want to look at the final notebook, it is available as `complete/4_1_ DataSelectionAndProjectionFinal`.

Data Generation

CoffeeCo is small and fortunately for us there are only a few main stores. To begin we'll prime a temporary SQL view called *stores* to represent our company's flagship stores (see Listing 4-1). If you are following along using the Zeppelin environment, go ahead and run the paragraph.

Listing 4-1. Generating the Stores Data and Temporary SQL View

```
%spark
case class Store(
    name: String,
    capacity: Int,
    opens: Int,
    closes: Int)

val stores = Seq(
    Store("a", 24, 8, 20),
    Store("b", 36, 7, 21),
    Store("c", 18, 5, 23)
)

val df = spark.createDataFrame(stores)
df.createOrReplaceTempView("stores")
```

We can see from the code in Listing 4-1 that we are simply generating a sequence (list) of stores. Each store opens and closes between 5am and 11pm on a 24-hour clock and has some information about occupancy limitation as well. With the data generation behind us, let's learn about the process of data *selection*.

Selection

The process of selection is arguably the most fundamental means of reducing the footprint of the data you are working with. This concept will be familiar to anyone with working knowledge of SQL. In a nutshell, selection enables us to *reduce the set of rows returned by a query* by way of a *condition*.

Let's start with the basics now and we'll look at the wide array of selection capabilities later in the chapter, when we get to "Selection Revisited." Say we wanted to find all the stores open on or after a specific time of day.

We can approach this problem by using a simple conditional query against the stores view. The code snippet in Listing 4-2 shows the query.

Listing 4-2. Returning Only the Rows that Match the Condition closes >= 22 via Simple Selection

```
%spark
val q = spark.sql("select * from stores where closes >= 22")
q.show()
```

There is only one answer to this question given that there is only one store that closes after 10pm (Store C).

Filtering

If the selection process feels to you a little like filtering, you'd be right. In fact, if you take a look at answering the same question using DataFrames, you'll see that we can use the filter or where function interchangeably.

Tip   When we select a column in a DataFrame, we have a few options for identifying the column. There are four distinct ways to provide the target column for the selection. The symbolic aliases ` and $ are implicit conversations that can be used by importing the implicit functions from the SparkSession:

```
df.where(df("closes") >= 22)
df.where(col("closes") >= 22)
df.where('closes >= 22)
df.where($"closes" >= 22)
```

Feel free to explore these options using the code snippet in Listing 4-3 from Zeppelin.

Listing 4-3. The where Clause Is Interchangable with the filter Function of the DataFrame

```
%spark
import org.apache.spark.sql.functions._
import spark.implicits._

val filter = df.filter($"closes" >= 22)
val where = df.where('closes >= 22)
filter.show()
where.show()
```

So, if filtering is functionally equivalent to selection and that process enables us to reduce the number of rows of data returned by a query, then you may be asking yourself what exactly is going on within the select * in our select * from stores? Well, funny enough, this process is known as *projection*!

Projection

If selection is the process of reducing the total number of rows returned by a query, then you can think of *projection* as the process of reducing the total number of columns returned by a query. This can be used as an optimization technique since there is a memory cost to returning all columns across all rows when only a subset of columns is really necessary to relay the results (or answers) provided by a query.

Let's look at an example of using projection and selection together using the data available within the stores view.

Say we want to find all stores where the minimum occupancy is greater than 20. In this case, we can assume we don't need to worry about when a store opens or closes, but rather we want to find the name of the store only (see Listing 4-4).

Listing 4-4. Reducing the Number of Columns Returned by Our Query Using Projection

```
%spark
// find all stores with an occupancy greater than 20
val pq = spark.sql(
  "select name from stores where capacity > 20")
pq.show()
```

The query in Listing 4-4 shows you how to use projection and selection together. The projection dictates which columns will be returned by the query, as seen in the `select` `name`, which directs Spark to return only the column labeled `name`. The selection portion of the query, which is a fancy filter or conditional predicate, dictates which rows meet the criteria to be returned by the query, as seen in `where capacity > 20`.

Let's see how we can build the same query using the DataFrame API directly (see Listing 4-5).

Listing 4-5. DataFrame Projection and Selection Using Functions

```
%spark
df
  .select("name")
  .where('capacity > 20)
  .show
```

The process of mixing projections and selections can be powerful in yielding performant transformations across your data. What you've seen so far are just the basics, but you can use this foundation as you look at some of the more powerful capabilities available to you when harnessing the full power of selection.

Exercise 4-1: Summary

Up until this point in the chapter, we've looked at the simple process of reducing the total number of rows returned by a query through the process of basic selection. We also looked at how to reduce the columns returned by our query by way of columnar projection. While many use cases fall into these kinds of simple filter and transform operations, such as basic reporting. We've only scratched the tip of the iceberg in terms of what is possible.

Selection goes well beyond just the conditional examples and empowers you to write transformative queries using many operators that span both traditional SQL (OLTP) and more analytical query (OLAP) processing through the use of common operators. As the chapter continues, we will explore the process of using joins, column, and expression aliases, as well as conditional select expressions and even nested queries.

Joins

In the wild, most datasets don't arrive in a perfect form to solve the problem you've been assigned. If and when you happen to stumble upon the perfect dataset that fits your needs to a T, drop everything and immediately thank the people responsible for thinking ahead and grasping how people would need to work with their data in the future! More likely, you will have to stitch data from a few different sources together in order to create the data representation needed to solve the problem at hand.

Joins are common within the data pipeline as a solution to combining data. These workflows fall under the umbrella of the ETL and can be used whenever you need to strategically combine and transform multiple sources of data into a single consolidated view that can be used to answer more targeted problems.

For example, say we were tasked with creating a job that generates the current available occupancy data for our coffee shops. For the sake of the exercise, let's say we already have a source of data that emits the number of occupied seats per coffee shop. We can use this data to join with our coffee `stores` data to create a new view that we can query to find which store can seat a variable sized party.

Later in the book, we'll be looking at streaming applications and pipelines and we will go over the process of emitting event data (such as the number of occupied seats per coffee shop) using Kafka and Redis Streams. We'll even walk through the process of reliably reading and composing new materialized views from those event streams!

Note   For the purposes of this exercise, we will be generating the occupancy data in a way that can also showcase the different common join styles available within Spark.

Exercise 4-2: Expanding Data Through Joins

When joining data, we commonly use one or more columns that can act as a join key (or selection expression) between the two datasets we want to combine. In this case, we will be using the coffee store name as our common key.

Create a new paragraph in Zeppelin and enter the code snippet from Listing 4-6. This creates the view you can use to join with the `stores` data.

Listing 4-6. Generating the Occupancy View for Our Joins

```
%spark

case class StoreOccupants(storename: String, occupants: Int)

val occupants = Seq(
  StoreOccupants("a", 8),
  StoreOccupants("b", 20),
  StoreOccupants("c", 16),
  StoreOccupants("d", 55),
  StoreOccupants("e", 8)
)

val occupancy = spark.createDataFrame(occupants)
occupancy.createOrReplaceTempView("store_occupants")
```

The pattern of generating data in Spark should start to feel familiar at this point. We start by creating a new temporary view based on our occupancy DataFrame in Listing 4-6. It is worth pointing out that the StoreOccupants class has some unknown stores, identified as d and e. These are only here to showcase the various join capabilities within Spark. Moving on, let's explore joining these two data sources (stores and occupancy) using the following join flavors: *Inner, Right, Left, Semi, Anti,* and *Full.* This helps us understand how the different styles of joins work. Selecting the correct type of join is key to significantly reducing the amount of energy and effort you need to apply when materializing a new view of data. For each join type, you can create a new paragraph in the Zeppelin note to work through things live.

Inner Join

The inner join is the simplest to understand and just so happens to also be the default join operation in Spark (given this is usually how people want to join data). The inner join works by selecting only the rows that meet the join selection criteria across both sides of the data being joined.

Spark SQL Inner Join

```
%sql
select * from stores
inner join store_occupants on stores.`name` == store_occupants.`storename`
```

DataFrame Inner Join

```
%spark
val inner = df
  .join(occupancy)
  .where(df("name") === occupancy("storename"))

inner.show()
```

The result of our join operation is a new DataFrame that combines all the columns of our two data sources where the join criteria is met. In this case, that's where there is a matching store name across both data sources.

The tabular output of the inner join shows how the composition works.

Results of Our Inner Join Operation

```
+----+--------+-----+------+---------+---------+
|name|capacity|opens|closes|storename|occupants|
+----+--------+-----+------+---------+---------+
|   a|      24|    8|    20|        a|        8|
|   b|      36|    7|    21|        b|       20|
|   c|      18|    5|    23|        c|       16|
+----+--------+-----+------+---------+---------+
```

Inner joins simply ignore all rows that don't have a matching join condition. Next up, let's look at the right join.

Right Join

The right join, or right outer join, returns all rows from the right-side data source explicitly joining all rows where the selection criteria is met with the left side of the data. When and the data doesn't match, it will insert null values instead.

Spark SQL Right Join

```
%sql
select stores.*, store_occupants.`occupants` from stores
right join store_occupants on stores.`name` == store_occupants.`storename`
```

DataFrame Right Join

```
%spark
// df is our stores data
val rightJoined = df
  .join(occupancy,
    df("name") === occupancy("storename"),
    "right")

rightJoined.show()
```

The result of our join operation is again a new DataFrame that combines all the columns of our two data sources, but instead of skipping the missing data between the two data sources, we have the following output.

Results of Our Right Join Operation

name	capacity	opens	closes	storename	occupants
a	24	8	20	a	8
b	36	7	21	b	20
c	18	5	23	c	16
null	null	null	null	d	55
null	null	null	null	e	8

Right joins can be used when you want to preserve the details of the missing data between two data sources. The target dataset, in this case occupancy, dictates the total number of rows returned by the operation. Given we have a higher number of rows in our right-side data (occupancy) than we do in our stores data (5 rows to 3), the missing values will be added to the left side as null columns.

Let's look at using the left join now, which is literally just the left-side equivalent of the right join.

Left Join

The left join, or left outer join, returns all rows from the left-side data source explicitly joining all rows where the selection criteria is met with the right-side side of the data. When the data doesn't match, it will insert null values instead.

Spark SQL Left Join

```
%sql
select stores.*, store_occupants.`occupants` from stores
left join store_occupants on stores.`name` == store_occupants.`storename`
```

DataFrame Left Join

```
%spark
// df is our stores data
val leftJoined = df
  .join(occupancy,
    df("name") === occupancy("storename"),
    "left")

leftJoined.show()
```

Given the output of the right join operation, what do you think the output of the left join will be? If you guessed that it would be *just three rows*, good job!

Results of Our Left Join Operation

```
+----+--------+-----+------+---------+---------+
|name|capacity|opens|closes|storename|occupants|
+----+--------+-----+------+---------+---------+
|   a|      24|    8|    20|        a|        8|
|   b|      36|    7|    21|        b|       20|
|   c|      18|    5|    23|        c|       16|
+----+--------+-----+------+---------+---------+
```

As we saw with the right join operation, the data that is backing the target data source is critically important to the results of the operation. In this case, we have the three rows in the `stores` data against the five rows in the `occupancy` data, and we can only ever have the same number of rows as the target data (three) in our result.

Semi-Join

The semi-join, or left semi-join, filters the set of rows returned on the left side of the join, by finding all rows on the right side that match the selection criteria. Essentially, the join is done only as a means of selecting which results to return from the left side of the join. This process works by finding the intersection of the join selection expression and returning only the intersecting rows for the left side of the join. I like to think of this as a selection proxy filter, since it uses another data source as its own selection criteria.

Consider the following use case to understand this concept better. Say we want to identify coffee shops that have fewer than 20 seats and store them in a new view as boutiques. People enjoy different sized shops, and I for one would rather be in a smaller, quieter shop.

We have two options for doing this. We can either create a new view by selection, or we can use what is known as an *inline view*. Let's look these processes now.

With the goal in mind of creating a new data source (and to showcase using semi-joins) we have a few options available to achieve this goal:

- We can use an inline table to manually create the data to use as our right-side source for the semi-join.

- We can create a new view using selection across an existing data source.

Create an Inline Table Called Botiques

```
%sql
create or replace
temporary view boutiques as (
  select * from VALUES ("c") as data(boutiquename)
)
```

Using an inline table is a simple way to create a temporary data table that can be used for a slew of different purposes:

- Test a new idea when you find yourself thinking "wouldn't it be great if we had this data...". This can save you time when you are building applications since you don't need to create any formal classes or even come up with column names.

- Simplify SQL selection statements by adding synthetic data that can assist in filtering or in complex selection.

- Comes in handy when exploring new sources of data or when you want to add some additional default columns, which can be done using a full join against the temporary inline view.

In our case, we can use this table to help filter our data using a semi-join. However, when considering our use case, we already have nice source-of-truth data located in our stores view that we could use for this purpose as well. The nice thing about reusing an existing view is that you don't have to manage yet another source of truth; you can instead just apply a filter where capacity < 20 and it becomes easy for others to understand the concept of the boutiques view.

Creates the Botiques View Through Selection and Projection Aliasing

```sql
%sql
create or replace
temporary view boutiques as (
  select stores.`name` as boutiquename
  from stores
  where capacity < 20
)
```

Now that we have created our boutiques view, all we have left to do is apply our semi-join operation.

Spark SQL Semi-Join

```sql
%sql
select * from stores
semi join botiques
on stores.`name` == botiques.`botiquename`
```

DataFrame Semi-Join

```scala
%spark
// semi-join
val botiques = spark.sql("select * from botiques")
val semiJoin = df
  .join(botiques,
    df("name") === botiques("botiquename"),
    "semi")

semiJoin.show()
```

The result of the operation is that we find the one coffee shop that is a boutique coffee shop. It just so happens to be the one row from our `stores` data that intersects with the `boutiques` data. Let's look now at the reverse of the semi-join, which is the anti-join.

Anti-Join

The anti-join, or left anti-join, is the opposite of the semi-join. It uses the set difference between the left- and right-side data, based on the join selection expression, to reduce the rows returned for the left-side data.

Spark SQL Anti-Join

```
%sql
select * from stores
anti join boutiques
on stores.`name` == boutiques.`boutiquename`
```

DataFrame Anti-Join

```
%spark
val boutiques = spark.sql("select * from boutiques")
val antiJoin = df
  .join(boutiques,
    df("name") === boutiques("boutiquename"),
    "anti")

antiJoin.show()
```

The result of this operation is that we find the two coffee shops that are not boutiques.

Semi-Join and Anti-Join Aliases

While semi-joins and anti-joins allow you to use another data source as your selection filter, there is a more understandable way of declaring this same intention. You can use the IN and NOT IN operators, respectively, rather than using the semi-join or anti-join, to achieve exactly the same result. In fact, Spark will actually just use a semi-join or anti-join behind the scenes.

Using the IN Operator

Using the IN Operator as a Semi-Join

```
%spark
val inOper = spark.sql(
"""
select * from stores
where stores.`name` in (
  select boutiquename from boutiques
)
"""
)
inOper.explain("formatted")
```

Spark will take the intention of your IN condition and will convert it to use a semi-join behind the scenes. Given that the semi-join and anti-join may not be commonly known join types across your team, it is always better to optimize for maintainability of our code, and this sometimes means going with common conventions (or you can always document what you are doing in comments).

The output obtained by calling explain on the inOper DataFrame shows the steps that Spark takes behind the scenes to convert the Spark SQL expression into a series of transformations and actions. The engine handling things behind the scenes is the Catalyst engine, which we investigate more in Chapter 6.

Spark Execution Plan: Shows the Formatted View of the Steps Spark Takes to Parse, Optimize, and Generate a Plan of Execution

```
== Physical Plan ==
* BroadcastHashJoin LeftSemi BuildRight (4)
:- * LocalTableScan (1)
+- BroadcastExchange (3)
   +- LocalTableScan (2)

(1) LocalTableScan [codegen id : 1]
Output [4]: [name#0, capacity#1, opens#2, closes#3]
Arguments: [name#0, capacity#1, opens#2, closes#3]
```

```
(2) LocalTableScan
Output [1]: [boutiquename#248]
Arguments: [boutiquename#248]

(3) BroadcastExchange
Input [1]: [boutiquename#248]
Arguments: HashedRelationBroadcastMode(List(input[0, string, true])),
[id=#557]

(4) BroadcastHashJoin [codegen id : 1]
Left keys [1]: [name#0]
Right keys [1]: [boutiquename#248]
Join condition: None
```

Learning to read the Spark execution plans can help when you are optimizing complex queries, but for now you can simply use it to discover the behind-the-scenes processes at work when you use Apache Spark. Now back to joins.

Negating the IN Operator

Go ahead and create another paragraph to see what is happening when you switch from using the IN to NOT IN.

Using the NOT IN Operator as an Anti-Join

```
%spark
val notInOper = spark.sql(
"""

select * from stores
where stores.`name`
not in (
  select boutiquename from boutiques
)
"""
)
```

We've looked at various techniques for joining data sources in Spark. Let's conclude this journey now looking at the full join operation.

Full Join

The final join operation we'll explore is called the *full join*. This operation takes both data sources into account and will join the data where the selection criteria is met, and then will fill in nulls on either side where there is no match within the two datasets.

To showcase how the full join operation works, we will be adding some additional records to our stores DataFrame (df), for the purpose of having missing data across both the left and right DataFrames (see Listing 4-7).

Listing 4-7. Using Full Join and Union to Combine Your Datasets

```
%spark
// create 2 new stores
val addStores = spark.createDataFrame(Seq(
  ("f", 42, 5, 23),
  ("g", 19, 7, 18)))
  .toDF("name","capacity","opens","closes")

val fullJoined = df
  .union(addStores)
  .join(occupancy,
    df("name") === occupancy("storename"),
    "full")

fullJoined.show()
```

The code represented in Listing 4-7 showcases how to do a full outer join. It also shows off another kind of join that Spark has to offer—the *union*. There is also another trick that snuck into the top of this code example, and that is a third way to create a DataFrame. We've seen the process of using a case class, how to use an inline table, and now how to use a sequence of tuples. The inline table and the use of tuples with explicit column names are similar in style.

Unions   Unions are like joins except that both data sources need to have the same schema (matching column names and order). This process can be achieved through projection and column aliasing across one or more DataFrames (using the union or unionAll function) before creating an aggregate temporary view.

Given the effort required to coerce the underlying schemas of potentially many different yet related datasets, Spark 3 introduces the `unionByName` method so you can ignore any missing columns: `df.unionByName(addStores, allowMissingColumns=true)`

The full outer join is the most expensive of the joins explored with respect to complexity. This is because it needs to sort both the left-side and right-side data sources, in order to do a *sort-merge-join* operation across the data sources. Let's take a look at the output of the full join operation to get an idea of how it works.

The Output After the Full Join Across the stores and store_occupancy Data

```
+----+--------+-----+------+---------+---------+
|name|capacity|opens|closes|storename|occupants|
+----+--------+-----+------+---------+---------+
|   g|      19|    7|    18|     null|     null|
|   f|      42|    5|    23|     null|     null|
|null|    null| null|  null|        e|        8|
|null|    null| null|  null|        d|       55|
|   c|      18|    5|    23|        c|       16|
|   b|      36|    7|    21|        b|       20|
|   a|      24|    8|    20|        a|        8|
+----+--------+-----+------+---------+---------+
```

The output shows that the rows have been sorted in a reverse order. More important is the fact that we have not skipped or ignored any of the missing data between the two sets. Using the left, right, or full outer joins enables you to understand what is missing between your two datasets, and then it is up to you to decide how you want to move ahead. This is an important concept since it can be used tactically to measure corrupt or missing data from an upstream data producer and tracked as a metric when monitoring the frequency of missing data in a data pipeline.

Exercise 4-2: Summary

This exercise explored the many styles of joins available within Apache Spark as a means of providing you with strategies for joining and transforming data. Next, we put together everything we learned to solve an interesting seating problem related to our coffee shops.

Putting It All Together

Let's pull together what we've learned in this chapter to solve the initial problem of finding seating availability for a party of a specific size (based on the available occupancy data we have at our fingertips). We can break down the steps necessary to solve this problem into the following phases:

1. Compute the current available seats per store.

2. Find the stores that have at least enough seats for the party.

3. Return the store name that has availability.

We have almost all the pieces in our toolbox to be able to achieve this goal. You may be wondering how we compute the available seats per store?

Exercise 4-3: Problem Solving with SQL Expressions and Conditional Queries

First off, we know we can join the `stores` and `store_occupants` data to get the `capacity` and `occupants` columns, but we will still need to subtract these two columns in order to use the result in our selection. Lucky for us, there just so happens to be a fairly simple solution to this problem. We can use an *expression* to achieve this goal. For this exercise, you can continue along in the same note inside of Zeppelin.

Expressions as Columns

Using an expression column can be done by using the alias operator (`as`). We can also use the simple minus operator (`-`) in conjunction with aliasing to achieve this goal. Let's solve the first part of this task by computing just the availability within our stores.

SparkSQL: Use a Column Expression to Generate the Current Availability

```
%sql
select name, (capacity-occupants) as availability
from stores
join store_occupants
on stores.`name` == store_occupants.`storename`
```

The result of this operation is a new DataFrame with the column's name and availability.

```
+----+------------+
|name|availability|
+----+------------+
|   a|          16|
|   b|          16|
|   c|           2|
+----+------------+
```

Now we are getting close to a solution. The next step will be to simply select the rows where the availability is greater than or equal to that of the party size. We've created these kinds of conditions before, so let's see what happens if we try to find availability for a party size of four.

```
%sql
select name, (capacity-occupants) as availability
from stores
join store_occupants
on stores.`name` == store_occupants.`storename`
where availability > 4
```

When you attempt to run this query, you will see that Spark is unable to resolve the availability column.

```
org.apache.spark.sql.AnalysisException: cannot resolve '`availability`'
given input columns: [stores.capacity, stores.closes, stores.name, store_
occupants.occupants, stores.opens, store_occupants.storename
```

The reason that Spark cannot resolve this column is because it doesn't exist until after the expression is first resolved. There are two approaches to solving this dilemma:

- Use an inner query
- Change the expression column to a selection expression

Using an Inner Query

An inner query allows us to resolve the column named availability, by returning the column through projection for use within an outer query. This just requires us to create a wrapping select statement where we apply our conditional selection logic.

Use an Inner Query to Fully Resolved Columns, Including Expressions, in an outer Query

```
%sql
select name, availability from (
  select name, (capacity-occupants) as availability
  from stores
  join store_occupants
  on stores.`name` == store_occupants.`storename`
) where availability > 4
```

The inner query pattern allows us to use the resulting projected columns within the scope of the outer query. This is necessary if we want to return the availability column, but if we only care about the name of the store with availability, then we can greatly simplify our query by making a slight adjustment to the query syntax.

Using Conditional Select Expressions

There is a much better way to solve this problem. We just need to change how we express the availability filter, from a known (resolved) column to a select expression. This also happens to remove the need for an inner query as well.

Using a Conditional select Expression to Filter the Results

```
%sql
select name from stores
join store_occupants
on stores.`name` == store_occupants.`storename`
where (capacity-occupants) > 4
```

The optimized query itself is tighter and easier to understand after moving the position of the expression and removing the inner query. For completeness, lct's also look at how to solve this same problem using pure DataFrame transformations (see Listing 4-8).

Listing 4-8. Finding Seating Availability at the Coffee Shops Using Pure DataFrame Transformations

```
%spark
import spark.implicits._

val partySize = 4

val hasSeats = df
  .join(occupancy, df("name") === occupancy("storename"))
  .withColumn("availability", $"capacity".minus($"occupants"))
  .where($"availability" >= partySize)
  .select("name")

hasSeats.show
```

The DataFrame transformation shown in Listing 4-8 joins the occupancy data with the stores DataFrame (df), and then introduces the withColumn function on the DataFrame. This operation is functionally equivalent to the process you looked at before when using the inner query to fully resolve the availability column. In this case, we are just being more declarative. The result being a series of functional transformations that read almost exactly like the three steps outlined earlier.

Exercise 4-3: Summary

This exercise showed you how to create dynamic (derived) columns based on SQL expressions, and how to use nested (inner) queries with conditionally query derived columns from an outer query. This enabled us to return a listing of coffee stores with available seating for an arbitrarily sized group of people.

Summary

This chapter introduced you to the concepts of data projection and selection. We began with the basics and honed our skills along the way through practical hands-on examples. You learned to apply conditional filters, and how to reduce the number of columns in your DataFrame. You learned a variety of tricks to create DataFrames, from case classes to inline tables, and even sequences of tuples. Together we learned to join data across a wide variety of join styles, from inner joins to unions and even the semi-joins and anti-joins. We finished up by pulling together concepts from all across the chapter in order to answer a basic reservations query: "Which stores currently have availability for a party of four people." These skills are all transferable to the day job of any data engineer and you'll find yourself using most of these techniques on a daily basis.

The next chapter introduces you to working with external databases using JDBC. You'll be setting up your local container environment to work with MySQL 8, and the efforts from Chapter 5 will continue into Chapter 6, as you are introduced to using the Hive Metastore along with the Spark SQL Catalog.

CHAPTER 5

Bridging Spark SQL with JDBC

In the last chapter, we looked at common patterns and techniques for harnessing the powerful core functionality available to us when transforming data using Spark SQL and the DataFrame APIs. While we certainly covered a lot of ground, we purposefully skipped over some of the more exciting capabilities available to us under the Spark SQL umbrella. Along that line, wouldn't it seem to only make sense that we should be capable of connecting to and working directly with remote databases from the comfort of Apache Spark SQL? Additionally, wouldn't it also be advantageous to use SQL's strongly typed semantics when reading data into Spark? Couldn't we somehow also marry these rich type systems (inherent to Java/Scala) with both SQL and the strong internal typing mechanics of Apache Spark itself? Luckily, that is exactly what you will learn to do in this chapter.

You will learn to seamlessly read and write data between Spark and any JDBC-compatible RDBMS database (such as MySQL, PostgreSQL, Microsoft SQL Server, Azure SQL Database, Oracle, and others). You'll learn to natively load and transform data from external database rows into Spark DataFrames and then write back to the source-of-truth database as well.

Overview

This book is all about working with data using Apache Spark. In order to extend our reach and cover as much ground as strictly necessary, we will sometimes find ourselves working outside of Apache Spark in order to get a 360-degree view of how a process works. In this chapter, you learn how to spin up, bootstrap, and work with MySQL on Docker, in order to understand the full end-to-end bridge between Spark and MySQL (or really any JDBC compatible database). This Docker-based database will be reused again

© Scott Haines 2022

S. Haines, *Modern Data Engineering with Apache Spark*, https://doi.org/10.1007/978-1-4842-7452-1_5

in order to power your Hive Metastore in the next chapter. For some of you, this might be an initial crash course into database administration, but I promise it won't be painful.

It is my hope that through these exercises, you'll gain a clear understanding of exactly how Spark connects to and powers applications backed by traditional OLTP databases and even non-traditional cloud-based distributed databases using JDBC as a bridge to these relational database mapping services (RDBMS).

This chapter is broken into three fundamental sections and exercises:

- MySQL on Docker crash course

- Connecting the RDBMS world with Spark SQL using JDBC

- Continued exercises

So, without further ado, let's spin up our development environments and get our hands dirty!

MySQL on Docker Crash Course

You'll begin the chapter by familiarizing yourself with MySQL on Docker.

Starting Up the Docker Environment

Exercise Content   You can find the materials for this chapter under ch-05/ of the book's GitHub https://github.com/newfront/spark-moderndataengineering. Just start the environment using the run script and head on over to http://localhost:8080.

```
cd ch-05/docker && ./run.sh start
```

After starting the Docker environment with the run.sh start command, you'll notice the output has a reference to mysql now. The output is shown for clarity in Example 5-1.

Example 5-1. *The Output After Starting the Docker Runtime Environment for the Chapter Exercises Now Includes a Reference to MySQL*

```
ch-05/docker/run.sh start

SPARK_HOME is set. location=/Users/me/install/spark-3.0.1-bin-hadoop3.2
1c9d09c28d3d   mde          bridge     local
docker network already exists mde
Creating mysql ... done
Creating zeppelin ... done
Zeppelin will be running on http://127.0.0.1:8080
```

This additional service was added with a small change to the chapter's `docker-compose-all.yaml` file. Let's look at the Docker compose service configuration to see how MySQL was added.

Docker MySQL Config

The configuration contained in `ch-05/docker/docker-compose-all.yaml` now includes a section for `mysql` under the `services` area, which is shown in Listing 5-1.

Listing 5-1. The mysql Service Definition from the Docker Compose Configuration

```
services:
  mysql:
    image: mysql:8.0.23
    container_name: mysql
    command: --default-authentication-plugin=mysql_native_password
    restart: always
    volumes:
      - ${PWD}/data/mysqldir:/var/lib/mysql
    environment:
      - MYSQL_DATABASE=default
      - MYSQL_USER=dataeng
      - MYSQL_PASSWORD=dataengineering_user
      - MYSQL_ROOT_PASSWORD=dataengineering
```

```
healthcheck:
  interval: 5s
  retries: 10
ports:
  - 3306:3306
networks:
  - mde
hostname: mysql
```

This new configuration allows us to run MySQL alongside Apache Zeppelin within our local Docker environment. The configuration overrides the default behavior of the MySQL container on startup, thereby achieving the following:

- Creates a database named `default` on startup

- Generates the `dataeng` `MYSQL_USER`

- Provides a password for the `dataeng` user `MYSQL_PASSWORD`

- Grants all permissions to the `dataeng` user

- Declares the container port 3306 (the default MySQL port)

This config also adds the `mysql` hostname to the shared Docker network (mde). This allows our Zeppelin service to access the MySQL database from within the Notebook environment (across container boundaries).

Tip   Having the MySQL container running locally can come in handy if you want to bootstrap, or rebuild, an environment consisting of tables and data from another environment, such as a staging or production one (redacted data or sanitized if it is coming from production). You can simply import the SQL definitions for the tables and any data necessary to add some rows to each table.

Optionally, you can also add bootstrap SQL files to `CREATE` tables and views and to `INSERT` data to a prime local test environment. These useful scripts can be checked into GitHub or your favorite version control system, assuming the data doesn't contain any personally identifiable information (PII). By doing so, you allow other engineers or analysts to import the data definitions (DDL) and example data to use when they're getting started with an initial dataset.

Exercise 5-1: Exploring MySQL 8 on Docker

This first exercise walks you through some administrative tasks in your new MySQL container. You'll begin by creating a table and populating some rows using plain old SQL from the MySQL shell directly in the MySQL Docker container.

Working with Tables

To create and populate a new table inside the MySQL Docker container, you follow these four steps:

1. Connect to the Docker container.

2. Authenticate to the MySQL shell as the database user `dataeng`.

3. Create a new table in the MySQL database named `customers`.

4. Insert some initial rows into the `customers` table.

Our first objective is to connect to the `mysql` Docker container, and then authenticate yourself onto the `mysql` command line, so you can create and prime the `customers` table.

Note   There is a note titled "ConnectingToMySQL" in the Apache Zeppelin instance that you started just a moment ago. You can use the note as a hands-on guide to walk you through the entire process of connecting to Docker, authenticating to the MySQL shell, creating the `customers` table, inserting data into the `customers` table, and doing some simple `selects` to ensure the rows are correct.

Connecting to the MySQL Docker Container

Open a fresh terminal session. From this session, you can directly connect to your `mysql` container using the following `docker exec` command.

```
docker exec -it mysql bash
```

Your terminal prompt will change to `root@mysql:/#` when you are logged in to the container. Next, you need to connect to the actual MySQL database through the `mysql` shell.

Using the MySQL Shell

From within the mysql container, you'll use the mysql shell to authenticate your user and connect to the database.

```
mysql -u dataeng -p
```

Tip The password for the dataeng user is in the ch-05/docker/docker-compose-all.yaml and is also shown in Listing 5-1. Just look for the servi ces:mysql:environment:MYSQL_PASSWORD, and you'll see the password is dataengineering_user.

Once you're connected, you will see some output telling you about the MySQL version, followed by a mysql> command prompt. This means you are ready to rock and roll!

The Default Database

There are two databases created by the container startup process. There is the information_schema database and the default database. You can switch to the default database since that is where you will be building your table definitions. The use command is how you switch database contexts.

```
mysql> use default;
```

Now you are ready to create your table definitions.

Creating the Customers Table

The customers table is a simple table storing customer information. This table represents a basic registered customer within CoffeeCo, and is shown in Listing 5-2.

Listing 5-2. The Create Table Syntax for the customers Table

```
CREATE TABLE IF NOT EXISTS customers (
  id VARCHAR(32),
  created TIMESTAMP DEFAULT CURRENT_TIMESTAMP,
  updated TIMESTAMP DEFAULT CURRENT_TIMESTAMP,
  first_name VARCHAR(100),
```

```
  last_name VARCHAR(100),
  email VARCHAR(255)
);
```

Now the only thing left to do is insert some customer records. We'll revisit this table schema again toward the end of the chapter.

Inserting Customer Records

Now that we have defined our `customers` table, all that's left to do is add a few records so we'll have something to work with. Type the SQL command in Listing 5-3 and press Enter to execute the command.

Listing 5-3. Adding Sample CoffeeCo Customers Using the INSERT Syntax

```
INSERT INTO customers (id, first_name, last_name, email)
VALUES
("1", "Scott", "Haines", "scott@coffeeco.com"),
("2", "John", "Hamm", "john.hamm@acme.com"),
("3", "Milo", "Haines", "mhaines@coffeeco.com");
```

Let's do a quick sanity check to make sure things look alright before we exit the MySQL shell and move over to the Zeppelin.

Viewing the Customers Table

The customer records we just created can be easily viewed using a simple `select` query. For simplicity's sake, we are only projecting the `id`, `first_name`, and `email` columns. Go ahead and follow along from the shell to see the rows stored in your new table.

```
mysql> select id, first_name, email from customers;
+------+------------+----------------------+
| id   | first_name | email                |
+------+------------+----------------------+
| 1    | Scott      | scott@coffeeco.com   |
| 2    | John       | john.hamm@acme.com   |
| 3    | Milo       | mhaines@coffeeco.com |
+------+------------+----------------------+
3 rows in set (0.00 sec)
```

Now that you've confirmed the table exists and has some rows, you can exit the mysql shell.

Tip   Selecting a specific database context isn't strictly necessary, as you can always prefix your queries with the database selector. If you don't want to always run the use {database}; command, you can still query the customers table by adding the database prefix.

```
select * from default.customers;
```

This pattern lets you query from multiple database sources.

Exercise 5-1: Summary

So far, this exercise has taught you how simple it is to spin up, connect, and work with MySQL using Docker. This process enables you to compose consistent environments locally so there are no gotchas after you deploy. This means you can run sanity checks, test, or try new things, all without the need to spin up and maintain a permanent environment.

During the next exercise, you'll learn to connect to MySQL, fetch table (schema) definitions, read, query, and write data using Apache Spark and Java Database Connectivity (JDBC).

Using RDBMS with Spark SQL and JDBC

Connecting to a Relational Database Management System (RDBMS) from Spark requires two things, *a JDBC compatible driver* and the *database properties*. For the sake of simplicity, I may refer to RDBMS as databases interchangeably.

The JDBC SQL driver (JAR) can be added directly to your application (compiled) or dropped into the Spark JARs directory (included as part of a custom distribution, requires administrative support), or fetched on-demand. The database properties, including the hostname (mysql), RDBMS listening port (3306), database name (default), and credentials (user/pw). Together Apache Spark can make it easy to work with data stored in external databases.

The process of managing dependencies in your application can be handled at compile time and even updated or overridden at runtime. For example, the following are three common ways to handle dependencies with Spark:

1. Provide the necessary JARs as a direct dependency in your application using maven or sbt to be compiled alongside your application JAR (this can be referred to as building a *fat* JAR).

2. Download the required packages at runtime using the built-in runtime dependency support baked into Spark. This means using the `--packages` option when submitting an application or spinning up the `spark-shell`.

3. Reference a local or remote location where the application can locate additional dependencies using the `--jars` option when submitting an application or spinning up the `spark-shell`.

In the following section, we look at managing dependencies and providing additional configuration to the `spark-shell`. You can use these deploy patterns to augment the behavior of your application at runtime. Along the way, you'll learn to handle dependencies, augment configurations, and importantly how these techniques can be used to create *composite command-line applications*.

Managing Dependencies

We haven't built or compiled a Spark application locally at this point in our journey. Instead, we've relied on the `spark-shell` and the Apache Zeppelin notebook environment to dynamically compile and run these partial applications on our behalf. In later chapters, you'll be using the Scala build tool (`sbt`) to build, package, and manage your Spark applications dependencies, but rather than taking a long segue now, we'll look at other options built into Spark in an exercise that connects to your local MySQL from the `spark-shell`.

Note   MySQL should be running on localhost on port 3306. You can test if port 3306 is live using the net connection (nc) command from the terminal.

```
> nc -z localhost 3306
Connection to localhost port 3306 [tcp/mysql] succeeded!
```

Providing additional JARs to your Spark applications at runtime can be beneficial. You will learn how this capability enables you to augment the behavior of an application by mixing in additional dependencies. While runtime modification of your applications can lead to inconsistent behavior, it is a strategy that is often useful for testing and debugging, as you don't need recompile and re-release your application to change some behaviors. You'll learn that this process is worth a deeper dive, so we will be doing just that in this next section.

Exercise 5-2: Config-Driven Development with the Spark Shell and JDBC

Let's say for example that we've been asked to write a generic application that relies in principle on user, or environment, provided configuration, to connect to a remote database, in a reliable way, for the purpose of doing on-demand ad hoc database queries. In the process of doing our research, we figure out that Apache Spark already includes this basic functionality and can connect to any JDBC compatible database, as long as we include the appropriate dependencies.

We have come up with the following plan of attack:

1. Define our configuration parameter names. This will create a common config interface that can be shared in the documentation.

2. Provide common, supported database drivers (connectors) for use in the application. These can be pulled in automatically without requiring other engineers to worry about additional dependency management. However, if they wish to bring their own connectors as well, that can be done using the `--packages` option when starting the application.

3. Prepare and safeguard the application by handling any other necessary steps that will help others simply use the tool without needing to become an expert in Spark.

Tip   I am a firm believer in the "baby steps" method of learning. As we learn to walk, often we will trip and fall, but we can learn from our mistakes along the way. The same can be said when building reliable modules or shared libraries. If you provide guard rails that don't create roadblocks, but instead provide a smooth experience that can be adapted, if necessary, by the users of the library, all parties will be happy campers. For example, you can add sensible defaults to your code for the benefit of others.

We'll look at building this behavior into a simple `spark-shell` flow. We will identify the configuration needed, write a simple Scala file (shim), and look at how to use local and remote dependency management for our MySQL connectors. The two MySQL connectors we will be using have been added to the exercise material and are conveniently located in the `jars` directory in the chapter exercise materials at `/ch-05/docker/spark/jars/`.

Configuration, Dependency Management, and Runtime File Interpretation in the Spark Shell

To begin, let's take a closer look the `spark-shell` and learn about the available options for modifying the runtime behavior of this dynamic Apache Spark application. Given we are using the `spark-shell` as the starting point of this exercise, you can take advantage of the following parameters when starting up this Spark application, including those to augment the runtime configuration, remote and local dependency management, and dynamic class compilation and interpretation.

Runtime Configuration

Using the runtime property `--conf "key=value"` on the Spark Shell pushes additional config into the immutable SparkConf for use within the application at runtime. The value of any configuration added using the runtime conf properties will override any value of the key being provided. So be careful and document available runtime overrides.

Local Dependency Management

Using `--jars "<file.jar>,<file.jar>"` adds additional JARs into the application classpath on startup. The expectation is that the JARs are located on the local file system, or available to fetch from s3a or HDFS.

Runtime Package Management

Using the `--packages "<groupId:artifactId:version>,"` property provides Spark with the ivy parameters required to download one or more dependencies at runtime. You can use the runtime package manager to fetch libraries on-the-fly. Remember that not all packages will be available when you want them, so use this trick when reliability isn't your top priority, since it can lead to broken promises in production.

Dynamic Class Compilation and Loading

Using the `-i <file>` parameter interprets and loads any Scala files into the context of the session before returning control of the `spark-shell` to you. This option exists directly on the Scala REPL, and therefore is also part of the `spark-shell` since the Spark REPL extends the default Scala REPL.

Now, given you are looking to connect to MySQL from within Spark, you'll need to create some additional config options, based on the Spark JDBC connector properties, to be passed into the shell program at runtime.

The Spark JDBC (`DataFrameReader`) configuration options are shown in Table 5-1.

Table 5-1. *Spark JDBC DataFrameReader Configuration Table*

Config	Description	Example
url	The JDBC connection URL. This string is comprised of the *JDBC* protocol, host or server name, port, and database.	`jdbc:mysql:// 127.0.0.1:3306/default`
dbtable	The optional table to load.	`customers`
query	The optional query.	`select * from customers`
driver	The `ClassName` of the JDBC driver.	`com.mysql.cj.jdbc.Driver`
user	The database user accessing the db.	`dataeng`
password	The password for the user accessing the db.	`*********(secret)`
numPartitions	The optional maximum number of partitions to use to split up a series of conditional queries over your database table. This defaults to a single connection [TODO – check the true default].	`4`
partitionColumn	The column to use to bucket a query over the `numPartitions` value. This is required if you're using `numPartitions`. The column type must be `Integer`, `Timestamp`, or `Date`.	`created`
lowerBound	The lower bound to use when splitting a query. This creates a less than (<) condition on your behalf. This is required if you're using `numPartitions`.	`2021-02-16`
upperBound	The upper bound to use when splitting a query. This creates a less than (>) condition on your behalf. This is required if you're using `numPartitions`.	`2021-02-21`

Armed with the knowledge of what drives (literally!) the behavior of the *JDBC* connection, let's take some time to brainstorm some config names. We came up with the basic configuration for this application, as shown in Table 5-2.

Table 5-2. *Basic Application Runtime Configuration Options*

Config Name	Default Value
spark.jdbc.driver.class	com.mysql.cj.jdbc.Driver
spark.jdbc.host	localhost
spark.jdbc.port	3306
spark.jdbc.default.db	default
spark.jdbc.table	customers
spark.jdbc.user	dataeng
spark.jdbc.password	None. Passwords should never exist in your code!

Okay. Things are starting to feel more formalized now. We know how to configure our connections and have created fallbacks (defaults) to make our application easier to use. Including specific application config options, and fallbacks that can act as guardrails (or training wheels), making working with your library code enjoyable.

The startup script is shown in Listing 5-4.

Tip   The following code snippets expect that your local working directory is the root of the Docker directory within the chapter's exercises folder. `/ch-05/docker/`

Listing 5-4. Starting the Spark Shell with Locally Provided JARs, Additional Runtime Configuration, and Scala Files to Interpret and Load

```
$SPARK_HOME/bin/spark-shell \
  --conf "spark.jdbc.driver.class=com.mysql.cj.jdbc.Driver" \
  --conf "spark.jdbc.host=127.0.0.1" \
  --conf "spark.jdbc.port=3306" \
  --conf "spark.jdbc.default.db=default" \
  --conf "spark.jdbc.table=customers" \
  --conf "spark.jdbc.user=dataeng" \
  --conf "spark.jdbc.password=dataengineering_user" \
  --packages=mysql:mysql-connector-java:8.0.23 \
```

```
--jars=spark/jars/mariadb-java-client-2.7.2.jar \
-i examples/connect-jdbc.scala
```

This command starts the `spark-shell` from Listing 5-4 and does a lot of additional work behind the scenes, in a consistent order, enabling you to augment the runtime behavior of your `spark-shell` driven Spark application. These same capabilities are also available to you when you use the `spark-submit` command (which you will see more of in the following chapters).

Figure 5-1 shows the order of events when starting the `spark-shell`.

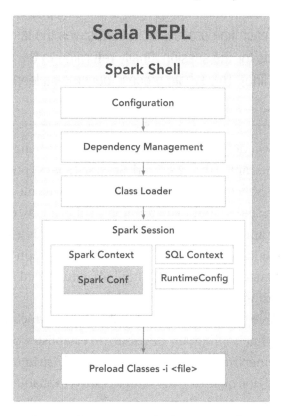

Figure 5-1. *The series of steps that execute before control is returned to you within an instance of the Spark Shell*

The Spark Shell extends the Scala REPL, enabling you to use Apache Spark specific parameters and mix in Scala REPL specific commands as well. By the time the Spark Shell is instantiated, all runtime configuration declared using the `--conf` option has been added to the `SparkConf` and `SparkContext`, all dependencies have been downloaded and resolved using the `--packages` option, and the JARs and relative class paths are loaded, including any additional local JARs that need to be added to the application classpath as defined using the `--jars` option. Lastly, after the `spark-shell` is initialized, as a final step, any additional Scala files included using the `-i <file>` option are loaded due to the behavior of the Scala REPL.

Note   In case you were curious, REPL stands for **R**ead, **E**val, **P**rint, **L**oop. In the case of the Scala REPL, this little console program reads each line of text (as a standard in), evaluates line by line (dynamically compiling and storing class information, variables, etc.), and then as long as no exceptions are thrown, prints the output (if any) for each line processed. It then saves the last line (or lines) to the stack of commands for the session, finally returning to the first read state (waiting for your next line). This whole process becomes a loop from R -> E -> P and back to R.

Armed with the knowledge from Figure 5-1 regarding the behavior of the `spark-shell`, we can reliably guarantee that the Spark Session (`spark`) provided by starting up the `spark-shell` can reliably be referenced by any of the Scala files (included using `-i`) within the context of the `spark-shell` environment. This order of events is what enables you to create composite applications from the `spark-shell`. It is a similar process to how Spark code is dynamically interpreted within each paragraph in Zeppelin. Each paragraph is read, evaluated (dynamically compiled), and loaded into the context of the Spark application context without you having to lift a finger!

Let's look at the source code from `connect-jdbc.scala`, which is shown in Listing 5-5. It includes what is minimally necessary to establish a connection to our MySQL database (container), and in short order, enables us to iteratively query the database and test our ad hoc queries. This source code is available unabridged inside the `examples` directory in `ch-05/docker/examples/connect-jdbc.scala`, including configuration fallbacks to default config values where applicable.

Listing 5-5. The Source Code of connect-jdbc.scala

```
// spark: SparkSession (from spark-shell)

// assign from spark conf
val jdbcDriver = spark.conf.get("spark.jdbc.driver.class")
val dbHost     = spark.conf.get("spark.jdbc.host")
```

```
val dbPort    = spark.conf.get("spark.jdbc.port")
val defaultDb = spark.conf.get("spark.jdbc.default.db")
val dbTable   = spark.conf.get("spark.jdbc.table")
val dbUser    = spark.conf.get("spark.jdbc.user")
val dbPass    = spark.conf.get("spark.jdbc.password")

val connectionUrl = s"jdbc:mysql://$dbHost:$dbPort/$defaultDb"

// load the remote SQL table
val df = spark
.read
.format("jdbc")
.options(Map[String, String](
    "url" -> connectionUrl,
    "driver" -> jdbcDriver,
    "dbtable" -> dbTable,
    "user" -> dbUser,
    "password" -> dbPass
  )
)
.load()
```

This partial application shown in Listing 5-4 relies directly on Spark's core configuration, and the Spark Session created when starting the spark-shell. This is all conveniently accessible to your code at runtime because of the order in which it is loaded *after* the spark-shell has completed all the work necessary to bootstrap the application (see Figure 5-1).

If you haven't tried to run the code yet, do so now, and then we will dive deeper into how we were able to read the configuration from within the scope of this partial application from Listing 5-5.

Spark Config: Access Patterns and Runtime Mutation

The Spark configuration is divided into two separate classes at runtime. The first is the main immutable configuration, the SparkConf. This read-only object is accessible from the SparkContext, which can be accessed using the SparkSession. This core application config is assembled from various locations when the application starts up:

1. Default or internal configuration provided by the standard Spark distribution (either built by you, or downloaded as an official build).

2. Spark Environment level overrides defined in $SPARK_HOME/conf. An example is available at /path/to/ch-05/docker/spark/conf/ spark-defaults.conf.

3. Application Submission overrides using the --conf <key=value> option on application start.

4. Application Internal configs. These configs are specified directly when constructing the Spark Session. Certain configs like the spark.sql.warehouse.location.dir and others can only be set once for the lifetime of an app.

The resulting SparkConf is immutable, and it is used to construct the driver application's SparkContext.

The second configuration is the runtime configuration stored in the RuntimeConfig object. This is a mostly mutable copy of the fully immutable config available in SparkSession. Figure 5-1 is a good reference for visualizing the duality of the Spark config types.

Viewing the SparkConf

Given the SparkConf is immutable, we can only read the values within the object. You can dump the active configuration using the spark-shell that you started in Listing 5-4. From the spark-shell, you can use the active SparkSession (spark), to get to the SparkContext, and finally the SparkConf itself, which you can use to print the configuration as a debug string. This process is shown in Listing 5-6.

Listing 5-6. Accessing and Printing the Immutable SparkConf Using the spark-shell

```
scala> spark.sparkContext.getConf.toDebugString
spark.app.id=local-1613941330294
spark.app.name=Spark shell
```

```
spark.driver.host=192.168.1.18
spark.driver.port=52829
spark.executor.id=driver
spark.home=/Users/me/install/spark-3.0.1-bin-hadoop3.2
spark.jars=file:///Users/me/git/newfront/spark-moderndataengineering/ch-05/
docker/jars/mariadb-java-client-2.7.2.jar,file:///Users/me/.ivy2/jars/
mysql_mysql-connector-java-8.0.23.jar,file:///Users/me/.ivy2/jars/com.
google.protobuf_protobuf-java-3.11.4.jar
spark.jdbc.default.db=default
spark.jdbc.driver.class=com.mysql.cj.jdbc.Driver
spark.jdbc.host=127.0.0.1
```
*spark.jdbc.password=*********(redacted)*
```
spark.jdbc.port=3306
spark.jdbc.table=customers
spark.jdbc.user=dataeng
spark.master=local[*]
spark.repl.class.outputDir=/private/var/folder...
```

You may have noticed that the configuration automatically redacted the value of `spark.jdbc.password`. This is one of the guardrails set in place by the Spark engine, and the value of the `redaction` regex can be extended or replaced using the property named `spark.redaction.regex`. By default, Spark automatically redacts any configuration properties matching `secret|password|token`. If you have other patterns or naming conventions for secrets, you can just override the `spark.redaction.regex`. To look at more of the specifics you can look Runtime Environment settings in the official Spark docs (`http://spark.apache.org/docs/latest/configuration.html#runtime-environment`).

Accessing the Runtime Configuration

You can use the SparkSession object (the `spark` variable in the `spark-shell`) to get (read) and set (write) values of the RuntimeConfig. Try this for yourself, from within your active `spark-shell`, by using Listing 5-7 to read the value of `spark.jdbc.driver.class`.

Listing 5-7. Get the Runtime Config Value with Fallback

```
spark.conf.get("spark.jdbc.driver.class", "com.mysql.cj.jdbc.Driver")
```

This access pattern (with fallback where applicable) is used throughout the partial application `connect-jdbc.scala`. We read and assign the JDBC config values into convenience variables, and then construct the JDBC `DataFrameReader`.

You can update most values using the runtime configuration by calling set on the SparkSession. Using the active `spark-shell`, try setting the `spark.jdbc.driver.class` value using the code from Listing 5-8.

Listing 5-8. Dynamically Setting Values of the RuntimeConfig

```
spark.conf.set("spark.jdbc.driver.class", "org.mariadb.jdbc.Driver")
```

By getting and setting values, you can update the behavior of your application at specific junctures or, to say that a little differently, within *specific stages* of execution.

Note   We'll see more of the `RuntimeConfig` along our journey to the book's end, but it is worth pointing out that each SparkSession can modify a subset of the immutable SparkConf without changing the default behavior of the Spark application. This is because the SparkContext retains the default configuration (as defined on initial application start), thus ensuring that there is a separation of concerns between one or more instances of the SparkSession.

Moving back to the running `spark-shell` (launched in Listing 5-4), let's look at the process of iterative development with the Spark Shell.

Iterative Development with the Spark Shell

From within your active `spark-shell`, try calling `printSchema` or executing the `show()` action on the DataFrame identified as `df`. You can even try using a selection and limit clause, which is shown in Listing 5-9.

Listing 5-9. Interacting with a JDBC Backed DataFrame Abstracts Away the Complexities of Connecting to a Remote (External) RDBMS

```
df
  .select("id", "first_name", "email")
  .limit(3)
  .show()
```

The results of calling show on the JDBC backed DataFrame yields the first three entries in the customers table from your MySQL database.

```
+---+----------+--------------------+
| id|first_name|               email|
+---+----------+--------------------+
|  1|     Scott|   scott@coffeeco.com|
|  2|      John|  john.hamm@acme.com|
|  3|      Milo|mhaines@coffeeco.com|
+---+----------+--------------------+
```

Now let's learn to describe views and tables.

Describing Views and Tables

You have learned to use JDBC to connect to your MySQL docker container. Wouldn't it make sense that we could describe the schema of the table by doing a simple SQL style describe? In theory, Spark should be referencing the backing relation—in this case our customers table so how about you try this out. Follow along using the code fragment in Listing 5-10.

Listing 5-10. From the spark-shell, Creating the Customers View and Describing the Table

```
df.createOrReplaceTempView("customers")
spark.sql("desc customers").show()
```

You will see the following output after calling `show` on the `desc customers sql` command.

```
+----------+---------+-------+
| col_name|data_type|comment|
+----------+---------+-------+
|        id|   string|   null|
|   created|timestamp|   null|
|   updated|timestamp|   null|
|first_name|   string|   null|
| last_name|   string|   null|
|     email|   string|   null|
+----------+---------+-------+
```

If this process seems a bit convoluted, that is because it is. If we already have access to a database, then why would we need to first know about the existence of the customers table beforehand in order to do a describe? How can a process like this even scale? How would one discover the location or schema of this table without access to tribal team-based knowledge?

We'll simplify the process of data discovery in the next chapter using the Hive Metastore, but for now remember what this feels like. It should feel like an unnecessary roadblock in your path to data discovery. But let's first finish this exercise by using our JDBC connection information to write back to our `customers` MySQL table.

Writing DataFrames to External MySQL Tables

To complete the full roundtrip from MySQL to Spark and back again, we need to go ahead and write back to the database. You are probably thinking, we need to create a few more records in order to do that, and yes, we definitely have to increase the number of records in our table.

Generate Some New Customers

You have become good at generating data inside of Zeppelin and the Spark Shell and even directly using the MySQL shell. While we could go ahead and continue to use the MySQL shell to add some more customers, we wouldn't have the joy of doing so directly from Spark.

We can reuse the schema of the current customer table, which we described just a moment ago. For reference, you can run `df.printSchema` or `df.schema.toDDL` to get a quick reference of the data structure. You will see the following schema.

```
root
 |-- id: string (nullable = true)
 |-- created: timestamp (nullable = true)
 |-- updated: timestamp (nullable = true)
 |-- first_name: string (nullable = true)
 |-- last_name: string (nullable = true)
 |-- email: string (nullable = true)
```

Now here comes the first roadblock. In order to add more rows to the database, you'll need to create a DataFrame matching the schema of the `customers` database table. In this case, that means we need to come up with a way of generating the good ol' *timestamp*.

If you remember back in Chapter 3, we discussed schemas and the `StructType`. We also went over the Spark data types. As a refresher, the internal Scala type used for a timestamp is a `java.sql.Timestamp`. If each row of data contains the appropriate Scala type, then Spark can do an implicit conversion to the internal `TimestampType` for us, and we can go on living our lives.

The code snippet shown in Listing 5-11 shows the code necessary to create our new customers. (The `ts` and `time` functions are provided to simplify the customer creation.) We give Spark exactly what it needs to be able to construct and write more customers!

Tip   Remember that you can paste multiple lines at a time into the `spark-shell` using the `:paste` command. When you are done pasting in your code, pressing Cmd+D will run everything you pasted in place.

Listing 5-11. Creating a DataFrame to Represent Additional Customers

```scala
import org.apache.spark.sql._
import java.sql.Timestamp
import java.time._

// spark: SparkSession (from spark-shell)
assert(spark.isInstanceOf[SparkSession])

def ts(timeStr: String) = Timestamp.valueOf(timeStr)
```

```
def time = Timestamp.from(Instant.now())
// create some new customers
val records = Seq(
  Row("4",ts("2021-02-21 21:00:00"),time,"Penny","Haines","penny@coffeeco.com"),
  Row("5",ts("2021-02-21 22:00:00"),time,"Cloud","Fast","cloud.fast@acme.com"),
  Row("6",ts("2021-02-21 23:00:00"),time,"Marshal","Haines","paws@coffeeco.com")
)

val newCustomers = spark.createDataFrame(
    spark.sparkContext.parallelize(records),
    customers.schema
)
```

The code from Listing 5-11 creates a simple helper function called ts that takes a simple DateTime string, for example 2021-02-21 21:00:00, and converts it into an instance of a Timestamp class, which Spark can implicitly convert to its internal TimestampType. In addition to this function, there is a simple function that can be used to get the current time, called time.

Next, we generate three new customers and create a new DataFrame by wrapping an RDD of our customer records, along with the schema that we lift from the temporary view customers, from Listing 5-10.

For those of you keeping track, this is now the fourth method we've encountered for creating DataFrames!

All that is left is to write our new records to the external customers table.

Using JDBC DataFrameWriter

We can reuse most of the variables defined by connect-jdbc.scala, which are available to us through the spark-shell since we initially loaded the connect-jdbc.scala into the same context. Now let's learn to write back to the customers table using the code fragment from Listing 5-12.

Listing 5-12. Creating an Instance of the JDBC DataFrameWriter and Attempting to Save the Rows Defined by the DataFrame newCustomers to the Customers MySQL Table

```
newCustomers
  .write
  .format("jdbc")
```

```
.options(Map[String, String](
  "url" -> connectionUrl,
  "driver" -> jdbcDriver,
  "dbtable" -> dbTable,
  "user" -> dbUser,
  "password" -> dbPass
)
)
.save()
```

If you have been following along with the exercise since the beginning of the chapter, you will have just seen an error pop up.

```
Table or view 'customers' already exists.
SaveMode: ErrorIfExists.
```

This error is thrown because Spark is defensive by default (another guardrail). When we go to save our three new columns, we could have actually overwritten the database table or added duplicates of the same rows. Let's thank our lucky stars since we were just saved by SaveMode.

SaveMode

Table 5-3 details the modes dictated by the SaveMode enum.

Table 5-3. *SaveMode Options on the DataFrameWriter*

SaveMode	Description
Append	If the data or table already exists, then the contents of the DataFrame are expected to be appended to the existing data.
ErrorIfExists	If the data already exists, an exception is thrown.
Ignore	If the data already exists, the save operation is expected to not save the contents of the DataFrame or change the existing data.
Overwrite	If the data or table already exists, the existing data is expected to be overwritten by the contents of the DataFrame.

Modify the code from Listing 5-12, and manually specify the *save mode* and try to save again. The change is shown in Listing 5-13. The full example is located in the Zeppelin notebook ("5-2_MySQLAndSpark"). Just pop over to `http://localhost:8080` to load the examples.

Listing 5-13. Appending New Rows to the Customers Table

```
newCustomers
  .write
  .format("jdbc")
  .mode("append")
  .options(Map[String, String](
    "url" -> connectionUrl,
    "driver" -> jdbcDriver,
    "dbtable" -> dbTable,
    "user" -> dbUser,
    "password" -> dbPass
  )
)
.save()
```

This time you will not see any exceptions thrown. You can even query the temporary customers view created earlier (in Listing 5-10) using the following query.

```
spark.sql("select * from customers").show
```

Magically, your new rows show up without having to refresh anything. Why is that? This is because the temporary `customers` view is just a promise to eventually go and get the data via our JDBC connection. In fact, if you connect back to the MySQL Docker instance, `docker exec -it mysql bash`, and log back into the `mysql` shell, you can query the `default.customers` table and you'll see your new rows happily sitting inside the `customers` table in our default database.

Note   The source code for writing to MySQL via JDBC is available in the chapter exercise materials under /ch-05/docker/examples/save-jdbc.scala. You can use the Scala REPL :load function if the process of using :paste has gotten a little unruly. Just :load the file. The end result will be similar to using the -i <file>, which you learned about in Figure 5-1.

You have now officially completed the full roundtrip from MySQL to Spark and back again! Pat yourself on the back and take a step back to consider all you've just accomplished. Considering we initially set out to create a command-line tool that could enable users to run ad hoc SQL queries and have done almost everything other than that, we should probably finish that up now before moving on.

Using all you've learned so far in this exercise, you can extend the previous connect-jdbc.scala example to allow the partial application to take a SQL query via the command line. Use the application launch command from Listing 5-14 to guide you on creating your command line SQL application. The code for this final exercise is located in the examples directory at /path/to/ch-05/docker/examples/jdbc-adhoc-query.scala.

Listing 5-14. The Command -ine Driven Shell Script for Querying a Table

```
$SPARK_HOME/bin/spark-shell \
  --conf "spark.jdbc.driver.class=com.mysql.cj.jdbc.Driver" \
  --conf "spark.jdbc.host=127.0.0.1" \
  --conf "spark.jdbc.port=3306" \
  --conf "spark.jdbc.default.db=default" \
  --conf "spark.sql.query=select id, first_name, email from customers where
  email LIKE '%coffeeco%' ORDER BY email ASC LIMIT 3" \
  --conf "spark.jdbc.user=dataeng" \
  --conf "spark.jdbc.password=dataengineering_user" \
  --jars=spark/jars/mariadb-java-client-2.7.2.jar,spark/jars/mysql-
  connector-java-8.0.23.jar \
  -i examples/jdbc-adhoc-query.scala
```

The application configuration properties from Listing 5-14 replace the spark.jdbc.table config property, introduced by connect-jdbc.scala, with spark.sql.query. This change takes into account the way the JDBC `DataFrameReader` works in Spark. You can pass in one of the following two connection properties—`dbtable` or `query`—but you can't pass in both. For the full list of parameters, return to Table 5-1.

Try creating the working Spark partial application to run the `spark-shell` based application from Listing 5-14. Keep in mind that in the real world, you would want to limit the database user actions to read-only and depending on the table size you will also probably want to add default limits to the query. You wouldn't want someone to trigger a query like `DROP DATABASE `default``.

Tip   Any application started with the `spark-shell` command will wait (or hang) until you manually exit the application. This is due to the nature of the loop in the REPL. If you want to exit the `spark-shell` automatically, as a last step in your application, you can use `sys.exit(0)` to exit the program normally. If there is a problem executing, say the query was malformed, then the `spark-shell` will throw an exception and exit. For command-line applications it is standard to use a zero for success and a non-zero value for non-successful operations. Think of this like HTTP status codes.

Exercise 5-2: Summary

This exercise covered a lot of ground, introducing you not only to using JDBC with Apache Spark, but also to the power of iterative development with composite files and the `spark-shell`. While using the composite application pattern on the `spark-shell` is not intended to power production applications, it does enables you, as an engineer, to work through a problem as a series of logical steps, with nearly immediate feedback from within the `spark-shell`, or other interactive development environments like you've seen with Apache Zeppelin. This process of iteratively decomposing a problem into its logical parts is critical for data engineers, software engineers, or really anyone working on any complex problem solving.

In its purest form, we literally separated the concerns between logical Docker container edges in Exercises 5-1 and 5-2. You learned to create tables directly at the database (MySQL Docker container), and to manually insert rows to create a base MySQL table for reading and writing from Apache Spark. You learned to use the `spark-shell` to your advantage, adding dynamic runtime dependencies and custom configuration. You can say you even started down the path of writing reusable code, even if that code was broken up across a few separate files within a composite application powered by the `spark-shell`.

Continued Explorations

Note   This section is optional and covers some more advanced processes. You can skip this section and come back later too.

The chapter exercises all assumed we had to work with the poorly defined `customers` database table schema. *Yes it wasn't very good!* We can do better. Consider the fact that wc can currently add almost unlimited duplicates to our database. That really isn't a good design, especially when MySQL supports constraints. I offer up these continued explorations as additional food for thought, and to go into a some more advanced areas, including table truncation and column-level deduplication techniques.

Note   The note titled "5_2_MySQLAndSpark" can be run from the Zeppelin UI. This contains additional steps and extended explanations for reading and writing using the JDBC connector (`http://localhost:8080/#/notebook/2G1SGWZBW`).

Good Schemas Lead to Better Designs

What if we had the option of creating a better schema for our `customers` table? Could it have made our lives easier? The `create table` syntax for the `bettercustomers` is shown in Listing 5-15.

Listing 5-15. The Modified Create Table Syntax for the bettercustomers Table

```
CREATE TABLE IF NOT EXISTS `bettercustomers` (
  `id` mediumint NOT NULL AUTO_INCREMENT COMMENT 'customer automatic id',
  `created` timestamp DEFAULT CURRENT_TIMESTAMP COMMENT 'customer
  join date',
```

```
`updated` timestamp DEFAULT CURRENT_TIMESTAMP COMMENT 'last record
update',
`first_name` varchar(100) NOT NULL,
`last_name` varchar(100) NOT NULL,
`email` varchar(255) NOT NULL UNIQUE,
PRIMARY KEY (`id`)
);
```

Right off the bat, you might be thinking that having a UNIQUE constraint on email might be a good idea. No? Maybe you are thinking of the fact that we now have automatically "monotonically" increasing row numbers and that's a step in the right direction? Maybe it is the fact that we have column-level comments in our table definition? All these answers are in fact correct. With a well-defined table definition, we can rely on the database to handle pieces of our core business logic, like storing the next customer ID and safeguarding against duplicate data. This can free up more time to focus on solving other problems in our applications.

Using the Zeppelin note ("5_2_MySQLAndSpark"), you can walk through the following example. Starting with fixing the customers table. This includes how to create and write initial rows into the bettercustomers table using only the non-default columns of the table schema and more. This means that only the first_name, last_name, and email columns are necessary.

Note   You'll have to create the bettercustomers table using the MySQL console for these additional exercises. Follow the directions in the notebook, or use what you've learned in the chapter, to create the new table using the table definition from Listing 5-15.

Write Customer Records with Minimal Schema

The paragraph shown in Listing 5-16 shows a simplified way of adding customers to your bettercustomers table. The code flow is based on Listing 5-11 and removes the need to manually specify timestamps. This technique is more advanced and explicitly ignores specific columns, which could always change, causing problems in the future. You can run this code using Apache Zeppelin as per the section instructions.

Listing 5-16. Using Database Table Constraints and Default Values Can Enable You to Write Sparse Rows to a Remote Database Directly from Spark, Cutting Down on Bandwidth Transfer Costs

```
%spark
import org.apache.spark.sql.types._
import org.apache.spark.sql._

val simpleFieldNames = Set("first_name", "last_name", "email")
val simpleSchema = StructType(
    betterCustomers.schema.fields.filter(sf => simpleFieldNames.
    contains(sf.name)))

val saveMode = "append"

// create simplified Rows!
val rows = spark.createDataFrame(
  spark.sparkContext.parallelize(
    Seq(
      Row("Nanna","Haines","nhugs@coffeeco.com"),
      Row("The","Rock","djohnson@coffeeco.com"))),
  simpleSchema)

rows
.write
.format("jdbc")
.mode(saveMode)
.options(Map[String, String](
  "url" -> connectionUrl,
  "driver" -> jdbcDriver,
  "dbtable" -> "bettercustomers",
  "user" -> dbUser,
  "password" -> dbPass
))
.save()
```

Now isn't this exciting! What is equally exciting is how problematic this query can become. There is nothing like a good old thorn in the side to hamper the hard work and effort of what feels like a job well done.

Here is the honest truth—the query you just ran *is a disaster waiting to happen*. Well, why is that you ask? The problem stems from the fact that the SaveMode value is sitting around in a variable. One of the things we love to do as engineers is to simplify problems and many times that means writing core components in a more config driven, or library-based approach (very much like how the Spark community wrote the JDBC connector). With the simple use case from Listing 5-16, this means someone could naively change the value of *SaveMode* as overwrite, which in the case of the JDBC DataFrameWriter will drop the current table and rewrite it using the contents of the writing DataFrame. This would be a disastrous result given the time and effort we just spent defining and creating the table definition.

Furthermore, the new table would be replaced instead with the three columns (first_name, last_name, and email) from the rows DataFrame. Essentially, we set out to make our lives easier, in this case inserting new rows using our minimal schema, but we've potentially just enabled a critical flaw and disaster waiting to happen for any critical production database table. Yikes! What is worse is that only the two records from Listing 5-16 would exist in our database afterwards.

The good news is that if your intention really is to overwrite the bettercustomers table, you can keep the existing table definition, by adding the truncate option on the JDBC DataFrameWriter. This option maintains the original table syntax from Listing 5-15. The updated JDBC configuration is shown in Listing 5-17.

Listing 5-17. Using Truncate to Preserve the Table Definition

```
rows
.write
.format("jdbc")
.mode(saveMode)
.options(Map[String, String](
  "url" -> connectionUrl,
  "driver" -> jdbcDriver,
  "dbtable" -> "bettercustomers",
  "user" -> dbUser,
  "password" -> dbPass,
  "truncate" -> "true"
))
.save()
```

Now that is better. In the next exploration, you'll learn how to deal with the problems introduced by the poorly defined `customers` schema. You can rely on *truncation* to solve the next problem as well. However, there is an interesting catch. So, read on and good luck!

Deduplicate, Reorder, and Truncate Your Table

For this continued exercise, you will need to come up with a way to deduplicate, reorder, and truncate the records stored within the `customers` database table. This exercise teaches you how to deal with fixed and problematic schemas and data in general. For this use case, you can't simply modify the table schema for the `customers` database table. See there is a slight problem with the *save + append* operation when applied to the original `customers` table schema in Listing 5-2. Given *there are no constraints declared* on the table schema, Spark will happily continue to write duplicate rows again and again, if you simply command it to. There is no pre-baked method for Spark to ignore duplicate rows when writing into the database itself using just the JDBC connector and zero RDBMS based constraints.

As you saw in the *"Good Schemas Lead to Better Designs"* exploration, duplicate record rejection can be handled with constraints. However, the simple `customers` table can have unlimited duplicates. To get an idea of just how simple it is to create many duplicates, try running `:load examples/save-jdbc.scala` a few dozen times in your active `spark-shell` you started, in Listing 5-4, to create a ton of duplicates.

Now to solve for this problem. You can use the following techniques to help you along your way.

Drop Duplicates

Using the `dropDuplicates` function, can conditionally remove duplicate rows from within your DataFrame. This will take only the first row identified by its unique column value and ignore additional duplicates. For example: `df.dropDuplicates("id")`.

In the case where you want to have more control over which of the duplicates you select, you can mix a grouping clause and the distinct min/max operators to for example, select only the row with the earliest creation date, and the latest update date. This technique of preprocessing with aggregate functions can be helpful when reading and processing Change Data Capture (CDC) logs. You can look at the Zeppelin paragraph titled Select Distinct and Sort for an interactive example.

Sorting with Order By

Using the ORDER BY clause to sort data using the temporary customers view, or using the sort function directly on your DataFrame, enables you to reorder the records in a dataset. For example: df.sort(asc("id")).

Truncating SQL Tables

Truncating a table clears all committed records, which means all rows that have been written to the database's write-ahead log (WAL) and acknowledged by the RDBMS. It is like calling DELETE FROM customers instead of DROP TABLE customers which can be better than dropping the physical table if you must. You were introduced to the truncate table JDBC option in Listing 5-17.

Remember that dropping the full table will can break other applications that rely on the table.

Caution   Use table truncation with caution, since when you are doing a full table *overwrite,* you may just find yourself deleting everything in the table by mistake.

Stash and Replace

Think about how you could store a current snapshot of the table before doing the actual overwrite if you need to overwrite an entire table, for deduplication or to fix corrupt entries. Apache Spark enables you to store a persisted snapshot of a dataset using the saveAsTable function on the DataFrame. This method uses the Hive Metastore and distributed file system to store a reliable (persistent) backup. The persistent table can then be used to save your distinctly sorted table before truncating the physical table itself and replace it with the stashed (persistent) records. You'll learn how to use the Hive Metastore in the next chapter.

Note   The full solution to the problem is in /ch-05/docker/examples/ deduplicate-reorder-truncate.scala. I recommend trying things out and exploring possible solutions before looking at the answers. Happy trails!

Summary

This chapter taught you how to complete a full data roundtrip from MySQL to Spark and back again, just leaning on native JDBC drivers, the JDBC `DataFrameReader`, and `DataFrameWriter`. If you compare the process you learned for writing composite applications using the `spark-shell` to what you've experienced with Apache Zeppelin, then it's probably clear that composites (partial applications) provide yet another avenue for rapid development. One that enables you to again reuse the same Docker-based development environment to choose your own data engineering adventures!

In the next chapter, you'll learn about using the Apache Hive Metastore. You'll be setting up, configuring, and running the Metastore backed entirely by our Docker-based MySQL database. You will see how the Metastore simplifies the process of data discovery and enables data exploration within the context of your Spark applications. You will see how the Hive Metastore introduces new ways of working with database tables, and how you can even protect your source-of-truth data stores and provide documentation and shared table definitions between multiple Spark applications.

Are you ready?

CHAPTER 6

Data Discovery and the Spark SQL Catalog

Being able to connect and work with the data systems and services that are included in most companies' modern tech stack is a critical skill for data engineers. Lucky for you, Spark provides the mechanisms to work with and transform data so you can take action and solve problems, instead of writing and maintaining yet another piece of custom infrastructure code. By relying on the core capabilities of Spark, you learned to harness the power of JDBC to interoperate with data stored in a traditional database. Accessing any JDBC-compatible RDBMS enables you do write your SQL queries once, which means you're not burdened with supporting separate applications with different business logic.

This chapter continues the lessons learned in Chapter 5. This will be our first look at using data lakes and relational databases together to populate the contents of the *Spark SQL Catalog*. This process will enable you to annotate your data sources with rich metadata about the tables that your Spark applications produce. This contextual information details the input and output formats for your data pipelines and can also be extended to auto-generate the *data lineage* graph or can be used as a general data discoverability service internally and even as an external *data catalog*. This is not just invaluable to you and your team, but also to any other member of the data organization looking to reuse or analyze the data your Spark jobs emit.

We conclude this chapter by looking at how Spark optimizes your transformative flows, by extending the earlier exercises using Spark's strongly typed datasets. This final exercise introduces you to how Spark encodes the intentions of your application code and magically generates and optimizes the logical phases of execution necessary to run applications from start to finish, all through the internal Catalyst Optimizer.

153

© Scott Haines 2022
S. Haines, *Modern Data Engineering with Apache Spark*, https://doi.org/10.1007/978-1-4842-7452-1_6

This chapter is broken into the following areas:

- The Hive Metastore and Spark SQL Catalog

- The Catalyst Optimizer

- Spark datasets

Exercise Content   You can find the materials for this chapter in the `ch-06/` folder of the book's GitHub at `https://github.com/newfront/spark-moderndataengineering`. Just start up the environment using the run script and head on over to `http://localhost:8080`.

`cd ch-06/docker && ./run.sh start`

Data Discovery and Data Catalogs

In the previous chapter, we explored how easy it can be to work with JDBC connections to your database (RDBMS) from within Apache Spark. While we were able to connect to the database we spun up for the exercise, there was one main issue that could prove to be an issue down the road. That hindrance is namely this. Without prior knowledge of the location of a database, and knowledge of what kind of data is represented in the physical tables of that database, as well as how that data is represented by its underlying table schema, the simple act of data discovery is all but nonexistent. But data discovery is actually the key to success for data-driven companies, and as a data engineer, this should really matter to you.

Why Data Catalogs Matter

Data discovery may not seem like an immediate concern for us given that the data we are currently working with, within the context of the book, represents only a small number of customers in a fictitious company's initial customer database, as well as a small number of coffee brands and some basic metadata (the roast levels as expressed by their boldness), a small number of stores, and some information about the foot traffic to each store during the day. However, in the real world, data discovery problems are a much larger concern. Let me paint that picture.

Data Wishful Thinking

As any company grows, small teams naturally expand as new teammates are hired on. As teams continue to grow, it is only natural at a certain point to fork off and divide into smaller teams again, with more specific scopes and areas of focus. New teams, made up of prior members of original teams, continue this cycle of growth and division as the company continues to grow organically. For a while, teams can rely on tribal knowledge, or under documented processes learned over time at the company, and cross-team communication enables teams to continue to share tribal knowledge, data locations, table schemas, and raw data formats directly.

For some reason many startups, and even some very mature companies, have fallen for the idea that in the cloud age, teams are able to be more agile, move fast, and even break things, to keep up with the blinding pace of innovation. In doing so, it is easy for engineers and PMs to skip, or turn a blind eye to, the ever-important step of thorough documentation and software testing. The unspoken side-effect of this is that it is easy to watch tech debt pile up, becoming skeletons in the closet. A necessary risk for the continued forward progress of the company! Unfortunately, history tends to repeat itself and this high technical debt can lead to processes and patterns that quickly become unsustainable.

As companies grow, acquire new companies, turn over engineers, reorganize departments and as other cross-team/cross-initiative data platforms emerge, the old ways of relying on tribal knowledge, or expecting teams to wait for a core team member to respond to a Slack message, or join a video conference, or even physically show up, to walk a new team through an archaic process becomes detrimental to the well-being of the company and the people working for it. Clear documentation and established ways of working cross-functionally is of even higher importance when dealing with data.

For data teams, there is a large cost and hinderance to progress, when relying on manual communication of data schema changes, or the age-old game of telephone (you know Bob told Alice who told me that Team X is making a change to their data format). Unreliable processes quickly crack under pressure, and as a side-effect, reduce what used to be core, reliable data into unreliable, untrustworthy data. Under communicated changes to the schema of one database table, or the raw format of records within the data lake, will cause production outages, as teams may not even know who relies on their data or who is subject to breaking upstream changes. Outages mean people get mad and the blame game begins. This is a tipping point, a fork in the road, and one of the reasons that data platforms and centralized processes and other "ways of doing things" emerge as companies get to a certain size.

Data Catalogs to the Rescue

Luckily, there are standard ways of doing things that exist outside of the internal context of a company, and like most design patterns, people follow them because they can save time and potential heartache of getting burned when things come crashing down. The picture I painted before was a worst-case scenario of what can happen when the trust in core data layers within the company break down. The fact of the matter is that data documentation is as important as any API documentation. These metadata-based data definitions represent and act as a centralized contract for how, when, and why to consume and use the data stored in a table, and one of the most widely used, standard ways, of achieving this is using the Apache Hive Metastore.

The Apache Hive Metastore

The *Hive Metastore* was originally constructed to provide important metadata that could be used to simplify how various teams worked with the distributed data that lived inside their data lakes. These pseudo-table definitions mapped key missing information, including the schema, partitions, encoders, and decoders needed to interoperate with the data stored in these distributed SQL tables.

This metadata was used at runtime to provide the glue needed for Apache Hive to run distributed SQL-like queries over the data stored in HDFS. Rewinding back to Chapter 1, we discussed the emergence of Distributed File Systems, data pipelines, and ETL. Once upon a time, Apache Hive was a godsend to the data engineers who were writing and maintaining ETL jobs using MapReduce.

Big Data is really just a technique for storing, working with, and querying massively partitioned and fully distributed data that doesn't fit in the traditional SQL database. Hive provided a mechanism to treat these distributed file systems in the same way that people were accustomed to treating traditional databases, as tables with a schema. This enabled data engineers to focus on the pipelines that ingested data into HDFS, the representative data structures and schemas of their data as tables, and the data lineage information all rolled up into the table definition of each distributed Hive table.

This table metadata exists in the Hive Metastore and provides the keys to describe the who, what, where, and why of each Hive table. In addition, the metastore enables teams with access to the metastore to discover new tables and avenues to reuse the data available in this distributed data warehouse. All this without having to rely on a lot of cross-team communication, lengthy meetings, and sharing tribal knowledge to begin to be productive.

Metadata with a Modern Twist

Today the Hive Metastore is still alive and well, but just serving a different purpose than originally intended. It is now typical to run the metastore without ever intending to use Hive itself. This change can be thought of as an evolution that coincides with other changes across the new modern data tech stack. The rise in popularity of cloud-based distributed object stores, such as Amazon S3 and Azure Blob Storage, have seen the industry shift focus away from Hadoop and HDFS.

A similar migration has also taken place with companies moving away from many of the older Hadoop technologies like Hive itself. In Chapter 2, you learned how Spark was born from the best of what made MapReduce successful. While changing the core execution model to reuse computations in-memory and allow for multiple iterations over the same data without the need to explicitly go back to the source-of-truth data, it was just a natural technological evolution.

Analogous to the concept of recycling, the ecosystem simply reused components of MapReduce, but with a twist. The same can be seen with respect to Spark SQL and Hive. Spark SQL is more than capable of operating as a fully-fledged SQL engine, and as you will soon experience first-hand, it can also take advantage of the Hive Metastore. While it can interoperate with Hive, it can also do just fine on its own.

Exercise 6-1: Enhancing Spark SQL with the Hive Metastore

The Hive Metastore provides a single source of truth for describing the location, data encoding (e.g., Parquet, ORC), columnar schema, and statistics of the tables recorded for the purpose of democratizing data for use by data engineers, analysts, and machine learning engineers and data scientists alike. Using the Metastore, folk can find datasets without the need for asking around or going on wild goose chases!

The next section takes you through the five simple steps required to get the Hive Metastore up and running. Like the previous chapter's exercises, you can run everything using the chapter source material to follow along live.

Bootstrap Option   There is a bootstrap option available for the `run.sh`. If you simply want to bootstrap this environment and not worry about all the twists and turns, I promise I will not be offended. The Apache Zeppelin note titled "6_1_HiveMetaStore" starts with *Bootstrap the Environment,* which will walk you through the commands and setup. You can skip ahead to the Spark SQL Catalog and come back to this section for more information regarding initializing the database tables, user permissions, and Spark configuration.

Configuring the Hive Metastore

There are a few steps required to configure the Hive Metastore, so let's begin by getting things to work in the context of Spark. Fortunately, these steps are fairly straightforward:

1.   Create the `metastore` database.

2.   Grant access to the database to your MySQL user.

3.   Create the database tables.

4.   Configure the Spark environment to access the metastore.

5.   Provide the Hive dependencies for the Spark environment.

This shouldn't feel like too many steps. In fact, Spark is capable of running the Hive Metastore without the need to run any additional services (or containers). The only real requirement is that you have the database and tables initialized and have granted access to the database to your Spark `dataeng` user.

Create the Metastore Database

To simplify the process of initializing the Hive Metastore tables after you finish creating the actual metastore database, the table definition SQL files have been provided inside of the chapter's `exercise` directory. The `docker file copy` command has been included in the run.sh. The command-line parameter hiveInit will copy the necessary files into the running mysql container on your behalf. Simply execute the command in Listing 6-1.

Listing 6-1. Start the Chapter Exercises' Environment and Initialize the Apache Hive Metastore

```
cd /path/to/ch-06/docker &&
  ./run.sh start &&
  ./run.sh hiveInit
```

The hiveInit command will run the following two Docker copy commands.

```
docker cp hive/install/hive-schema-2.3.0.mysql.sql mysql:/
docker cp hive/install/hive-txn-schema-2.3.0.mysql.sql mysql:/
```

Connect to the MySQL Docker Container

Use the following docker exec command to log in to the container context.

```
docker exec -it mysql bash
```

Authenticate as the root the MySQL User

After logging in to your local MySQL container, authenticate as the MySQL root user using the mysql shell.

```
mysql -u root -p
```

The MYSQL_ROOT_PASSWORD value can be found in ch-06/docker/docker-compose-all.yaml.

Create the Hive Metastore Database

Once you are inside the mysql shell, go ahead and create the metastore database.

```
mysql> CREATE DATABASE `metastore`;
```

Now that the database has been created, it is time to grant access to your Spark mysql user.

Grant Access to the Metastore

Grant the dataeng user access to the metastore database.

```
REVOKE ALL PRIVILEGES, GRANT OPTION FROM 'dataeng'@'%';
GRANT ALL PRIVILEGES ON `default`.* TO 'dataeng'@'%';
GRANT ALL PRIVILEGES ON `metastore`.* TO 'dataeng'@'%';
FLUSH PRIVILEGES;
```

Tip   It is common to create use case-specific users to govern access to your Hive Metastore database. Users should be granted either read/write or read-only access to the metastore, depending on the access patterns of the application. Many applications only consume data and copy the table data into different locations outside the context of the Hive Metastore. Remember that data access is a privilege!

You are done with our root user session for now. You should exit this session and log back in as the dataeng MySQL user.

```
mysql> exit
```

Create the Metastore Tables

We are almost done setting up the metastore. What we have to do next is create the underlying Hive Metastore tables themselves. There is a one-time setup cost for configuring the metastore database tables. Luckily, we can just import the table definitions into the mysql console.

Authenticate as the dataeng User

From the MySQL container context, change your authenticated MySQL user to dataeng.

```
mysql -u dataeng -p
```

Switch Databases to the Metastore

```
mysql> use metastore;
```

160

Import the Hive Metastore Tables

The SOURCE command can be used to easily read and apply the SQL table configuration commands for our version of Hive (v2.3.0). This file is available inside the mysql Docker container because you used the hiveInit command earlier in the setup process.

```
mysql> SOURCE /hive-schema-2.3.0.mysql.sql;
```

That's it. You are now the happy owner of a brand-new installation of the Hive Metastore. If you want to see which tables were initialized during this process, you can run the following commands from your current MySQL session.

```
use metastore;
show tables;
```

Running the show tables command should yield a laundry list of table names (more than 50). Feel free to look through the tables by using the DESCRIBE command on any of the tables for more context, or simply exit the console. For now, your metastore is ready for action. Next up, we will be adding the configuration needed for Spark to access the Hive Metastore.

Configuring Spark to Use the Hive Metastore

The Hive Metastore is traditionally configured through hive-site.xml. Spark will automatically look for this configuration file inside of the conf directory, located under the $SPARK_HOME directory. This file, and others, have been preconfigured for you and are available in the exercise files located within the ch-06/docker/spark/conf/ directory.

If you don't want to look at the configurations now, you can skip over this section and move to the Spark SQL Catalog instead.

Configure the Hive Site XML

The hive-site.xml file consists of a mere four properties. They provide Spark with the default connection properties to the Hive Metastore database that will be used in your application. These properties are shown in Table 6-1.

Table 6-1. *The Hive Metastore Configuration Options*

Property	Description
`javax.jdo.option.ConnectionURL`	Defines the JDBC connection URL to your metastore.
`javax.jdo.option.ConnectionDriverName`	Defines the JDBC driver `ClassName` to use to connect to the metastore.
`javax.jdo.option.ConnectionUserName`	Defines the metastore database username.
`javax.jdo.option.ConnectionPassword`	Defines the password of the metastore user.

There are many more configuration options available in the official Hive documentation, including different access patterns to the shared Hive Metastore. These include running the fully standalone Hive Metastore server (v3.x and above) as well as the *Hive Thriftserver,* and how to provide multiple thrift URLs for domain sharding, which helps to reduce congestion from too many connections to any single Hive Metastore instance. Since the metastore is stateless, meaning there is no physical state stored on the service itself, you can simply create multiple fronting services for the sake of load balancing the connections to the thrift server, and then onto the underlying metastore database.

Configure Apache Spark to Connect to Your External Hive Metastore

You just learned that Apache Spark uses the `hive-site.xml` file to connect to the Hive Metastore. What we didn't cover at the beginning of the exercise was that all required configurations are provided for you in the exercises `conf` directory. This directory is then mounted into the Docker environment, using the `docker-compose-all.yaml` file.

This means you don't need to copy any files to your $SPARK_HOME location. Open the Docker compose file `/ch-06/docker/docker-compose-all.yaml` and peek at `services:zeppelin:volumes` for more details.

From within the context of the `ch-06/docker` directory, the following three directories are volume mounted:

- /spark/conf – This directory includes spark-defaults.conf and hive-site.xml.

- /spark/jars – This directory includes all the MySQL JDBC driver JARs that we want to include locally when we spin up Spark.

- /spark/sql – This directory is a mounted so we can use it as the root directory for what is known as the Spark SQL Warehouse.

The spark-defaults.conf file is an optional file that can be used to override the default Spark configuration values for *all Spark jobs* that are run within the context of that Spark installation. For our current use case, we only care about the seven properties detailed in Table 6-2.

Table 6-2. *The Hive Metastore Configuration Properties*

Spark Property Name	Description	Example
spark.sql.warehouse.dir	The default location to read and write distributed SQL tables. This location can be located on the local file system and on any HDFS compatible file system.	hdfs:// metastores/ hive/ file:///spark/ warehouse/ s3a://your-bucket/hive/
spark.sql. catalogImplementation	Defines the backing SQL catalog for the Spark session.	hive
spark.sql.hive. metastore.version	Defines the Hive version for the metastore.	2.3.7
spark.sql.hive. metastore.jars	Defines the location of the Hive JARs to load into the Spark classpath. Spark will default to using the built-in JARs.	builtin, maven, or /path/to/ the/jars
spark.sql.hive. metastore.sharedPrefixs	The comma-separated class name prefix's to search for and load a specific version of the Hive Metastore.	com.mysql. cj.jdbc,com. mysql.jdbc,org. postgresql

(continued)

Table 6-2. (*continued*)

Spark Property Name	Description	Example
spark.sql.hive.metastore.schema.verification	If this is set to true, the Spark metastore version must match the value of VERSION in the Hive Metastore.	true
spark.sql.hive.metastore.schema.verification.record.version	If set to true, Spark will enforce record-level schema verification against what is published in the Hive Metastore.	true

The configurations described in Table 6-2 have been preconfigured in your `spark-defaults.conf` file from the chapter's exercises. For the sake of simplicity, we are using the standard built-in Hive JARs (v2.3.7 at the time of writing).

Using the Hive Metastore for Schema Enforcement

It is worth pointing out that the Hive Metastore can be used as a reliable arbiter for any schema disputes between applications writing into a distributed table. This means that bad data can't pollute your data lake. This is one of the many reasons for using the Hive Metastore. This option is configured using the `spark.sql.hive.metastore.schema.verification.record.version` setting (described in Table 6-2).

Note   You can upgrade the Hive Metastore version to use alongside Spark using the configuration from Table 6-2. Specifically, using the `spark.sql.hive.metastore.jars` and the `spark.sql.hive.metastore.version` properties. You can download the required Hive Metastore JARs from the official Apache Hive website to run an alternative version of the Hive Metastore.

SPARK DEFAULT CONFIGURATION

The default values, defined in the `spark-defaults.conf` file, enable us to `define-once` the specific cluster-wide settings we would like to automatically apply to any Spark application that is started in the context of the defaults. In our case, this means any application started in our Zeppelin environment. The default values can be explicitly overridden using application-level overrides, or by providing explicit configuration `--conf` values on application startup.

Using Spark Defaults to Protect the Cluster

You can think about the default configuration values as *important common settings* that are provided to your Spark applications for the benefit of the application runtime and to protect any other application running in the cluster.

For instance, if you are managing or running your own Spark clusters, you would most likely need to balance the cost of operations, and it is typical to deploy a fixed cluster size that considers the needs of the application(s) that will run on a daily or hourly basis. Additionally, you would also need to consider whether these are batch or streaming applications, and if the applications are IO bound or CPU bound. Then each application that is launched into your cluster can take a fixed or dynamic number of resources (min/max share) from the total cluster of resources. If all applications respect the boundaries of the other application resources, this strategy can work out just fine.

Providing Default Limits

Given each application can request a non-fixed number of resources from the shared cluster resources, cluster management can be difficult, to say the least. Especially when you expect each of the jobs in your data pipeline to run. It then becomes very important to be hyper-defensive when it comes to setting predefined limitations on the total number of resources an application can consume, since not every application will need the same number of compute cores or RAM to be able to run successfully. This is where the defaults come into play. The magic here is that app owners actually need to understand how some configurations work to override them, and in my experience, many applications can run just fine on a default small number of cores and RAM.

For example, if you consider a common t-shirt sizing approach (s,m,l,xl), a small application may consist of the following:

```
spark.cores.max 8
spark.executor.cores 1
spark.executor.memory 1g
```

This configuration would pin an application to eight compute cores and eight executors running with 1GB JVM size.

By making it easier for other application owners to claim a portion of the shared cluster, or in the case of a more dynamic environment like Kubernetes, there are a lot of configuration options that can be simply abstracted away into the defaults. Chapter 14 covers deploying applications using Spark Standalone and Chapter 15 covers deploying applications using Kubernetes.

Production Hive Metastore Considerations

The benefit of using an external Hive Metastore means that your table definitions will persist between Spark SQL sessions. If there is no `hive-site.xml` provided for the SparkSession, a local metastore will be created using Derby. The Derby-based metastore is not intended for production use cases and can only be used with a single SparkContext, which you should interpret as "one location."

It is worth also pointing out that the Hive Metastore isn't tied directly to MySQL. Most common flavors of RDBMS (MS SQL Server, MariaDB, Oracle, and Postgres) are also supported. For more information about configuring and using additional external RDBMS, you can view the Admin Manual online.

Lastly, for running production quality, highly available metastores, it is essential to protect yourself from any single points of failure, including the backing database itself. Running a SQL/JDBC-compatible, managed, cloud-based database, such as Amazon Aurora DB or Google's Cloud Spanner, provides good options for reducing potential failures incurred when operating databases without a seasoned DBA. Any fully managed RDBMS that has monitoring, backups, and automatic failover is recommended over running things in-house, if that isn't something you are equipped for.

Exercise 6-1: Summary

This initial exercise walked you through the process of bootstrapping the Hive Metastore to create an external, reliable, Hive Metastore. Now that you have the Hive Metastore bootstrapped, the `hive-site.xml` file configured, and the `spark-defaults.conf` file prepared, it is time to focus on the Spark SQL Catalog and learn how to use the external Hive Metastore to power a more reliable SQL experience.

The Spark SQL Catalog

The *Spark SQL Catalog* is Spark's internal SQL metadata manager. The functionality provided by the Spark SQL Catalog provide resources for working with databases, temporary or managed tables, views, functions, and more. You just learned about the Hive Metastore; the Spark SQL Catalog provides simple abstractions that enable you to interoperate with the Hive Metastore easily. As mentioned earlier, it is essential to the success of your data platform and organizational goals to ensure that the data you produce is self-documented in a way that makes it easier for other teams to fetch it on demand, without the need to schedule another meeting in order to understand what the column *wU35* or l2W mean (of course, these mean women under 35 years old and the last two weeks). Clearly code words don't make things simple for data teams. If each table definition is backed by a schema, and each column stores not just the data type, but also a brief comment, then other engineers working with your tables can use the Spark SQL Catalog to learn about your data in a self-service way.

Exercise 6-2: Using the Spark SQL Catalog

The Spark SQL Catalog is available directly off the SparkSession and offers many useful methods to introspect the available data residing inside of your data warehouse and data lakes.

> **Note**   Follow along with this exercise using the Zeppelin note titled "6_2_ SparkSQLCatalog." Just like in prior chapters, create a new paragraph within the note and use the `%spark` code blocks to build up the working note.

Creating the Spark Session

Let's learn to create the SparkSession from scratch. Using the code fragment in Listing 6-2, create a new paragraph in your Zeppelin note and run it to generate the SparkSession.

Listing 6-2. Create a New SparkSession with Apache Hive Support

```
%spark
import org.apache.spark.sql._
val sparkSessionHive = SparkSession.builder
  .master(spark.conf.get("spark.master"))
  .config(spark.sparkContext.getConf)
  .appName("hive-support")
  .enableHiveSupport()
  .getOrCreate()
```

The SparkSession we just created can be used to further explore the Spark SQL APIs. We will use the SparkSession now through a fun process of discovery. Given we have data in a database (our local MySQL database), the first area of exploration will be listing and viewing what is available.

Spark SQL Databases

Just starting out, we know we want to be able to look for and find what databases are available to use. This process can take advantage of our old friend SQL as we gain our footing and understand the SQL catalog capabilities.

Listing Available Databases

```
%spark
sparkSessionHive
  .sql("show databases;")
  .show
```

This method allows us to view the available databases. In our case, there is only the default database. You may have noticed a glaring piece of missing information. Can you spot it? Maybe ask yourself what default really means? Does this mean our MySQL database that we named default from the last chapter, or something different?

168

To find out more, you have to descend into the Spark SQL Catalog, as shown in Listing 6-3.

Listing 6-3. Using the Spark SQL Catalog to List Databases

```
%spark
sparkSessionHive
  .catalog
  .listDatabases
  .show(truncate=false)
```

The result of running the listDatabases block from Listing 6-3 is shown in Table 6-3.

Table 6-3. *The Default Spark SQL Database, aka the Warehouse*

name	description	locationUri
default	Default Hive database	file:/spark/sql/warehouse

This makes sense. When we configured the Hive Metastore earlier in the chapter, we only provided the JDBC information to connect to the Metastore, not to our actual MySQL database, which just so happened is also called default.

Finding the Current Database

The current database is equivalent to using the SQL use command. This just tells you which database is currently active within the Spark Session.

```
%spark
sparkSessionHive
  .catalog
  .currentDatabase
```

We've looked at listing and viewing the database stored within our local data lake. What happens if we want to create a new database in order to organize the tables underneath, say, a team name or business unit?

Using a database as a means to categorize the tables owned by an individual team is a nice way to efficiently view ownership across the metastore. There are many reasons that this pattern is important, but it mainly comes down to common file system best practices and design patterns.

Within the Hive Metastore, a database is a path prefix associated with a physical directory within the distributed "warehouse." The database construct is a means of restricting access to a collection of tables and identifying the ownership of a set of tables, or of a table itself, and it doesn't denote a separate physical database.

If all teams share the common default database location for their tables, then given enough time, the opportunity for a naming collision in the tables is high. What's worse, you could find your table being dropped accidently due to user error and naming collisions. Therefore, it is important to separate teams, or business units, by database name. Additionally, database prefixing can also help mitigate issues that arise due to read/write capacity problems.

Whether you are running on top of HDFS or using a cloud object store like S3 as the distributed file system for your data lake, scaling out to support higher demands with respect to read and write capabilities ultimately means file systems should be composable using path prefixes. With HDFS, you can set up a NameNode (think about this as the coordinator node, as it understands how the distributed file system routing works) to use federated routing, to route inbound requests to separate HDFS clusters based on the path prefix for the file system. You can utilize this technique to enabling smart scaling for large and small workloads (if you are running your own Hadoop cluster). With S3 and other cloud object stores, it is common to use base64 encoded prefixes, which enable dynamic sharding (or partitioning) within the S3 file system, where each path prefix can be looked up and resolved using the metastore or a resolution database.

For now, assume that the performance aspects of the metastore have been taken care of. You will learn how to create a database, which will map itself as a prefix-bound directory in the distributed file system.

Creating a Database

Creating a database in the Metastore can be done by following the example in Listing 6-4.

Listing 6-4. Using Spark SQL to Create a Database in the Hive Metastore

```
%spark
val dbName = "coffee_co_common"
val dbDescription = "This database stores common information regarding
inventory, stores, and customers"
val defaultWarehouse = spark.catalog.getDatabase("default").locationUri
val warehousePrefix = s"$defaultWarehouse/common"
spark.sql(s"""
CREATE DATABASE IF NOT EXISTS $dbName
COMMENT '$dbDescription'
LOCATION '$warehousePrefix'
WITH DBPROPERTIES(TEAM='core',LEAD='scott',TEAM_SLACK='#help_coffee_
common');
""")
```

The LOCATION property in the SQL query from Listing 6-4 ties the database to the directory, or path prefix, in the distributed file system that is being used to power the SQL table. The beauty of being able to define team- or organization-based databases using the Spark SQL Catalog and the Hive Metastore means that you can provide all the tools necessary for a team to track ownership of a specific database and the underlying tables. This simple step is useful and enables teams to act as responsible data stewards.

Loading External Tables Using JDBC

Now that you've seen how to create a new database, let's look at loading the external database table bettercustomers into our new coffee_co_common database. The first step is to provide Spark with the connection information to authenticate to our MySQL database. Then we can read and then write this table into our data lake. Use the code from Listing 6-5 to connect to the bettercustomers table (created in the continued explorations from Chapter 5).

Listing 6-5. Reading the External Bettercustomers Table Using JDBC to Create a Temporary View in Spark

```
%spark
val jdbcDriver = spark.conf.get("org.mariadb.jdbc.Driver")
val dbHost     = spark.conf.get("mysql")
val dbPort     = spark.conf.get("3306")
val defaultDb  = spark.conf.get("default")
val dbTable    = spark.conf.get("bettercustomers")
val dbUser     = spark.conf.get("dataeng")
val dbPass     = spark.conf.get("dataengineering_user")

val connectionUrl = s"jdbc:mysql://$dbHost:$dbPort/$defaultDb"

val betterCustomers = sparkSessionHive.read
  .format("jdbc")
  .option("url", connectionUrl)
  .option("driver", jdbcDriver)
  .option("dbtable", "bettercustomers")
  .option("user", dbUser)
  .option("password", dbPass)
  .load()

betterCustomers.createOrReplaceTempView("customers")
```

This initial step simply prepares for the eventual load of our customer's table. Given that Spark will not eagerly load the temporary view, you can ensure the table is accessible to the Spark Session. This is a nice way to interact with your session and ensure that all the data you expect to be loaded is in fact available.

Listing Tables

The Spark SQL Catalog enables you list and view all tables available to a particular Spark Session.

```
%spark
sparkSessionHive
  .catalog
  .listTables
  .show(truncate=false)
```

This will show you the current state stored in the Spark session catalog.

```
+--------+--------+-----------+---------+-----------+
|    name|database|description|tableType|isTemporary|
+--------+--------+-----------+---------+-----------+
|customers|    null|       null|TEMPORARY|       true|
+--------+--------+-----------+---------+-----------+
```

The `customers` table has a `tableType` of `TEMPORARY`, which just means it is only available in memory. If we stop our Spark Context (shut down Zeppelin), the table will no longer be available in the Spark SQL Catalog metadata. We would like this data to be available in our data lake and accessible using the Hive Metastore. So we need to load this data into our data lake (warehouse).

If you think back to the ETL flow examples from earlier in the book, this is just another type of ETL. In this case, we are simply loading the data into the data lake so that other data teams, or even your own data team, will have an easier time acting on this data.

Creating Persistent Tables

In the last chapter, we looked at using the `saveAsTable` method on the `DataFrameWriter`. We will be using this method again in order to write our table into our `coffee_co_common` database, all stored within the data warehouse.

```
%spark
val coffeeCoDatabaseName = "coffee_co_common"

sparkSessionHive.catalog
  .setCurrentDatabase(coffeeCoDatabaseName)

betterCustomers
  .write
  .mode("errorIfExists")
  .saveAsTable("customers")
```

Just like when we want to switch the context of the database using traditional SQL, we have to tell Spark which database to use, unless we want to default back to the `default` database. This can be done simply by using the `setCurrentDatabase` method on the Spark SQL Catalog. Once the context of our Spark Session is switched to using the `coffee_co_common` database, we are free to save our table into the SQL catalog.

If you go ahead and list the tables available in the SQL Catalog context, you'll see that this new table now shows up.

```
%spark
sparkSessionHive
  .catalog
  .listTables.show
```

```
+---------+----------------+-----------+---------+-----------+
|     name|        database|description|tableType|isTemporary|
+---------+----------------+-----------+---------+-----------+
|customers|coffee_co_common|       null|  MANAGED|      false|
|customers|            null|       null|TEMPORARY|       true|
+---------+----------------+-----------+---------+-----------+
```

Under the database column of the customers row, you'll see that the table was indeed stored in our new coffee_co_common database, and this table is now fully *managed* by Spark. This means that not only is the metadata managed (in this case, the data stored in the Hive Metastore), but also the underlying distributed table in the data lake.

Finding the Existence of a Table

The last thing that may come in handy is the ability to just check on the existence of a table in a specific database context in your Spark applications.

```
%spark
val dbName = "coffee_co_common"
val tableName = "customers"

sparkSessionHive
  .catalog
  .tableExists(dbName, tableName)
```

This can be used as a simple Boolean to see if the table you need to work with exists. This check can also be used as a sanity check if your application is expecting to write to a specific location for the first time. Ultimately, using the default save mode (errorIfExists) on your DataFrameWriter is safer than worrying about if your application is going to overwrite data in a shared location.

174

Okay, moving on. We will look at how the Metastore database is laid out and go through some exercises intended to help you learn enough about the Hive Metastore so you can go spelunking on your own with confidence. After this slight departure from the Zeppelin exercises, we will return to the Zeppelin notebook and look at reading (loading) data from these managed Spark tables. You will see how to enable data discovery options to your tables and table columns, and you will finish up this section by looking at how to manage table caching within your Spark application, as well as how to ultimately drop a table when it is no longer needed.

Databases and Tables in the Hive Metastore

To understand how the metadata for your database and table is managed, it helps to understand where the metadata is stored and how to find your way to the Hive Metastore database. You can use the following steps to view this record:

1. Open a new terminal session or add a new tab to your current context.

2. From there, create a new Bash session on the MySQL Docker container, and then log in to the MySQL CLI using the `secrets` file on the container.

3. Switch to the metastore database and query the TBLS table.

```
docker exec -it mysql bash
mysql --defaults-file=./secrets
mysql> use metastore;
```

Next, we will view the metadata for our new database and managed table.

View Hive Metastore Databases

The databases stored in the Hive Metastore are located in the DBS table. The available schema for the table is a simple `describe` away and can help to query it efficiently. In this case, you have only one entry, but in the real world there can be 100s to 1000s of database entries and finding your way can be a little complicated.

```
mysql> describe DBS;
+------------------+----------------+------+-----+---------+-------+
| Field            | Type           | Null | Key | Default | Extra |
+------------------+----------------+------+-----+---------+-------+
| DB_ID            | bigint         | NO   | PRI | NULL    |       |
| DESC             | varchar(4000)  | YES  |     | NULL    |       |
| DB_LOCATION_URI  | varchar(4000)  | NO   |     | NULL    |       |
| NAME             | varchar(128)   | YES  | UNI | NULL    |       |
| OWNER_NAME       | varchar(128)   | YES  |     | NULL    |       |
| OWNER_TYPE       | varchar(10)    | YES  |     | NULL    |       |
+------------------+----------------+------+-----+---------+-------+
```

Given we created a database called `coffee_co_common` from within Spark, we can use a simple equality query by `NAME` to find the database.

```
mysql> select * from DBS where NAME = 'coffee_co_common';

+-------+-----------------------------------------------------+-------------+
| DB_ID | DESC    | DB_LOCATION_URI     | NAME    | OWNER_NAME | OWNER_TYPE |
+-------+-----------------------------------------------------+-------------+
|     1 | This database stores ... | file:...warehouse/common | coffee_co_
common | zeppelin   | USER        |
+-------+-----------------------------------------------------+-------------+
```

The only row you really need to be concerned with for now is the `DB_ID` row, since this is the primary key. This will be used to find the managed customers table in the next section.

View Hive Metastore Tables

The Hive Metastore stores the table metadata in the `TBLS` table. That is a very meta sentence! Just like before, let's begin by describing the `TBLS` table schema so we understand which columns are available to us for our query.

```
mysql> describe TBLS;
+--------------------+--------------+------+-----+---------+-------+
| Field              | Type         | Null | Key | Default | Extra |
+--------------------+--------------+------+-----+---------+-------+
| TBL_ID             | bigint       | NO   | PRI | NULL    |       |
| CREATE_TIME        | int          | NO   |     | NULL    |       |
| DB_ID              | bigint       | YES  | MUL | NULL    |       |
| LAST_ACCESS_TIME   | int          | NO   |     | NULL    |       |
| OWNER              | varchar(767) | YES  |     | NULL    |       |
| RETENTION          | int          | NO   |     | NULL    |       |
| SD_ID              | bigint       | YES  | MUL | NULL    |       |
| TBL_NAME           | varchar(256) | YES  | MUL | NULL    |       |
| TBL_TYPE           | varchar(128) | YES  |     | NULL    |       |
| VIEW_EXPANDED_TEXT | mediumtext   | YES  |     | NULL    |       |
| VIEW_ORIGINAL_TEXT | mediumtext   | YES  |     | NULL    |       |
| IS_REWRITE_ENABLED | bit(1)       | NO   |     | b'0'    |       |
+--------------------+--------------+------+-----+---------+-------+
```

Now we can query the table we just created using Spark to see what information is stored at the TBLS level.

```
select TBL_ID,TBL_TYPE,TBL_NAME,SD_ID
from TBLS where DB_ID = 1;
```

```
+--------+---------------+-----------+-------+
| TBL_ID | TBL_TYPE      | TBL_NAME  | SD_ID |
+--------+---------------+-----------+-------+
|      4 | MANAGED_TABLE | customers |     4 |
+--------+---------------+-----------+-------+
```

After all of that, we have essentially retrieved mainly the information we already knew about TBL_NAME and the TBL_TYPE, but this did give us another clue, which is the TBL_ID. The TABLE_PARAMS table in the Metastore is where we will start to see the information that is needed for Spark to run.

As we continue down this rabbit hole, you are probably wondering where this ends. This journey will connect two more dots and then from there you can choose your own adventure, as the Metastore has a plethora of additional tables that all work together to enable a powerful discovery layer for your data.

Hive Table Parameters

The next table that is worth mentioning is the core parameters table, which is akin to the spark-defaults.conf file or other properties or configuration files. The difference is that it's stored in the database versus in a flat file in a distributed file system.

```
mysql> describe TABLE_PARAMS;
+-------------+--------------+------+-----+---------+-------+
| Field       | Type         | Null | Key | Default | Extra |
+-------------+--------------+------+-----+---------+-------+
| TBL_ID      | bigint       | NO   | PRI | NULL    |       |
| PARAM_KEY   | varchar(256) | NO   | PRI | NULL    |       |
| PARAM_VALUE | mediumtext   | YES  |     | NULL    |       |
+-------------+--------------+------+-----+---------+-------+
```

Using the TBL_ID from before, we can view the table properties of our customers table.

```
mysql> select * from TABLE_PARAMS where TBL_ID = 4;
+--------+---------------------------------+----------------------+
| TBL_ID | PARAM_KEY                       | PARAM_VALUE          |
+--------+---------------------------------+----------------------+
|      4 | comment                         | Loyal patrons of...  |
|      4 | numFiles                        | 1                    |
|      4 | spark.sql.create.version        | 3.1.1                |
|      4 | spark.sql.partitionProvider     | filesystem           |
|      4 | spark.sql.sources.provider      | parquet              |
|      4 | spark.sql.sources.schema.numParts | 1                  |
|      4 | spark.sql.sources.schema.part.0 | {"type":"struct"...} |
|      4 | totalSize                       | 1896                 |
|      4 | transient_lastDdlTime           | 1615756742           |
+--------+---------------------------------+----------------------+
```

This last table enables us to view the important configuration that is "automagically" applied to the managed tables by Spark. This includes all the information needed to instruct Spark to read our tables. It also simplifies the ways in which we can use the data, including simplifying how we load or read a table into Spark.

Switching gears, let's move back to our Zeppelin environment so we can load the data we saved into our managed table.

Working with Tables from the Spark SQL Catalog

Once the metadata is available to the catalog object on the SparkSession, you can use the `table` method to load data for use in your application.

```
%spark
val dbName = "coffee_co_common"
sparkSessionHive
  .catalog
  .setCurrentDatabase(dbName)

sparkSessionHive
  .table("customers")
  .show()
```

Well, now that is a breath of fresh air, isn't it? Aside from setting the current database in our session catalog context, there is no special handholding or guidance necessary when it comes to accessing a table. The Spark SQL Catalog abstracts away all of the business logic regarding querying the Hive Metastore. As we saw when manually querying the Hive Metastore, all of the metadata needed to introspect, load, and transform these distributed tables is essentially right at our fingertips.

Data Discovery Through Table and Column-Level Annotations

This section has been all about using the Spark SQL Catalog to enable you, your team, or really any data team to efficiently discover data. You've seen how the Hive Metastore plays a critical role in enabling teams to discover data stored in these hybrid data lake/ data warehouses. We looked at organizing data using the namespace pattern, which essentially creates a physical directory where table ownership is documented, along with all the tables' owners within a specific context. You learned how to query the myriad tables in the Hive Metastore directly, since mastery of the underlying technology used by you or your team plays a critical role in the health of any data system.

Lastly, we will wrap up our deep dive into the Hive Metastore by manually altering our table and column-level comments so that anyone using the data we produce will be able to fully understand our intentions. They won't have to look things up in a separate wiki or document archive, but rather can use the tools they are most familiar with.

Adding Table-Level Descriptions and Listing Tables

Table-level descriptions help document the intentions of a table. Imagine what would happen if a new team was looking to understand the customers table and they were met with the following.

```
%spark
sparkSessionHive
  .catalog
  .listTables
  .where($"database".equalTo("coffee_co_common"))
  .select($"name", $"description", $"tableType")
  .show()
```

```
+---------+-----------+---------+
|     name|description|tableType|
+---------+-----------+---------+
|customers|       null |  MANAGED|
+---------+-----------+---------+
```

This table could represent anything. What if there were multiple tables that could potentially be used, such as customers_current, customers_a, and customers_test? Now there are many places to look, and this means one of two options. The team or individual could either set up a meeting to ask you or your team which table to use, or equally as likely, especially if they're in a rush to generate some report, the team may flip a coin and use whichever table seems more up to date. They might even spend hours, days, or weeks combing through the data to figure things out before finally reaching out when they have a deadline looming.

We can prevent this problem with a single ALTER TABLE command to add TBLPROPERTIES to describe the customers table.

```
%spark
sparkSessionHive.sql(
"""
ALTER TABLE customers
SET TBLPROPERTIES (
  'comment' = 'Production Customers Data',
  'active' = 'true'
)
""")
```

Now when listing the table, the team would have instead seen the *Production Customers Data* descriptor, and this could help them determine that this is the correct table to use.

```
+---------+------------------------+---------+
|name     |description             |tableType|
+---------+------------------------+---------+
|customers|Production Customers Data|MANAGED  |
+---------+------------------------+---------+
```

Adding Column Descriptions and Listing Columns

Now that you've applied a description to the table itself, you can repeat this exercise by applying column-level descriptors. This process is similar to applying table-wide annotations using the TBLPROPERTIES, but instead we are targeting the comment field using ALTER COLUMN on the table.

```
%sql
use coffee_co_common;
ALTER TABLE customers ALTER COLUMN id
  COMMENT "Unique identifier for our Customer.";
ALTER TABLE customers ALTER COLUMN created
  COMMENT "timestamp when the account became active";
ALTER TABLE customers ALTER COLUMN updated
  COMMENT "timestamp when this record last changed";
ALTER TABLE customers ALTER COLUMN first_name
  COMMENT "The first name on the account";
```

```
ALTER TABLE customers ALTER COLUMN last_name
  COMMENT "The last name on the account";
ALTER TABLE customers ALTER COLUMN email
  COMMENT "customers email on file. Unique Constraint. Not to be used for
  Marketing";
```

With the columns updated in the Hive Metastore, it is now easy to understand each column. If any team needs to list the columns of a table now, they can easily do just that.

```
%spark
val dbName = "coffee_co_common"
val tableName = "customers"

sparkSessionHive
  .catalog
  .listColumns(dbName, tableName)
  .select($"name", $"description", $"dataType")
  .show(truncate=false)
```

The listColumns method of the Spark SQL Catalog has two variants. For our use case, given we have two tables loaded into the catalog context(a non-managed and managed table named customers), there is some ambiguity and Spark may end up providing the wrong information when listing the columns, even if we have the current database context pointing to the coffee_co_common database. From my experience, these kinds of use cases don't usually occur since your jobs will typically be split across a specific separation of concerns, with one batch job (or streaming job) writing data into the managed table, while the other job picks up from the managed table and continues to transform the data necessary to solve the data use case. Regardless, it is always better to test things first; remember the old adage, "measure twice, cut once."

The output now provides anyone with the necessary information to understand what each column is used for. Note: This output was truncated to fit on the page.

```
+----------+--------------------+---------+
|      name|         description| dataType|
+----------+--------------------+---------+
|        id|Unique identifier...|      int|
|   created|timestamp when th...|timestamp|
|   updated|timestamp when th...|timestamp|
```

```
|first_name|The first name on...|    string|
| last_name|The last name on ...|    string|
|     email|customers email o...|    string|
+----------+--------------------+---------+
```

We have explored most of the features of the Spark SQL Catalog, but we are not done just yet. There are a few more things that will nicely round out this deep dive. First, we have yet to discuss table caching, what it means to cache a table, how to clear a cache (to free up memory), how to update what is cached when another process has modified the table you are working with. Then comes the fun part, which is how to drop or delete a table. Given you should keep data around only as long as it needs to exist, it is more important than ever to keep your workspace clean.

Caching Tables

There are many reasons to cache that table you are working with. This can help to reduce the burden on the distributed file system (e.g., S3 or HDFS) by removing many additional trips back to the data lake in order to process complex queries. A complex query is any query that has to process and transform the same dataset more than once. In the case of reporting, or any analytical queries, it is common for example to use self-joins to handle de-duplication. Or, in the case of complex subsets, it can be easier to use semi-joins, like you learned earlier in the book when you looked at using joins.

Dropping Temp Views   In order to simplify things, we can drop the temporary `customers` view so that we only need to concern ourselves with the managed `customers` table from here on out.

```
%spark

sparkSessionHive

.catalog

.dropTempView("customers")
```

Now onto cache management in the catalog.

Cache a Table in Spark Memory

Armed with only the catalog on the Spark Session, we can use the `cacheTable` method to store the table data in memory (or if the table is too large, we can get to that).

```
%spark
val tableName = "customers"
spark.catalog.cacheTable(tableName)
```

That is all there is to it. Well, almost. We haven't looked at the Spark UI at this point in the book. We will look at just one of its features now, which is the Storage tab.

The Storage View of the Spark UI

You can open the Storage tab of the Spark UI at `http://localhost:4040/storage/`. If you haven't skipped ahead in the notebook, you will see a lovely blank screen, as shown in Figure 6-1.

Figure 6-1. *The Storage tab of the Spark UI*

Given that we haven't triggered an action, the data from the `customers` table, while being marked for cache, does not actually exist in Spark's storage. This is because there is no reason to eagerly run a full load on the table, given the application you are writing may only need to make a single pass over the data. In this case, Spark loves being lazy and does so to protect your application from the possibility of running into memory issues before starting to do any work.

Tip If you applied the mapping 127.0.0.1 spark to your /etc/hosts then you can also use `http://spark:4040/storage/` to view this tab. This process is covered in the chapter's materials, and an example can be found in /ch-06/docker/etc.

Force Spark to Cache

To force Spark to cache the data from the `customers` table, you need to call Spark into action. In this case, you can simply call `head` to return the first row located in the table.

```
%spark
val tableName = "customers"
sparkSessionHive.table(tableName).head
```

This is all that is needed. Now if you refresh the Storage UI, you will see your in-memory table. The reference UI is shown in Figure 6-2.

Figure 6-2. *Spark Storage UI with a list of cached RDDs*

That was fun, wasn't it? Now you have an alternative way to ensure that the applications you write are in fact behaving the way you expect them to. With the `customers` table living in the cache, it is simple to check the behavior when you want to remove a table from that cache.

Uncache Tables

When we cache data in the memory of our Spark applications, that means you are reducing the total usable memory allocated for object storage in your Spark applications. We won't go into the nitty gritty regarding how memory is managed, but each Spark application splits the available memory allocated into two spaces—one space is dedicated to the memory needed to run the application (from the JVM overhead, IO connections, etc.) and the other space is dedicated to object storage in the application (like we see when we cache our in-memory table). This is controlled by `spark.memory.fraction` and it defaults to a 60/40 split.

For now, let's look at the process of uncaching that table we cached.

```
%spark
val tableName = "customers"
sparkSessionHive.catalog.uncacheTable(tableName)
```

Clear All Table Caches

It is possible that there are many tables being cached in different places in your Spark application. If you want to start with a blank slate, you can use the simple clearCache catalog method.

```
%spark
sparkSessionHive.catalog.clearCache
```

This is easier than having to track down all the tables that are cached. Yes, you can iterate over all the tables available to the catalog, check the existence of their cache, and then optionally remove each cache, but sometimes it is better to just nuke everything.

Refresh a Table

In the excitement to clear all of the caches, we skipped over an important concept. When you cache a table in memory and the upstream data store is updated outside of your control (e.g., another application is writing to the table location, therefore invalidating your cache), then to ensure that you have the most up-to-date data, you can use the refreshTable method on the Spark catalog. This refreshes the metadata regarding the table, as well as the underlying files composing the table.

```
%spark
sparkSessionHive
  .catalog
  .refreshTable("customers")
```

Here is an interesting fact. If your Spark application is purposefully overwriting a table you have cached, it will also invalidate your local cache, invalidating the need to include any table refreshes in your own application logic.

Testing Automatic Cache Refresh with Spark Managed Tables

This is a fun one. There are a few steps necessary to fully showcase this use case:

1. Copy the contents of the `customers` table into a new managed table called `customers_temp`.

2. Read from this new managed table `customers_temp` into a reference variable called `tableCopy` and then cache the table.

3. Call `tableCopy.head` in order to cache off the table in local Spark memory.

4. Union the `tableCopy` with itself to duplicate the data.

5. Write into yet another managed table named `union`.

6. Now you are free to overwrite the initial `customers_temp` table without Spark informing you that you are trying to overwrite a table you are currently reading from (which it will do, by the way).

7. You have tricked the safety check and now have the `duplicates` loaded in `customers_temp`.

8. Interestingly enough, if you look at the Storage tab in the Spark UI (`http://localhost:4040/storage/`), you will see that Spark has purged the invalid cached table view.

You can follow along in the "6_2_SparkSQLCatalog" notebook under the section titled "Testing Automatic Cache Refresh in Spark" for this example, or without looking at the examples, try to leverage what you've learned about the Spark SQL Catalog in this chapter. In the end, you have to create two additional managed tables in order to trick Spark into overwriting another managed table. When you are done with this exercise, you can move onto the process of cleaning up the tables you just created.

Removing Tables

Dropping, aka deleting, tables can be done with the `DROP TABLE` command. Or, with a little more effort, you can also use the underlying Spark SQL Catalog Catalyst operations directly. These operations all fall under the traditional data manipulation language (DML) category of operations.

DATA MANIPULATION LANGUAGE

DML stands for Data Manipulation Language. This is a category of SQL operations encapsulated by the INSERT, UPDATE, and DELETE commands. Earlier in the chapter, when you set up the Hive Metastore, there was a brief note about read-only, write-only, and read-write access for Spark applications for the Hive Metastore.

Grants

The grants (permissions) that govern traditional RDBMS via DCL grants also apply to governing access for Spark applications that connect to the Hive Metastore. For data security best practices, you must control access to the data. Only the minimum set of privileges should be set, for example those that enable read-only operations for users like SELECT. Add capabilities based on the actual needs of the application.

For example, if our dataeng user needed to read basic Hive Metastore table metadata in order to read upstream table data, then transform, and lastly write that data to a new managed table using saveAsTable, we could have safely used a subset of the grant permissions for the user.

```
Read/Write Access
REVOKE ALL PRIVILEGES ON `metastore`.* TO `dataeng`@`%`;
GRANT CREATE, SELECT, INSERT ON `metastore`.* TO 'dataeng'@'%';
FLUSH PRIVILEGES;
```

```
Read-Only Access
REVOKE ALL PRIVILEGES ON `metastore`.* TO `dataeng`@`%`;
GRANT SELECT ON `metastore`.* TO 'dataeng'@'%';
FLUSH PRIVILEGES;
```

```
Full Access
REVOKE ALL PRIVILEGES ON `metastore`.* FROM `dataeng`@`%`;
GRANT ALL PRIVILEGES ON  `metastore`.* TO  'dataeng'@'%';
FLUSH PRIVILEGES;
```

This book doesn't cover all the ins and outs of security and data access management. However, some food for thought is to look into the availability of support for role-based grants for user data access. This powerful feature enables efficient governance of access based on common role names. Here is an example of read/write only access to the Hive Metastore.

```
Roles
CREATE ROLE readwrite;
GRANT CREATE, SELECT, INSERT, UPDATE, ALTER ON `metastore`.* TO readwrite;

Apply Roles
REVOKE ALL PRIVILEGES ON `metastore`.* FROM `dataeng`@`%`;
GRANT readwrite TO `dataeng`@`%`;
FLUSH PRIVILEGES;

View Grants
mysql> SHOW GRANTS FOR `dataeng`@`%`;
+--------------------------------------------------------+
| Grants for dataeng@%                                   |
+--------------------------------------------------------+
| GRANT USAGE ON *.* TO `dataeng`@`%`                    |
| GRANT ALL PRIVILEGES ON `default`.* TO `dataeng`@`%`   |
| GRANT `readwrite`@`%` TO `dataeng`@`%`                 |
+--------------------------------------------------------+
```

This can drastically reduce the amount of effort required to grant and revoke common access to your Hive Metastore.

Drop Table

Using DROP TABLE will do different things depending on if the table is managed or external. If the table is managed by Spark, the command will delete, entirely, all files across all partitions of the distributed table, as well as all metadata, and any annotations provided on the table. if the table is defined as being external, the DROP TABLE command will only remove the table metadata from the Hive Metastore, and the external table will be untouched.

Remember to try things locally, like we are now via Docker, so that you can test different commands out before running them against real datastores.

```
%sql
DROP TABLE customers_temp;
DROP TABLE unioned;
```

As you just experienced, it's trivial to drop a table using the SQL interpreter on the SparkSession. If your application tries to drop the same table twice, the operation will throw an exception. As discussed in the DML sidebar, role-based access or explicit user privileges can help to safeguard these shared datastores from accidental deletion.

Conditionally Drop a Table

In order to safeguard your application, it is wise to use conditions. The IF EXISTS condition is a Boolean operation and a false condition will short circuit or stop the processing of the command.

```
%sql
DROP TABLE IF EXISTS customers_temp;
DROP TABLE IF EXISTS unioned;
```

Last but not least, let's look at how to drop a table using the Spark SQL's Catalyst classes directly. This is the functional equivalent of the interpreted SQL DML query you just ran to conditionally drop the distributed customers table.

Using Spark SQL Catalyst to Remove a Table

This will be our first foray into directly using the Spark Catalyst classes; in the following section we will be looking more into the Catalyst Optimizer.

```
%spark
import org.apache.spark.sql.catalyst.TableIdentifier

val catalog = sparkSessionHive.sessionState.catalog
val tableName = "customers_temp"
val tableId = TableIdentifier(tableName)

catalog
  .dropTable(tableId,ignoreIfNotExists=true,purge=true)
```

Why have we gone through this trouble, you ask? Well, it turns out that there is no simple method to drop a table using the methods in the Spark SQL Catalog.

This is actually a good thing since this advanced technique requires you, the author of the application, to be very determined to drop tables using the DSL. There is a reason that `dropTable` doesn't exist in the Spark SQL Catalog after all. Alternatively, yes you can use the simple `DROP TABLE` from the Spark SQL interpreter, but let's imagine a world where that isn't possible. If people start misbehaving, you can restrict the operations available to the Spark applications running inside of your data platform using the role-based or user grants approach mentioned in the DML sidebar. Special privileges should be granted to the data stewards, or responsible maintainers of a particular table or database in your warehouse.

Exercise 6-2: Summary

Spark SQL and the Hive Metastore work better together, as a team, providing all of your Spark applications with a centralized metadata service for the data that is generally available in your organization. As you learned over the last section, there are many nuances to pick up in order to get a 360 view of this powerful form of data discovery. We looked at user-based access, how to move data from a traditional RDBMS (MySQL in our case) into the distributed data warehouse using simple methods on the SparkSession. We looked at paying things forward by annotating Hive Metastore database ownership, table columnar descriptions, and table name properties.

Moving on. It is time for a tour of the Spark Catalyst Optimizer. This is the glue between the interpreted SQL queries and DataFrame functionality you've been introduced to throughout this book.

The Spark Catalyst Optimizer

Behind the scenes of the functional transformations and powerful DataFrame operations we've encountered throughout the course of this book exists a powerful yet seemingly transparent engine. It's silently working to optimize and orchestrate each and every last job, stage, and task required by your application execution. Essentially, the Catalyst Optimizer parses your application code and converts it efficiently into a series of optimized steps, called an *execution plan*. See Figure 6-3.

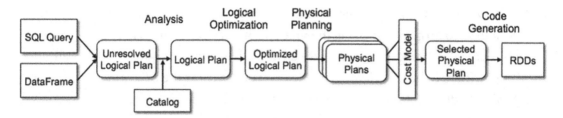

Figure 6-3. *The Catalyst Optimizer resolves the high-level intentions of your Spark application and converts each step, across a series of transformations, into a highly optimized, distributable, battle plan*

This process is broken into four distinct phases:

- Analysis

- Logical optimization

- Physical planning

- Code generation

To full appreciate the work being done by the Catalyst Engine, let's look at each of these distinct phases using the baked-in introspection layer of Spark. The following example analyzes a simple query against the parquet data stored in our data warehouse.

Note   Follow along using the Zeppelin note titled "6_3_CatalystOptimizer." This final note requires you to have gone through the full chapter exercises to produce the data required for the section.

Introspecting Spark's Catalyst Optimizer with Explain

You can use the explain method at any point up until an action is called on a DataFrame or by using the EXPLAIN clause in the Spark SQL interpreter.

```
%spark
sparkSessionHive
  .table("customers")
  .where(
    col("email").like("scott%")
```

```
  )
  .select($"id", $"email", $"last_name")
  .explain("extended")
```

This DataFrame query is functionally equivalent to the following SQL query.

```
%sql
select id, email, last_name from customers
where email like 'scott%'
```

Regardless of how you decide to craft your query, the Spark Catalyst Optimizer will begin at the same starting point, which is the query parsing. This first operation marks the beginning of the Catalyst journey from parsing, through optimization and planning.

Logical Plan Parsing

The initial phase of execution is responsible for converting the intentions of your query into a hierarchical tree.

```
== Parsed Logical Plan ==
'Project [unresolvedalias('id, None), unresolvedalias('email, None),
unresolvedalias('last_name, None)]
+- Filter email#5 LIKE scott%
   +- SubqueryAlias spark_catalog.coffee_co_common.customers
      +- Relation[id#0,created#1,updated#2,first_name#3,last_
name#4,email#5] parquet
```

This step is responsible for transforming and building up each query as a tree of relations and expressions on those relations. This first step lays out the work of identifying unknown columns (UnresolvedAlias) and, in the case of interpreted Spark SQL, you will see that the coffee_co_common.customers table, aka relation, is marked as UnresolvedRelation.

```
|== Parsed Logical Plan ==
'Project ['id, 'email, 'last_name]
+- 'Filter 'email LIKE scott%
   +- 'UnresolvedRelation [coffee_co_common, customers], [], false
```

This is where the logical analyzer comes into play.

Logical Plan Analysis

The Analyzed Logical Plan is tasked with entity resolution (UnresolvedAlias, UnresolvedRelation, and more) so that the columns and tables can be validated against the known schema and data source or sink locations.

```
== Analyzed Logical Plan ==
id: int, email: string, last_name: string
Project [id#0, email#5, last_name#4]
+- Filter email#5 LIKE scott%
   +- SubqueryAlias spark_catalog.coffee_co_common.customers
      +- Relation[id#0,created#1,updated#2,first_name#3,last_
         name#4,email#5] parquet
```

For each step in the optimizer's journey, a subset of specific tasks is checked off. In the above example, this is as simple as resolving columns and validating that the data source can provide the data requested by a query.

Unresolvable Errors

Unresolvable columns can happen due to a number of factors, such as column name spelling errors in freeform SQL queries. Spark will fail fast and tell you why the analysis phase failed with an AnalysisException.

You can change the query to throw an exception by adding a new, unresolvable field to your select.

```
%spark
customersTable
  .where(col("email").like("scott%"))
  .select($"id", $"email", $"last_name", $"unknown")
  .show()
```

This will trigger an exception, failing fast, before physically executing the actual plan.

```
org.apache.spark.sql.AnalysisException: cannot resolve '`unknown`' given
input columns: [spark_catalog.coffee_co_common.customers.created, spark_
catalog.coffee_co_common.customers.email, spark_catalog.coffee_co_common.
customers.first_name, spark_catalog.coffee_co_common.customers.id, spark_
```

```
catalog.coffee_co_common.customers.last_name, spark_catalog.coffee_co_
common.customers.updated];
'Project [id#0, email#5, last_name#4, 'unknown]
+- Filter email#5 LIKE scott%
   +- SubqueryAlias spark_catalog.coffee_co_common.customers
      +- Relation[id#0,created#1,updated#2,first_name#3,last_
      name#4,email#5] parquet
```

This short circuit at the analyzer phase is handy, especially when attempting to debug data-related issues. Next, after resolving all unresolved entities within the plan, Spark will move on to optimizing the query.

Logical Plan Optimization

At this phase of planning, Spark introduces optimizations on top of your query, for example by adding nullability checking and conversions to known static expressions like the StartsWith in the following optimized plan.

```
== Optimized Logical Plan ==
Project [id#0, email#5, last_name#4]
+- Filter (isnotnull(email#5) AND StartsWith(email#5, scott))
   +- Relation[id#0,created#1,updated#2,first_name#3,last_
   name#4,email#5] parquet
```

This phase of optimization converted the following:

```
Filter email#5 LIKE scott% became
Filter (isnotnull(email#5) AND StartsWith(email#5, scott))
```

Now onto the final planning phase. The physical plan.

Physical Planning

The physical plan is the last step before Spark converts your query into generated Java bytecode.

```
== Physical Plan ==
*(1) Project [id#0, email#5, last_name#4]
```

```
+- *(1) Filter (isnotnull(email#5) AND StartsWith(email#5, scott))
   +- *(1) ColumnarToRow
      +- FileScan parquet coffee_co_common.customers[id#0,last_
         name#4,email#5] Batched: true, DataFilters: [isnotnull(email#5),
         StartsWith(email#5, scott)], Format: Parquet, Location:
         InMemoryFileIndex[file:/spark/sql/warehouse/common/customers],
         PartitionFilters: [], PushedFilters: [IsNotNull(email),
         StringStartsWith(email,scott)], ReadSchema: struct<id:int,last_
         name:string,email:string>
```

This plan provides Spark with all the necessary information to generate a real plan of attack for reading, filtering, and loading the projected fields (ID, email, and last_name) into Spark as a DataFrame. The last and final piece of the puzzle is the code-generation component.

Java Bytecode Generation

This final step compiles and converts the physical plan into actual Java bytecode. This process is called *whole stage codegen*, and it optimizes your query by inlining all of your Java method calls into a single function optimized for the JVM rather than for a human (although some humans take pride in being able to read and fully grok the codegen).

```
%spark
customersTable
  .select($"id", $"email", $"last_name")
  .where(col("email").like("scott%"))
  .explain("codegen")
```

This will dump the partially human readable details for analysis.

```
Found 1 WholeStageCodegen subtrees.
== Subtree 1 / 1 (maxMethodCodeSize:369; maxConstantPoolSize:142(0.22%
used); numInnerClasses:0) ==
*(1) Project [id#0, email#5, last_name#4]
+- *(1) Filter (isnotnull(email#5) AND StartsWith(email#5, scott))
   +- *(1) ColumnarToRow
```

```
+- FileScan parquet coffee_co_common.customers[id#0,last_
name#4,email#5] Batched: true, DataFilters: [isnotnull(email#5),
StartsWith(email#5, scott)], Format: Parquet, Location:
InMemoryFileIndex[file:/spark/sql/warehouse/common/customers],
PartitionFilters: [], PushedFilters: [IsNotNull(email),
StringStartsWith(email,scott)], ReadSchema: struct<id:int,last_
name:string,email:string>
```

```
Generated code:
/* 001 */ public Object generate(Object[] references) {
/* 002 */    return new GeneratedIteratorForCodegenStage1(references);
/* 003 */ }
/* 004 */
/* 005 */ // codegenStageId=1
/* 006 */ final class GeneratedIteratorForCodegenStage1 extends org.apache.
spark.sql.execution.BufferedRowIterator {
/* 007 */    private Object[] references;
/* 008 */    private scala.collection.Iterator[] inputs;
/* 009 */    private int columnartorow_batchIdx_0;
...
```

Knowledge of these advanced features isn't necessary for most Spark engineers, but it is critical to understanding how Spark operates. Additionally, harnessing Catalyst can be invaluable for creating extensions to the Spark engine, such as when you would want to read and write data formats that are unknown/unsupported by the Spark core encoders.

Now that you know the steps that Spark goes through to optimize your application flow, we will wrap up this extensive chapter by looking at datasets.

Datasets

A *dataset* in a nutshell is a strongly typed and efficiently encoded collection of objects residing in native system memory, accessed and orchestrated by the Spark engine. While similar to a DataFrame conceptually, the Dataset API exposes typed capabilities, including the ability to create custom stateful aggregations (for streaming applications) and other strongly typed aggregations to be used with more traditional batch-based analytics.

When talking about *typed* versus *untyped,* all this means is that Spark is expressly aware, ahead of time, of the format and structure of the data, aka its *schema.* The other thing worth mentioning is that a dataset has an immutable fixed schema.

RDDs   Recall that discussed the RDD (resilient distributed data) objects earlier in the book, when discussing the Spark programming model. As a refresher, the RDD is the intelligent backbone for both the DataFrame and the Dataset. They store a logical map of the distributed (partitioned) metadata regarding the data lineage within an application—from the initial loading of data, from a reliable data source or an in-memory data structure, through each stage of transformation (mutation), leading up to a terminal action (trigger point) in your Spark application. We have seen actions in action every time we've called write, show, or collect.

While the RDD itself stores no physical object data. The Spark driver application's SparkContext is expressly aware of where the data resides in the memory and disk space of the Spark Executors assigned to the live runtime of the application. At each complete stage of execution, the resulting output data is stored across immutable data buffers, and a new step in the transformational data lineage is recorded. Although this process happens more or less transparently, we will look a little closer at this process throughout the book.

Now where were we? Datasets are strongly typed, immutable data collections that are encoded from JVM objects (Scala or Java classes). They play nicely with the rest of the Spark ecosystem since they are just a layer on top of the RDD. Because the dataset has an immutable, or fixed, schema, this means that Spark can take certain liberties when processing the data in this context of your application. It is also worth mentioning that the Python and R APIs cannot use datasets, which is a big benefit to using the Scala APIs.

Exercise 6-3: Converting DataFrames to Datasets

Let's begin with a practical example. Given you have access to the managed customer data from before, we can reuse that data for our use case. All we need to do now is generate a simple case class to act as a proxy to our customer data. Remember that we looked at Java type interoperability in the last chapter, in the process of creating additional rows to insert into our RDBMS. This same technique can be applied to a case class, which can interoperate implicitly with the catalyst rows.

Create the Customers Case Class

```
%spark
import java.sql.Timestamp
import org.apache.spark.sql.Dataset

case class Customer(
  id: Integer,
  created: Timestamp,
  updated: Timestamp,
  first_name: String,
  last_name: String,
  email: String)
```

With our class defined, we can now convert our DataFrame to a dataset.

Dataset Aliasing

There is a simple helper method, called as, which converts a DataFrame to a Dataset using encoders.

```
%spark
import org.apache.spark.sql._
implicit val customerEnc = Encoders.product[Customer]

val customerData: Dataset[Customer] = customersTable
  .as[Customer]
```

Datasets enable you to use standard functional programming in Scala or use the catalyst expressions depending on your use cases. This paradigm allows you the most flexibility in your data engineering tasks, since sometimes complicated tasks can't be converted into meaningful SQL operations. Falling back to core engineering chops can come in handy in these cases. For example, you can use mixed Catalyst and Scala operations.

Mixing Catalyst and Scala Functionality

We can create our own lambda functions that can be inlined and used to filter for an email string starting with a pattern, my first name for example.

```
customerData
  .filter(_.email.startsWith("scott"))
  .explain("formatted")
```

This method of filtering allows you to write specialized Scala filters that can operate directly using the Scala Customer class, as opposed to using pure Catalyst expressions (like we saw earlier in the chapter). This function will be called across each row, as Spark iterates over the internal Spark rows, encapsulating each batch. For each filter operation, the row will be converted from the internal Spark format, which is an optimized, native memory format called *Tungsten*, into a Java object that can be used with the Scala lambda function.

Caution   This conversion from native memory to JVM memory can cause garbage collection cycles that will eat away at the performance of your application, so take heed of this and remember not to ignore the catalyst native functions whenever possible. The `org.apache.spark.sql.functions` are optimized for efficiency and ship for free with Apache Spark. As a rule of thumb, excess garbage collection, not to mention support of lambda functions, can weigh down your Spark applications. For example, I've seen performance increases of 50-60% when switching from using lambda expressions to native SQL functions.

The same filter operation can be called using the typed Catalyst expressions easily.

Using Typed Catalyst Expressions

The following is functionally equivalent to the previous example, except that we are using a typed Catalyst columnar expression to do our `startsWith` filter.

```
%spark
customerData
  .filter($"email".startsWith("scott"))
  .explain("formatted")
```

This allows us to use predicate push down (optimization function), rather than requiring all parquet data to be read into memory first, and then filtered. You will look into more optimization techniques later on in your Spark journey, but using Datasets enables the best of both worlds, where you can leverage Spark or traditional Scala in a hybrid execution model.

Exercise 6-3: Summary

Using Datasets is often preferrable when building Spark Structured Streaming applications, as their ridged structure provides an explicit data contract at compile type versus running into problems at runtime. Additionally, over time, I've preferred using Datasets to define the ingress and egress (input and output) of the applications I write. This ensures that I have a solid API boundary for the data I produce. Given that it is simple to convert from a Dataset to a DataFrame (`ds.toDF`), applications can choose how they need to operate. However, being able to share these formats across the organization (either through the Hive Metastore or within a shared library of common data formats) helps you reuse data without the worry that data can change out from under an application.

I've also found that datasets can drastically speed up and improve the process of writing unit tests. With a little upfront effort, it becomes straightforward to write Spark applications using a more test-driven approach. This can save the long cycles of compilation and deployment, plus good tests enable engineers to easily debug any style of Spark application, from standard batch to streaming. We will look at the process of writing and testing a full end-to-end pipeline application in Chapter 7.

Summary

In this chapter, you learned all the ins and outs of using the Apache Hive Metastore alongside the Spark SQL Catalog. You discovered how easy it is to add important rich metadata about the ownership of hive databases, simple descriptors and definitions for the tables you produce, and how to provide rich column-level information so that through your actions you can enable easy data discovery. This simple act of improving data discovery can assist in paying things forward in your data platform environment. Now other teams can be empowered to seek out the data they need today to solve the data problems of tomorrow.

We then took a deeper dive behind the scenes and looked at the optimization pipeline Spark uses within the Catalyst Optimizer. We saw how it plans and transforms your query from a parsed plan to an analyzed plan, into an optimized plan and then on to physical plan. This series of events drastically improves the Java bytecode that is generated during the codegen stage, and ultimately, your application benefits directly from this careful compilation.

We finished the chapter looking at the process for converting a case class into a dataset using explicit encoding. Over the next few chapters, as you tackle building and running both batch and streaming pipeline jobs, you will lean on the information from these first six chapters in order to take a pragmatic approach to building the core components of a modern data engineering environment.

## CHAPTER 7

Data Pipelines and Structured Spark Applications

There is a central processing paradigm that exists behind the scenes and can help connect just about everything you build as a data engineer. The processing paradigm is a physical as well as a mental model for effectively moving and processing data, known as the *data pipeline*. We first touched on the data pipeline in Chapter 1, while introducing the history and common components driving the modern data stack. This chapter will teach you how to write, test, and compile reliable Spark applications that can be weaved directly into the data pipeline.

Data Pipelines

In Chapter 1, you learned about the origins of data pipelines and how the pipeline job evolved from the well-known extract transform load (ETL) design pattern. Conceptually, ETL enabled frameworks like MapReduce to grow in popularity, as complex problems could be defined as workflows, or a series of jobs, that divide and conquer difficult distributed data processing into digestible computations. Data pipelines establish a consistent processing paradigm for handling batch or stream processing. Given a data pipeline is simply a conduit for moving data reliably between systems, starting to think about data as it flows between points within a larger data topology using ETL processing can help to establish a clear mental model. As you become more familiar with complex data processing where data flows fan out, forks off from, or rolls up into one or more data-processing pipelines, your mental model can begin to shift to more of a graph of interdependent jobs being coordinated through workflow orchestration (which we'll discuss next chapter).

© Scott Haines 2022
S. Haines, *Modern Data Engineering with Apache Spark*, https://doi.org/10.1007/978-1-4842-7452-1_7

As an analogy, consider a data pipeline as a software-driven automobile or manufacturing assembly line. Each station along the assembly line does one job really well, and that makes sense in terms of synchronous data processing, where each job relies on the success of an upstream job to accomplish its predefined goals. Figure 7-1 shows an example of a more complex data pipeline, where some jobs need to be run synchronously, following the assembly line style of processing, and others can be run fully asynchronously.

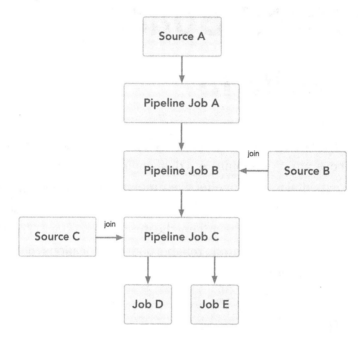

Figure 7-1. *Hybrid synchronous/asynchronous data pipeline*

Figure 7-1 shows that pipeline jobs A through C are dependent on each other completing in a specific synchronous order. This is typically the case of composition-style data processing. You are composing new data objects as a composite, or combination, of other data available to you. This is observed with the composition of the data sources A, B, and C. However, after the first three upstream pipeline jobs are complete, the downstream jobs, D and E, are independent and can run asynchronously.

Given the dynamic nature of data pipelines, new data-processing jobs can be added, inserted, or removed from the data-processing network as long as the changes to the data flow don't disrupt the downstream data requirements of any dependent jobs. While this flexibility provides an efficient mechanism to reuse data generated from upstream work, data lineage problems or data-processing blind spots can creep in, causing isolated problems. This can result in stale data, mainly due to broken plumbing (bad upstream jobs) or system-wide problems caused by data corruption from any critical point in the data network.

As the complexity of your data pipelines grow, you will see an increasing need for automation. At first, simpler solutions like crontabs and simple workflows triggered by webhooks and APIs will keep things running. However, as things expand and scale over time, it is common to migrate to more specialized workflow orchestration frameworks to manage the larger data-processing footprint. You learn how to use Apache Airflow with Spark in detail in Chapter 8, when we look at Workflow Orchestration. This chapter provides the blueprint to create reliable batch-based Spark applications that are ready to be dropped into Airflow.

Pipeline Foundations

Over the previous six chapters, you've been preparing for your journey to data pipelines as you've learned the important history, concepts, components, and foundations required by the modern data engineer. As a refresher, the first two chapters introduced and taught you about what has been and where things have gone and got you up and running with Apache Spark locally. Chapters 3 and 4 explored reading, parsing, transforming, and writing data, and discussed the importance of data documentation, strong types, and paying things forward for others.

With the introduction of the JDBC data source in Chapter 5, you learned to interoperate with your existing data stores using Apache Spark in order to work with external databases. Chapter 6 introduced the Spark SQL Catalog. You learned to power Spark SQL using your external Apache Hive Metastore. This can simplify the process of working with external or managed tables and can solve some of the common problems related to data discovery, data ownership, and data access.

Along the way, you've been learning to use Spark to solve data problems without needlessly getting sidetracked thinking about how to write applications that can be compiled, unit tested, documented, and released following standard software development best practices. You learn to apply common software engineering best practices to the work you do as a data engineer in this chapter.

Spark Applications: Form and Function

Until this point in the book, you've relied on Apache Zeppelin or the manual process of using the `spark-shell` to feed commands and process data interactively. These *interactive style applications* are just one of the many forms a Spark application can take. The other two forms the Spark application can take are the *batch-based application* and the *streaming application*. Looking at the similarities and differences between these three styles can help you decide what style of Spark application you want to write. This process is also a critical decision-making point in the journey from idea to production.

Interactive Applications

As the name denotes, interactive applications are alive in a sense, as they are waiting to read, evaluate, and process input from a user. This mechanism provides a nice immediate feedback loop for many different use cases, including data discovery and exploratory data analysis, as well as a tool to lean on when learning or trying something new. The Spark Shell and Apache Zeppelin (notebook environments) provide a gentle introduction to using Spark without needing to compile code and set up any IDEs. There are limits to those approaches though.

Spark Shell

The REPL that comes with Spark (aka, the `spark-shell`) was covered in detail in Chapter 5, as we explored how to create quick dynamic applications on top of the Spark runtime environment. Although you can build fairly complex routines with a mix of dynamic file loading and Spark configuration, the limiting factor is that these shell applications are not designed to be completely consistent or reliable for the prime-time needs of production systems. At best, this mode of operating is like a powerful interactive scripting environment.

Notebook Environments

We've used Apache Zeppelin throughout the book to simplify how we work with Apache Spark and Scala using notebooks. The notebook environment is a powerful tool. It brings together the simplicity of the `spark-shell`, in a consistent runtime, that can also be shared among your peers by passing notes or working together in real-time (hosted Zeppelin). Zeppelin enables ideas (heuristics and code flow) to be fully fleshed out, in a cascade of paragraphs that can be run as a synchronous flow (holistically) or individually, which is commonly the case when debugging a paragraph.

The icing on the cake (at least from my perspective) is the ability to mix inline documentation with interactive graphs and tables, alongside the data processed in each note. We return to Zeppelin for analytical processing and insight generation in Chapter 12.

There is also a healthy mix of pros and cons when using notebook environments on the job. While it can be straightforward to write a fairly sophisticated series of data-processing operations against both internal and external data sources, this ease of use also brings with it the potential to harm production systems. Consider a note that makes many relatively expensive queries to non-production data systems (in a non-production environment). While the intention of the engineer writing the notebook is not to cause harm, it is possible for these expensive queries to take down a service, or database, causing an incident if the same note is migrated to production or saved accidently.

Tip   Know the size of your data: Understanding the size of the data you are working with, not only the total number of records/rows, but also the required memory and IO (bandwidth) overhead. Understanding the size of the data you are processing is critical to operating healthy data applications. This is why many applications start out in the planning phase using notebooks and end up as compiled applications for stream or batch as a prerequisite for moving to production.

Batch Applications

Batch applications don't have to be any more complex than some of the simple paragraphs you've written inside of Zeppelin. The main difference is the process that goes into creating a compiled batch application as opposed to writing a quick note or paragraph in Zeppelin. We'll walk through the process of building a blueprint for batch-based applications in the next section of this chapter, but before we do, here are some things to keep in mind.

Like traditional software development best practices, writing solid, reliable data applications requires you to write code that includes tests (unit and automated), meets a team's or company's standard of excellence, goes through code review, and is released in a phase-based approach. It must also ensure that downstreams, as well as other dependent systems, are protected. It finally must ensure that each job runs in a performant way, with the proper metrics and application-level monitoring to ensure each job fulfills and meets agreed-upon SLAs.

When it comes to writing batch applications with Spark, there are two approaches that can help as you design your interfaces and configurations. Applications can be *stateless* or *stateful*.

Stateless Batch Applications

Stateless batch applications don't require state management. This approach to writing batch applications is intended to handle very specific tasks where all necessary configuration is provided to the application at runtime. This kind of app is well suited for traditional data pipeline jobs. For example, the pipeline job could be configurable to import all rows from an external table between a start and end date. Because there is a clear separation of concerns and no long-term state management, each run of the batch application can dictate its own boundaries. For instance, if you were required to backfill data, you could create a series of one or more jobs and trigger these jobs to run in parallel without any concern of cross-contamination (if the boundaries are specific, such as loading data partitioned by day). *Backfilling* is the process of loading missing data into a data source.

Tip   Design for Reuse: While you can create single serving applications, it is better to consider how your applications can be built for reuse driven by external configuration. Otherwise, have to repeatedly make minor changes to internal configuration, release, run, and repeat.

An example of a stateless batch application is a daily reporting job. Reporting jobs typically run against a specific subset of data, such as a daily or hourly metric or event rollups. The batch job configuration shown in Listing 7-1 shows how an example application could be configured to run externally.

Listing 7-1. External Configuration Allows Your Batch Applications to be Written Once and Used in Different Ways

```
$SPARK_HOME/bin/spark-submit \
  --class "com.coffeeco.reporting.DailyReport" \
  --master "spark://thehost:7077" \
  --deploy-mode "client" \
  --conf "spark.report.table.source=daily-summaries" \
  --conf "spark.report.table.sink=daily-reports" \
  --conf "spark.report.range.start=2021-03-01" \
  --conf "spark.report.range.end=2021-03-02" \
  ...
```

The opposite of the stateless batch application is the stateful batch application.

Stateful Batch Applications

Stateful batch applications can be designed to use any number of external systems to reliably manage their configuration and their application state, such as Apache Zookeeper or HDFS. However, an alternative trick is to use Spark Structured Streaming configured to run exactly once per trigger. This technique enables you to lean on the stateful semantics of the powerful Structured Streaming Engine to track and manage the effective state of everything, from the data source(s) to what data has been processed and fully committed, for each run of the batch job. All this without having to manually capture and manage the job metadata from batch to batch in your data pipeline. With more complex data-processing jobs, which include reading and joining data across multiple external data sources across a mix of batch and streaming sources, manually managing the external state can become unwieldly quickly.

Tip   If your stateful batch job will be run as a periodic batch job, used to reliably move data from a streaming source into a reliable data store like HDFS or S3, determine which aspects of the job can be reused to build generic base applications. Generic applications can be configuration-driven or be built on top of abstract classes (libraries) and shared among common interfaces. Base applications enable common data engineering problems to be solved once, and for the lessons learned to be reused without the need for long lead times from idea to production.

From Stateful Batch to Streaming Applications

Another common use case for stateful batch applications is the periodic processing of data from streaming sources, such as Apache Kafka. There is lot of metadata required to reliably track your application's position within a stream of data, across all distributed partitions of a given topic, that simply isn't well suited for stateless batch jobs.

Lastly, you can use stateful batch applications to help when performance tuning your applications, before turning them on for full-time streaming. By mixing external configurations with internal stateful processing, you can turn up the number of records processed per batch and test how the application performs across a mix of different CPU/RAM allocations. Applications that can be tested first in batch can be ramped up to their breaking points or configured for maximum throughput of data. When you are happy with the results, with the switch of the trigger, your application can run as a streaming job, picking up right where the last batch left off. You learn how to write reliable stateful batch jobs in Chapter 10, including the ins and outs of using external configuration to control the processing frequency and rates per micro-batch.

Streaming Applications

Apache Spark enables you to build applications that process unbounded data sources without the need to change your mental model too much from that of a batch processing one. Given that Spark Structured Streaming applications default to running as a series of micro-batches of data that is continuously processed, it is fairly straightforward to think about how to tackle things batch by batch. Spark enables you to work with streaming data across two main processing modes: *micro-batch processing* and *continuous processing*.

Micro-Batch Processing

This processing mode periodically checks a streaming data source (like Apache Kafka) and triggers a new batch to be processed. This means you can think about your application as working on periodic sets of finite data across an infinite stream of data.

Working with micro-batches of data is better suited for common analytical processing needs, like rolling up metrics across windows of time. Remember back to Chapter 4, when we looked at joining the current store occupants of each of our coffee shops to construct a way to figure out if there were enough seats for a party of a variable size. This kind of event processing is a great candidate for real-time, as these streams of data can quickly become irrelevant if not processed in a timely way. For instance, this data could be used to drive a mobile app that helps customers find the closest location to them that has available seating.

However, considering the stateful batch use case from earlier, this data is also interesting from a historical perspective, just not with the same fidelity. You may need to roll this data up into 15-to-30-minute intervals in order to enable other teams to generate daily/weekly/monthly performance reports on a per store basis.

Continuous Processing

Rather than working across micro-batches of data, continuous processing enables individual processing of each record, ensuring that each data point is processed in a stateful way, exactly once. This kind of processing can be used to enable real-time billing use cases, where you want to process each transaction individually (and durably). For billing it is especially important that each operation either succeeds or fails, at the event level, as opposed to having the possibility of at-least-once processing, which means you may double-bill a customer. In the case of a partial failure within a micro-batch, Spark will rewind and reprocess the slice of the dataset, which is a common pitfall with micro-batch stream processing and billing.

Note   We don't cover continuous processing in this book, but if you are curious, you can use the environment in Chapter 11 (Apache Kafka) to test things yourself. (At the time of writing, the only data source supporting continuous processing is Apache Kafka.)

Understanding the various operating modes that a Spark application can take is the first step to designing reliable data applications. The second step, is of course, actually writing the application. We'll go through an exercise now to prepare you for the application you'll be writing in the rest of the chapter.

Designing Spark Applications

As you know, data engineering is essentially the art of moving and transforming data efficiently and reliably, and to do our jobs well each job, or series of jobs we write to run in our data pipelines should be the result of a specific data need, encoded into a published data application that goes through the traditional software development cycle. The software development cycle is shown in Figure 7-2. It encapsulates the seven common conceptual phases of execution required to build and operate software successfully. The same process can be (and should be) applied to the planning and successful operations of any data engineering team or data organization.

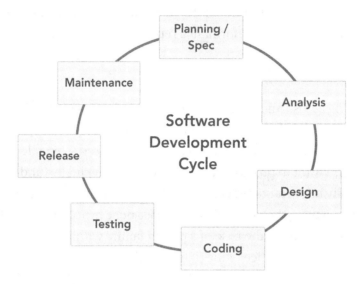

Figure 7-2. *The software development cycle*

Reliable data applications follow a traditional software development cycle. What you are planning to write, why you need to write it, who the intended audience is, and when things are required to be completed by—all these questions branch off from traditional specification-based planning and ideation. These plans lead to work that requires quick and careful analysis, before setting out to design and write the code required to accomplish a task. You may find yourself working inside a notebook environment like Apache Zeppelin during the planning, analyzing, and design phases of the project. You may also find yourself returning to these interactive environments to debug issues in production, or to quickly pull up a graph to share with your coworkers.

Tip   Interactive environments are pure gold for several reasons, but they can be invaluable for running quick research spikes. They help you find gaps early in the software cycle, before you end up spending time formulizing the code, writing unit tests, going through code reviews, and preparing for the production release.

We'll be working through an extended use case in this next section. The purpose of this exercise it to help you as you think about solving problems with data. Given that data engineers focus on data, learning to think abstractly about data and modeling data are essential skills. Data modeling is an essential skill for any data practitioner, and while conversations about taxonomies, naming conventions, and general theory are outside the scope of this book, this next use case will introduce you to the art of framing foundational data problems.

Use Case: CoffeeCo and the Ritual of Coffee

Not everyone enjoys coffee or tea. However, when it comes to rituals and community, there have been huge successes behind the business of coffee. Early on in this book, we touched upon some of the odds and ends of being a new data engineer on the job for a company called *CoffeeCo*. Consider this to be any of the major coffee institutions you routinely visit. Or, if you have a strong aversion to coffee, you are equipped with the tools to process and transform this data into something that's more to your liking (and I hope we can still be friends).

Throughout this chapter and the rest of the book, you'll be focused on writing Spark applications to facilitate and coordinate the data needs of CoffeeCo. You'll be extending the basic primitives from earlier chapters, like our model for the coffee shops, inventory, customers, and customer purchasing behavior. You'll be using this data while writing purpose-driven applications to drive the CoffeeCo business just like you would on the job. These lessons can be applied to just about any work you'll come across when working horizontally across a data platform, or vertically as a data engineer focused on a specific feature or business unit. The lessons learned can be used to solve common problems applied to most businesses.

Consider yourself promoted. You are now the founding data engineer at CoffeeCo. Your first task is to build the core data models that can be used to represent many facets of the CoffeeCo business. To do that, you'll need to understand more about the business.

Thinking about Data

If you look from a high enough vantage point, you'll notice that there is a common thread among most customer-facing businesses. That being the fact that businesses have some *product* to sell, *places* to sell their product (either online or at brick-and-mortar shops), and *customers* who purchase their products. It is like the notion of a noun: Business revolve around people, places, and things.

Thinking about the data needs of a company, which aspects of the business need to be captured to make informed decisions, and how do we extract meaning from the data? These can be monumental tasks for all new data operations. Tactical strategies differ from place to place, but you can use one mechanism that can be profoundly helpful, and it's taken directly from product management and the process of user stories. To simplify it, we will use the ever-creative name, *data stories*.

Data Storytelling and Modeling Data

No matter where you work, and how you work, all jobs revolve around the notion of tasks. This is the process of decomposition; you break down large ideas or chores into smaller tasks that can be completed without the need to think about everything at once. This is a skill learned over time. Many engineers work for many years learning how to think about a problem, dissecting it into functional sub-components, and reusable classes and interfaces. These skills can easily transfer to how we think about data problems, since data engineering is closely linked to traditional software engineering. It just takes a little practice. Luckily, you'll be walking through the high-level data modeling process now.

Exercise 7-1: Data Modeling

The process of storytelling can be a useful tool for understanding the data relationships necessary to break up a complex idea like a company. In this case, our company is CoffeeCo. We will break it into a hierarchy of *ideas, entities, categories,* and *relationships* used to construct the data models. The process can even be useful for strategic planning and decision making when it comes to what data to capture and integrate. Depending on your role at the company, you may be helping to drive a lot more outside of your role as a data engineer. Understanding the big picture or how things fit into place (with respect to data) is a valuable skill to have. Data modeling, like many things, begins with a story.

The Story

CoffeeCo is focused on providing people with a unique coffee drinking experience, custom tailored to their individual needs and tastes. We start off with the best quality coffee sourced through a global network of coffee growers who are like family to us. These critical relationships ensure that we have a steady stream of seasonal product that grows naturally and organically. As an added benefit, it ensures that each variety of coffee comes from a well-known region that supports the growing requirements and natural flavor profiles we are looking to achieve. Using state-of-the-art personalization systems, we enable our customers to rate all aspects of the coffee experience, from the way that their coffee is roasted, to the variety of bean, favorite farms and producers, and even which shops are their favorite. Each cup, every visit, and each individual experience increases their customers' love of their coffee as well as the community they build around it.

Breaking Down the Story

You may have found that your brain has already started the process of breaking down and digesting the CoffeeCo story. You may have started to separate specific relationships and consider names and overarching entities that could be used to convey a specific high-level concept, such as vendor(s), location(s), and customer(s). You may have skipped past the data modeling entirely and focused instead on what the data can be used for and how. This is all good, as it means the story has your attention. Let's look first at breaking out the specific entities, and then we can move on to how the data can be used.

Extracting the Data Models

At a high level, you can use the story to dictate the relationships between the core entities that support the business. Essentially, CoffeeCo provides each *customer* with a *product*. Products are sourced from *vendors*. Each vendor operates within one or more specific *locations*. Products are sold across *stores* and each product has availability at one or more *locations*. The customer experience can be tracked through *ratings* across physical products consumed by each customer. The experience in each store can be graded by *customer ratings*. Ratings as a construct can be used to generate a personalized experience that adapts and learns over time for each customer.

Once you've extracted the high-level entities, you can break down each type further by telling their story. As you will see, this process can yield a good data model directly from a story.

Customer

A *customer* is associated with an account identifier. Each account is uniquely identified based on a randomly assigned UUID that is either generated online at account creation time or alternatively made available to a customer within at the time of purchase in a store via a QR code. Customer accounts become active when they finish registering and linking their account identity with the friendly mobile application. This requires their first and last name as well as their email address.

```
case class Customer(
  active: Boolean,
  created: Timestamp,
  customerId: String,
  firstName: String,
  lastName: String,
  email: String,
  nickname: String,
  membership: Membership,
  preferences: Preferences
)
```

This customer data model is used to plan baseline information as well as relationships, such as in the case of the *membership* and *preferences*. These are placeholder relationships to the customer and can be composed later in the data modeling journey.

Store

Customers buy coffee products at the various store locations. Each store is associated with a unique identifier, and each location has additional information regarding the date it opened, the physical location (such as the address, longitude, and latitude), and the hours of operation. Each store also has a maximum occupancy limit and additional amenities, such as outdoor seating, drive-thru, and many others. In thinking about the differences between one store and another, there is a daily menu. This is the available coffees, prices, and all that jazz.

```
case class Store(
  alias: String,
  storeId: String,
```

```
  capacity: Int,
  openSince: Timestamp,
  openNow: Boolean,
  opensAt: Int,
  closesAt: Int,
  storeInfo: StoreInfo,
  location: Location
)
```

Just like with the customer data model, there are breadcrumbs for a follow-up exercise for the StoreInfo, Location, and Menu data models. We visit these some more in the next section.

Product, Goods and Items

Each product is a unit of good to be sold. In this case, there are more items that can be sold than just the coffee. Each type of product sold has a unique identifier, a friendly name or alias that can be used as a human recognizable name, and a unit cost. Given you can sell a cup, carafe, instant coffee, freshly roasted coffee, and other bulk items, you need to think abstractly here. Given each product sold is on the menu, and given Scala has a core Product class, we will give this model the name Item.

```
case class Item(
  itemId: String,
  availability: Availability,
  cost: Int,
  name: String,
  points: Int,
  vendor: Vendor
)
```

Given that items have general availability, seasonal availability, and even geographic availability, we can model the items' availability as a sub-resource of the Item case class. The same is true with the Vendor information. This allows us to work with the data and allows the data model to evolve over time.

Vendor

CoffeeCo works with various vendors from all over the world. The vendors provide goods and services, like the farms that grow and produce seasonal varieties of coffee, all the way to the manufacturers of cups and trinkets for sale in each store. We will only be working with coffee-related data in this book, but it is worth thinking about how data models can be expanded in the future, so that you aren't having to juggle tons of different data models that end up being more or less the same thing.

```
case class Vendor(
  active: Boolean,
  name: String,
  location: Location,
  vendorId: String
)
```

You may have noticed that each vendor has one or more products, and that each product only has one vendor. This lets you decide how you want to join data while processing a specific pipeline. Consider that each vendor may also operate across many locations, so tying a vendor to a single location can cause problems. You may want to add the notion of an *operation*. Vendors can have many operations that exist across different locations, producing different products. I'll leave this as food for thought.

Location

The location is an abstract concept that deals with a point of origin. Consider that businesses operate at specific locations, either online or in the real world, and the same goes with the Location data model. The data model includes information common to many locational needs.

```
case class Location(
  locationId: String,
  city: String,
  state: String,
  country: String,
  zipcode: String,
```

```
  timezone: String,
  longitude: String,
  latitude: String
)
```

Location data is useful for correlating trends and patterns of customer behavior within a system. In the case of CoffeeCo, this locational data can be used to understand which vendors produce certain styles of coffee that a customer particularly enjoys. This data can also be used to understand which stores provide the best experience across most customers and whether the physical location plays a critical role. Sometimes a mixture of outdoor seating, a sunny day, and good cross-traffic is all it takes to brighten someone's day.

Rating

The last data model from the CoffeeCo story is that of a *rating*. Given a rating is issued by a customer for a particular store or item, we can track this information generally using a rating system. For simplicity, we can start with something basic as many systems do. Remember you can always add complexity later.

```
case class Rating(
  created: Timestamp,
  customerId: String,
  score: Int,
  item: Option[String],
  store: Option[String]
)
```

The data model for ratings is broken down into a flexible format that enables a framework for ratings. This includes support for product- or store-based ratings that are scored as individual customer ratings.

Exercise 7-1: Summary

The process of turning stories into data has been around for as long as people have been painting in caves. If you think about it, a picture tells 1,000 words (well, not every time, but you get the idea), and well-defined data models can be used to capture data in the same way, either at a high level of resolution or simply as an outline of what was. This exercise was intended to help you conceptualize the data modeling process and to think about data composition.

From Data Model to Data Application

Now is the time you've all been waiting for. It is time to start writing your first major Spark application. This process will teach you how to architect and write testable application code that you can use as a blueprint for the myriad apps you'll write down the line. The source code for this exercise is available in the chapter materials.

Exercise Materials   You can find the materials for this chapter in the `ch-07/` folder of the book's GitHub resources at `https://github.com/newfront/spark-moderndataengineering`.

Every Application Begins with an Idea

You have been asked to build out a new data pipeline that will ingest customer ratings and join each rating on the customer's table, essentially doing a glorified ETL. This data can then be further processed downstream to create more complex datasets, such as each customer's preferences (or a taste profile, or a store profile).

The Idea

After a little brainstorming, you come up with a simple data model for these rating events and decide to test things out quickly using the tools you currently have at your disposal. What are these tools, you ask? Well, you have you IDE (I am partial to IntelliJ) and the ability to reuse your local Docker environment to move fast and test your ideas.

However, you need a plan of attack. You need one so that you can timebox the amount of work needed to vet your idea. This is where the application blueprint comes into play. If you think about the work you've done within the Zeppelin environment up until now, then you can think of this next step as a way of solidifying your notebooks or workflows into a compiled application.

Exercise 7-2: Spark Application Blueprint

The Spark application blueprint is the framework, or boilerplate code, needed to create an application. This includes the general layout, configuration, and build settings required to compile and assemble a fully functioning Spark application.

As a general rule of thumb, each of your Spark applications should be laid out the same way every time. This creates a consistent way of doing things and gives you and everyone writing applications a similar frame of reference for where to look for configuration files, as well as the expectations for how to build and run the app.

Default Application Layout

Open the project code located in ch-07/app. You will see the directory structure and configuration files in Listing 7-2.

Note   Feel free to skip ahead to "Common Spark Application Components" if you've been writing Java and Scala applications for a while. This next section is an overview of the application directory structure.

Listing 7-2. The Spark Application Directory Structure

```
app/
  conf/
    local.conf
  project/
    build.properties
    plugins.sbt
  src/
    main/
      resources/
        application.conf
        log4j.properties
      scala/
    test/
      resources/
```

```
      application-test.conf
    scala/
  build.sbt
  README.md
```

The nested directories and files shown in Listing 7-2 are typical of any application written in Scala and should be familiar to anyone who has written Java applications. Let's go over the structure, starting at the root level.

README.md

This is your chance to inform and guide other engineers who will inevitably be working on applications you worked on down the line. It is important to document the requirements to build and run your application, as well as any additional information that can be of use to future engineers. If you use GitHub then your project will automatically render this file when someone goes to check out your code.

build.sbt

This is your application's *build definition*. The build.sbt file guides the build, providing variables and rules, including which version of Scala and Spark to compile against. It also assembles a fat JAR (rolls up all of your dependencies into one JAR, which simplifies how you deploy your application). This is leans on additional configuration files living inside the project directory in the project's root. Together this provides you with the tools you need to build and release your applications.

conf

The conf directory is where you store your *external* application overrides. This is a common pattern for supplying the correct configuration for multi-environment overrides, such as your development, staging, performance, and production environments, as well as for regional overrides. Lastly, it is useful to supply a local testing configuration, which can be used when running your application in your local development environment.

Multi-Environment Configuration

```
conf/
  local.conf
  dev.conf
  stage.conf
  prod.conf
```

project

If you are new to `sbt`, the `project` directory is your one-stop shop for configuration. The `build.properties` tracks which version of `sbt` you want to use., as well as any declared plugins and resolvers. Plugins add capabilities to your `sbt` builds, while the resolvers tell `sbt` where to find dependencies, such as Apache Spark and others.

src

The sources (`src`) directory is where your application code lives. This is your application root. The directory structure in Listing 7-3 shows the application source code and resources.

Listing 7-3. The Spark Application Source Code and Runtime Resources

```
src/
  main/
    resources/
      application.conf
      log4j.properties
    scala/
      com.coffeeco.data/
        config/
          Configuration.scala
        SparkApplication.scala
        SparkBatchApplication.scala
        ...
  test/
    resources/
      application-test.conf
    scala/
      ...
```

Now that you have a general overview of the application layout, let's go over the common components you can use in your applications.

Common Spark Application Components

Each application you write will have common components that can be abstracted away for reuse. This ensures that there is a simple wireframe for doing initialization and setting up (bootstrapping) the application runtime. In this chapter, you're introduced to the following three common components:

- `Configuration.scala`. This component is intended to be reusable across many Spark applications and is a general means of providing layered configuration for an application. Unless a config value needs to be overridden at runtime, it is encouraged to provide the lion's share of your application config through the `application.conf` or within any environment or regional override conf. This way, all required configuration can be distributed alongside the application through your CI/CD pipelines.

- `SparkApplication.scala`. This component is intended to be used in conjunction with the `Configuration.scala`. When generating the `SparkSession` for your Spark applications, there may be additional default behaviors that you want all Spark applications to inherit. For instance, monitoring the application performance, operational metrics, as well as problems resulting in task failure. Having eyes and ears on your application ensures that you know when each app is no longer running in a healthy state. We cover monitoring and revisit the `SparkApplication` trait again when we look at "The Road to Production" in Chapter 12.

- `SparkBatchApplication.scala`. This trait extends the `SparkApplication` trait and is used specifically for batch-based applications.

Let's begin by looking at the `Configuration` object.

Application Configuration

The Configuration object provides the Spark application with a customizable means of generating explicit settings to supply to the core SparkConf on application initialization. There are a few different ways that your application arrives at the state of its current configuration. In Chapter 5, you looked at how to access and modify the SparkConf using the conf parameters on the spark-shell, and in Chapter 6, you learned a little more about the spark-defaults.conf file, which handles the cluster-wide Spark configuration. Now we are introducing one more way, which we cover in detail.

Let's start by looking at the Configuration.scala source code (shown in Listing 7-4 and available in the online chapter contents).

Listing 7-4. Common Configuration Helper

```scala
object Configuration {
  private lazy val defaultConfig = ConfigFactory.load("application.conf")
  private val config = ConfigFactory.load().withFallback(defaultConfig)

  config.checkValid(ConfigFactory.defaultReference(), "default")

  private lazy val appConfig = config.getConfig("default")
  lazy val appName: String = appConfig.getString("appName")

  object Spark {
    private val spark = appConfig.getConfig("spark")
    private val _settings = spark.getObject("settings")
    lazy val settings: Map[String, String] = _settings.map({ case (k,v) =>
      (k, v.unwrapped().toString)
    }).toMap
  }
}
```

The Configuration code shown in Listing 7-4 is built on top of the popular Typesafe Config library. This library plays nicely with Apache Spark and enables your application to create layers of config, which can be lazily loaded, evaluated, and mixed into the runtime configuration for any of your Apache Spark applications. Let's take a peek at the format of the application.conf to understand how the config is generated.

Application Default Config

The default application config is compiled along with your Spark application during app assembly. It provides a guaranteed common base of configs for *any* instance of your application. Think of this as a first concrete layer of config foundation. The defaults are compiled along with your application and can be found in the app source code under `ch-07/app/src/main/resources/application.conf`. See Listing 7-5.

Listing 7-5. Application Default Config (Partial Config Shown)

```
default {
  appName = "spark-event-extractor"
  spark {
    settings {
      "spark.sql.session.timeZone" = "UTC"
      "spark.sql.catalogImplementation" = "hive"
      "spark.sql.hive.metastore.version" = "2.3.7"
      "spark.sql.hive.metastore.jars" = "builtin"
      "spark.sql.hive.metastore.sharedPrefixes" = "..."
      ...
    }
  }
}
```

The `application.conf` is loaded and parsed by the `ConfigFactory` object in Listing 7-4. It's loaded into the `default` named `config`. The process from Listing 7-4 also checks the config's existence and validity, ensuring that a corrupt object doesn't exist at the `default` path. If the config is corrupt, the parsing of the config will fail with a `ConfigException` and your application will fail in turn. *Recall that failing immediately is better than failing later at runtime.*

Once the config has been parsed, your custom configuration is available at the named default config.

```
private lazy val appConfig = config.getConfig("default")
```

From this point in Listing 7-4, you have access to the application name `appName` and the map contained under `default.spark.settings`. This is really all you need at this point, since you want to have a common way of shipping the application config alongside the application code.

This base set of configuration values can also be modified at runtime using an external config file. This will effectively load the `application.conf` from your compiled application resources and then overlay (or override) the application defaults on application start. You will see this pattern used when you have regional specific configurations or settings based on the runtime environment common with development, staging, and production environments.

Runtime Config Overrides

Being able to augment your application using additional external configs enables many capabilities, from simply adding regional or environment overrides to more advanced automation like automatically augmenting an application CPU core and RAM profile . Regardless of the case, from the simple to the cutting edge, using explicit overrides enables you to easily change your application without having to recompile and rerelease your application. Consider the following use case.

Say you are required to run a reporting job across two separate time zones. If the application default specifies `"spark.sql.session.timeZone"="UTC"` then you could simply override the default `application.conf` and seamlessly handle this use case with a simple override file named `timezone-pst.conf`. Now you may have additional requirements, but in the most simple use case, you can run a consistent job in a specific time zone.

```
default {
  spark {
    settings {
      "spark.sql.session.timeZone" = "America/Los_Angeles"
    }
  }
}
```

Additionally, you could also achieve the same override using runtime configuration overrides in your Spark submit (like you learned in Chapter 5).

```
--conf "spark.sql.session.timeZone= America/Los_Angeles"
```

Note   In Chapter 8, you learn how using Airflow can add more control to use cases like the time zone shifting reporting job. Each DAG represents a workflow in Airflow and ensures that runtime overrides can be set in stone. That means your downstream dependencies are never left looking for the data you are supposed to be producing. Consistency and reliability are your top priorities as a data engineer.

With the `Configuration` object out of the way, it is time to see how this is used alongside the `SparkApplication` trait to provide a seamless Spark initialization.

Common Spark Application Initialization

The `SparkApplication` trait is a simple module that provides a common way to initialize your Spark application. It leans on the `Configuration` object from the last section to streamline the generation of the immutable `SparkConf`, which is then feed into the construction of the `SparkSession`, which is the entry point into the Spark runtime.

So how exactly is this accomplished? Look at the trait shown in Listing 7-6.

Listing 7-6. The SparkApplication Trait

```
trait SparkApplication extends App {

  val appName = Configuration.appName

  lazy val sparkConf: SparkConf = {
    val coreConf = new SparkConf()
      .setAppName(appName)
    // merge if missing
    Configuration.Spark.settings.foreach(tuple =>
      coreConf.setIfMissing(tuple._1, tuple._2))
    coreConf
  }

  lazy implicit val sparkSession: SparkSession = {
    SparkSession.builder()
      .config(sparkConf)
```

```scala
      .enableHiveSupport()
      .getOrCreate()
  }

  lazy val validationRules: Map[()=>Boolean, String] = Map.empty

  def validateConfig()
  (implicit sparkSession: SparkSession): Boolean = {
    if (validationRules.nonEmpty) {
      val results = validationRules.foldLeft(List.empty[String])(
        (errs: List[String], rule: (()=>Boolean, String)) => {
          if (rule._1()) {
            // continue to next rule
            errs
          } else {
            // if the predicate is not true, we have a problem
            errs :+ rule._2
          }
        })
      if (results.nonEmpty)
        throw new RuntimeException(s"Configuration Issues Encountered:\n
        ${results.mkString("\n")}")
      else true
    } else true
  }

  def run(): Unit = {
    validateConfig()
  }
}
```

Listing 7-6 shows the source code for the SparkApplication trait. This trait, or mixin, extends the Scala App trait, which is included in the core Scala library. The App trait provides functionality to quickly turn any object into an executable, which just so happens to be a seamless starting point for initializing our Spark applications (since the executable means that we also have a main class). The other reason we extend the App trait is because it delays initialization for our Spark application. This means we have guarantees that our Spark application will be evaluated after all other static classes and objects are initialized.

In this first pass of the SparkApplication trait, you'll probably notice that there is nothing too complicated happening. What you are doing is ensuring that any application that mixes in this trait has access to a common interface of commands for config validation, and generation of the SparkSession and SparkConf, and that all applications have a run method. Simple in theory, right?

Building an entry point to run each Spark application now only requires a little setup to get started. You can essentially begin writing every new application without having to worry about common boilerplate and supporting many ways of doing similar things. Rather you now have a common framework for running applications holistically. The scaffolding being done behind the scenes is similar to how the spark-shell and Apache Zeppelin are initialized.

The last common component to look at is the SparkBatchApplication trait, which is a minor extension to the SparkApplication trait. It provides you with a consistent set of methods for creating batch applications.

Dependable Batch Applications

The trick to building common components is to start small and work smart. Also, it is worth pointing out that common components should be the end result of more than one application doing exactly the same thing. Many of the best libraries are built by solving common problems, so let nature run its course and get feedback and input from folks on your team, in your community, or online. Then watch how things evolve. Finally, ensure that your libraries simplify a problem and introduce guardrails only when they are necessary.

With that in mind, Listing 7-7 is an example of starting small and ensuring that all batch applications have some guardrails in place.

Listing 7-7. The SparkBatchApplication Trait

```
trait SparkBatchApplication extends SparkApplication {
  lazy val saveMode: SaveMode = SaveMode.ErrorIfExists
  def runBatch(saveMode: SaveMode = saveMode): Unit

  final override def run(): Unit = {
    super.run() // will call validate config
    runBatch()
  }
}
```

Any application inheriting from the `SparkBatchApplication` trait will need to define the `runBatch` method and add any config validations to get started. The trait also helps guide applications that inherit from it to be defensive by default. While the `saveMode` isn't enforced at the trait level, implementing the `runBatch` method requires application engineers to think about the `saveMode` (even if they weren't before). This isn't forcing batch jobs to do the right thing, but it provides a guide about what is important to think about for each application.

Thinking about this another way, the `run` method, which is overridden from the `SparkApplication` trait, is marked as `final` in the `SparkBatchApplication` trait. This restricts any implementing class from overriding the `run` method again, so in other words, engineers only need to concern themselves with the context of the `runBatch` method. Trying to work around the interface will fail. This forcing function enables you (as a library author) to do interesting things, including powering generic application launchers that require only a `run` method. In this way, whether you are launching a batch job or a streaming job, you only need to concern yourself with a single method. You should also provide the appropriate guardrails so you leave enough flexibility that people will continue to use your libraries!

Follow-up Exercise   It is only natural to provide these helper libraries as a common library. This way, all of your team/company applications can be built using the same common set of utilities, rather than leaving it up to each new application to reinvent the wheel. This can become tiring fast. If you are feeling up to it, after this chapter is over, look at extracting these common Spark components into a separate project.

Exercise 7-2: Summary

This second exercise was a tour of the Spark application layout (directory structure) and the three common traits provided to help make writing Spark applications simpler. Now that you've been introduced to the common Spark application components, it is time to write the actual Spark application.

Connecting the Dots

Equipped with the knowledge of how to lay out your Spark application, the common application initialization components, and the interfaces (traits), the next step is to write the application. As a precursor to writing your first streaming Spark application, this application will instead be based on the humble batch job. This process will reinforce earlier exercises, such as the work done with the containerized database tables and the distributed file system-based tables. It builds a batch job that can handle joins between two data sources.

Note   This process will help as you transition to working with the exciting concept of *unbounded tables* and *unbounded streams* of structured data, which are covered in Chapter 10 (after a mental model-building chapter on streaming data systems in Chapter 9).

Application Goals

The goal of this application is simple: Read all customer ratings from one managed Spark table and join each rating against the customer data stored in the distributed customer's table by `customerId`. Lastly, write the results of this joined data into a new table location.

Additionally, since this chapter is also leading toward workflow orchestration, we want to run this application as a pipeline job. This means that the driver application will require input in the form of runtime config for an input (or a source table) and an output (or a destination table) when submitting the application to run. This is an example of a *stateless batch-based Spark application,* as it requires external influence to fulfill the tasks at hand.

Let's work from the outside in, starting with modeling our events, designing the runtime config, determining what the `spark-submit` command will look like that launches the application, and then moving from the main application class all the way down the stack. If you are ready, let's begin by building the `spark-event-extractor` app.

Exercise 7-3: The SparkEventExtractor Application

First things first. This application is supposed to take a customer rating event. So we need to come up with a proposal for such an event.

The Rating Event

The customer rating event needs to provide our application with the time of the event (this will become more important down the line for streaming applications), the customer's unique identifier, a rating, and any other metadata we can use to further enhance the rating event in the application.

The Result of Brainstorming the Customer Rating Event

```
{
  "created": "2021-04-18 16:28:13.198",
  "eventType": "CustomerRatingEventType",
  "label": "customer.rating",
  "customerId": "CUST124",
  "rating": 4,
  "ratingType": "rating.store",
  "storeId": "STOR123"
}
```

The initial customer rating event is defined as a simple JSON object. JSON data is simple to serialize and can be a good way to test ideas without a lot of upfront overhead. After defining the initial event definition, we can model a simple Scala case class to assist with serialization and deserialization.

CustomerRatingEventType

The customer rating event allows us to define a wrapper class that can support customer ratings across stores and items, as shown in Listing 7-8.

Listing 7-8. Data Modeling with Case Classes Can Save Time

```
case class CustomerRatingEventType(
  override val created: Timestamp,
  override val eventType: String = "CustomerRatingEventType",
```

```
  override val label: String,
  override val customerId: String,
  override val rating: Int,
  override val ratingType: String,
  override val storeId: Option[String] = None,
  override val itemId: Option[String] = None
) extends CustomerEvent with RatingEvent
```

The beauty of modeling structured data definitions with case classes is that it's a flexible way to write unit tests and work with the DataFrame and Dataset APIs in Spark.

Tip   You can view all of the data models for this chapter in the `models` directory, inside the application root `ch-07/app/src/main/scala/com.coffeeco.data.models`.

Now that you've defined the data model for this customer rating event, it is time to move on to configuring the application.

Designing the Runtime Configuration

Next, we need to come up with the names for the configuration that can be fed to the application at runtime. Following the convention of namespacing our Spark config as `spark.{app.name}.{key.name}`, we'll use the following names:

- `spark.event.extractor.source.table`: Provides the name of the source, or upstream data table.

- `spark.event.extractor.destination.table`: Provides the name of the destination, or the final output table.

- `spark.event.extractor.save.mode`: Provides a way to optionally overwrite data if it exists in the destination table location.

Remember, we can always add more override configurations to the application as we go, but it is good to start off with only the bare minimum needed for the functional requirements of each Spark application. I also find it to be helpful at this planning phase to understand how things will look when we are ready to launch the application.

Planning to Launch

Writing the spark-submit command first can help you picture how you'd like your application to behave when you finally go to launch it. This can help you "gut-check" your config names and solicit feedback regarding your intentions. The spark-submit command, including the new runtime config settings, is shown in Listing 7-9. (The full spark-submit command is shown in the README.md file in the ch-07/app/ directory.)

Listing 7-9. The spark-submit Command for the Stateless Batch Application

```
$SPARK_HOME/bin/spark-submit \
  --master "local[*]" \
  --class "com.coffeeco.data.SparkEventExtractorApp" \
  --conf "spark.event.extractor.source.table=bronze.customerRatings" \
  --conf "spark.event.extractor.destination.table=silver.customerRatings" \
  --conf "spark.event.extractor.save.mode=Overwrite" \
  --driver-java-options "-Dconfig.file=conf/local.conf" \
  target/scala-2.12/spark-event-extractor-assembly-0.1-SNAPSHOT.jar
```

Starting with the spark-submit command may feel counterintuitive, but in practice, it is helpful to work from the outside in. This is a bit like the test-driven development (tdd) mindset, since to write tests for your application, you must first understand what the application needs to do. So, where are we so far?

Application Recap

You've learned what is required to assemble Spark applications, including the general application layout and structure and the build directives. You were introduced to the idea of reusable common component helpers, and all of these lessons can now be bundled into a general Spark application blueprint. Additionally, you were introduced to the idea behind the app and have a loose set of goals for what it must do, what the runtime configuration will look like, and what the spark-submit command will look like that will be used to launch the app.

This is what I refer to as all of the outside concerns needed to build and run the app, except for the performance tuning, instrumentation, deployment, and monitoring steps, which will be covered in detail at the end of the book. For now, the next step is to move down to the next layer and start writing and wiring up the actual classes.

Assembling the SparkEventExtractor

As we know, each Spark application consists of a main class, which initializes the SparkSession and one or more jobs that are run as a series of stages and tasks. These are orchestrated through the SparkSession at runtime. In keeping with tradition, we will build our application in exactly the same fashion. Starting off with the main class, which is called the SparkEventExtractorApp, and continuing down through the layers of the application architecture.

Tip   Open ch-07/app/main/scala/com.coffeeco.data/ SparkEventExtractorApp in your favorite IDE as we walk through things.

SparkEventExtractorApp

The main class is using a naming convention based on the name spark-event-extractor. This pattern was used earlier when we defined the runtime configuration that would be fed into the app (spark.event.extractor.source.table). Whether you like convention-based approaches to naming or not, this can help take the burden of naming things off your mind.

While you can technically have multiple main methods, all nestled neatly into a single application, each time you deploy an application you can only pick a single main class.

```
$SPARK_HOME/bin/spark-submit \
  --class "com.coffeeco.data.SparkEventExtractorApp"
  ...
```

So, we begin again at the outset and build the app, taking advantage of our SparkBatchApplication trait. For starters, implement the methods required by the trait.

```
object SparkEventExtractorApp extends SparkBatchApplication {
  override def runBatch(saveMode: SaveMode): Unit = {}
}
```

This starting point will guide us on our way. Let's start at the top and move down. This means we can start by validating the Spark configuration needed to properly run our application.

Validate the Spark Configuration

This process is twofold. First we create some constants to encapsulate our three expected runtime config values. Second, we need to test that validity to assume that our application is prepared to execute.

Add a new child object called Conf inside the app and add some constants to this object.

```
object SparkEventExtractorApp extends SparkBatchApplication {
  object Conf {
    final val SourceTableName: String =   "spark.event.extractor.
    source.table"
    final val DestinationTableName: String = "spark.event.extractor.
    destination.table"
    final val SaveModeName: String = "spark.event.extractor.save.mode"
  }
  ...
}
```

There is nothing special going on here. We are just ensuring that we have marked the configuration constants as final. This will allow you to reuse the application constants throughout your source code or in your unit tests. Next, let's add variables to store these three runtime values.

```
lazy val sourceTable: String = sparkSession
  .conf
  .get(Conf.SourceTableName, "").trim

lazy val destinationTable: String = sparkSession
  .conf
  .get(Conf.DestinationTableName, "").trim

override lazy val saveMode: SaveMode = {
  sparkSession.conf.get(Conf.SaveModeName, "ErrorIfExists") match {
    case "Append" => SaveMode.Append
```

```
    case "Ignore" => SaveMode.Ignore
    case "Overwrite" => SaveMode.Overwrite
    case _ => SaveMode.ErrorIfExists
  }
}
```

Now we have reliable constants and deterministic fallbacks in place for your important application configs. The next step is to identify the configuration rules that will validate the application configuration on startup. Listing 7-10 includes `validationRules`, which is an immutable map of functions (predicates) and helpful error messages for malformed configurations.

Listing 7-10. Validating the Application Config Using Validation Rules

```
override lazy val validationRules = Map[() => Boolean, String](
  (()=> sourceTable.nonEmpty) -> s"${Conf.SourceTableName} can not be an
  empty value",
  (()=> destinationTable.nonEmpty) -> s"${Conf.DestinationTableName} can
  not be an empty value",
  (()=> sourceTable != destinationTable) -> s"The source table ${Conf.
  SourceTableName}:$sourceTable can not be the same as your destination
  table ${Conf.DestinationTableName}:$destinationTable.",
  (()=> sparkSession.catalog.tableExists(sourceTable)) -> s"The source
  table ${Conf.SourceTableName}:$sourceTable does not exist"
)
```

The `validateConfig` method from the `SparkApplication` trait (see Listing 7-6) provides the mechanics for processing the `valiationRules` when the application is initializing. The object encapsulating the rules is a simple map taking a lambda function (as a predicate) and the exception information if a rule is broken and the application fails to initialize.

The validation logic (see Listing 7-10) is cut and dried. It does the following:

- Checks that the `sourceTable` and `destinationTable` values are not empty. Given that the configuration falls back to being an empty string, this is an effective way to test if the variables have been initialized by real config values.

- Checks that the `sourceTable` and `destinationTable` are not the same table. Since reading and writing to the same table is most likely not the use case we are looking to solve here.

- Checks for the existence of the `sourceTable` in the `spark.sql.catalog`.

If any of these validation fail, we throw a runtime exception, and the execution of the application is halted. The output provides the information needed for anyone looking to run the batch application to fix their mistakes and move on with their lives. Useful self-service troubleshooting pays itself forward in strides when other engineers need to redeploy an application and are unsure of the status of the runbook. Rather everything they need is provided by the application.

The last thing we need to do is implement the `runBatch` method.

Write the Batch Job

The `runBatch` method is the heart of the actual application. Given the goal of this application is to join customer ratings with their customer records and then write these records to a secondary table, all you have to do is wire up that request.

Let's look first at the final implementation (shown in Listing 7-11) and then work backward from there.

Listing 7-11. Implementing runBatch to Run the SparkEventExtractor

```
override def runBatch(saveMode: SaveMode): Unit = {
  SparkEventExtractor(sparkSession)
    .transform(sparkSession.table(sourceTable))
    .write
    .mode(saveMode)
    .saveAsTable(destinationTable)
}
```

The `runBatch` method is being used as a runtime delegate. What does this mean exactly, you ask? In practice, it is much easier to test your application logic when you can separate the concerns of application initialization from the underlying application logic. This way, you can use traditional software-testing frameworks, mocking frameworks, and your favorite IDE to assist in debugging and stepping through your code. The other reason for this separation of concerns is to handle application serialization. Let's look at the Spark delegation pattern.

Spark Delegation Pattern

Given that Spark is a distributed processing engine, this ultimately requires application code to be serializable so it can be shipped around the cluster efficiently. This doesn't mean that everything needs to be needlessly serializable though.

We can rely on the fact that your core application (`SparkEventExctratorApp`) acts as your Spark driver. As a refresher, this means it's solely responsible for providing the immutable `SparkConf`, initializing the `SparkContext`, allocating executor instances, parsing, generating the execution plan of your logical application, and lastly scheduling work to be done across the executors. This whole process is delegated to the Spark framework, and it is also nicely abstracted away inside the `SparkApplication` trait.

This means that only the `SparkEventExtractor` needs to be serializable. Herein lies the heart of the Spark delegation pattern. Let's look at it in action now.

The high-level architecture of the `SparkEventExtractorApp` in Figure 7-3 is a visual representation of the application's runtime operation. So, what does the class itself look like?

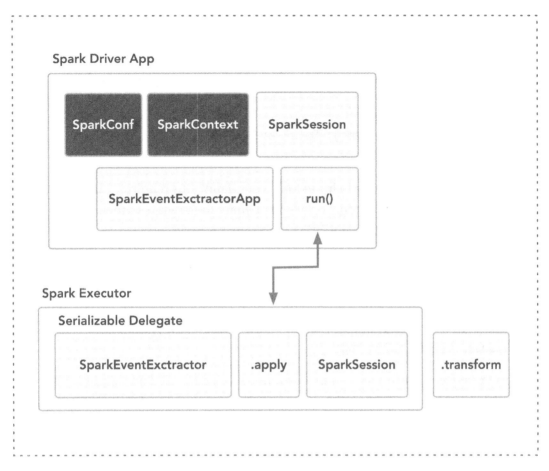

Figure 7-3. *The separation of concerns between the SparkEventExctractorApp (driver) and the work being delegated to the SparkEventExctractor class*

SparkEventExtractor

Let's start by looking at the class definition. Then we'll dive deeper down. The class signature in Listing 7-12 shows the Spark Delegate class.

Listing 7-12. Spark Delegation Pattern: The SparkEventExtractor Delegate Class

```
case class SparkEventExtractor(spark: SparkSession)
  extends DataFrameTransformer {}
```

The class is designed as a case class. This immutable construct provides inherent serialization, as well as an autogenerated companion object. The companion object has exactly one instance that is lazily generated, and the beauty here is that means Spark can

241

rely on a lazy chain of delegation. Given that the SparkSession is our only prerequisite for constructing the class itself, these details can be again offloaded onto Spark. The class also extends a trait called DataFrameTransformer. This simple trait is designed to provide additional structure to your application, mainly just a transform method.

```
trait DataFrameTransformer {
  def transform(df: DataFrame): DataFrame
}
```

The DataFrameTransformer helps to force your hand, thus requiring our class to implement the transform method. So, let's implement the transform method now, using what we know about the application requirements to do so.

```
import spark.implicits._
override def transform(df: DataFrame): DataFrame =
  df
    .filter($"eventType".equalTo("CustomerRatingEventType"))
    .transform(withCustomerData)
```

This transform code filters out any unsupported events based on simple string comparison and reduces the possible events down to CustomerRatingEventType. We can now safely pass the events to our withCustomerData transform method. Let's go ahead and implement that now.

```
def withCustomerData(df: DataFrame): DataFrame = {
  if (spark.catalog.tableExists("customers")) {
    spark.table("customers")
      .select(
        $"customerId",
        $"firstName",
        $"lastName",
        $"email",
        $"created".as("joined")
      )
      .join(df, usingColumn = "customerId")
  } else throw new RuntimeException("Missing Table:customers")
}
```

Now we have a fully functioning application, for the most part. Sure, there are plenty of opportunities to improve the code, such as adding layers of schema enforcement or expanding the runtime configuration to include the table name for the customers table (since hard-coding things can be problematic). But the important thing is that we have a proof of concept ready, and we are one step closer to getting this application out the door.

Compiling the Spark Application

At this point you can go ahead and compile your shiny new application using sbt.

```
> sbt clean assembly
```

Assuming you are using the provided source code from the chapter, you will see that in addition to your application building successfully, you also ran a whole suite of unit tests against the Spark application.

```
[info] ScalaTest
[info] Run completed in 12 seconds, 279 milliseconds.
[info] Total number of tests run: 4
[info] Suites: completed 1, aborted 0
[info] Tests: succeeded 4, failed 0, canceled 0, ignored 0, pending 0
[info] All tests passed.
```

Exercise 7-3: Summary

This first batch application was an introduction to many layers of nuance and many years of lessons learned writing Apache Spark applications. While the application doesn't do a ton of things now, what it does do exceptionally is set you up for future success. As we move into stateful batch and Spark Structured Streaming in the chapters to come, we will continue to fall back on the lessons learned in architecting this first batch application.

What you'll learn next is quite possibly the most important step in the Spark application development lifecycle. Just like with traditional software development, testing your code isn't simply a "nice to have," it's a requirement of the job. Furthermore, in the case of Spark applications, testing your code ensures that you don't waste valuable time and money debugging your code at runtime. I also find it invaluable to write tests like you'll do in Exercise 7-4, as you will be running Apache Spark in the test environment to power your unit tests, which opens the doors to rapid application development, better debuggability, and ultimately trustworthy applications that just work.

Testing Apache Spark Applications

Learning to properly test software is an artform. I'm sure you've heard the saying that it can take you much longer to write unit tests than it takes to write the application itself. This isn't necessarily true in all cases, but it can certainly feel like it when you are starting out. I've found in practice that the right frameworks, libraries, and design patterns make all the difference in the world. Let's look at adding the right dependencies now.

Adding Test Dependencies

The build.sbt located in the application root directory describes all the library dependencies for our application, including libraries that are only used for testing. Open the build.sbt file and scroll down to the section starting with libraryDependencies to see all of the dependencies.

```
libraryDependencies ++= Seq(
  ...
  "org.apache.spark" %% "spark-sql" % sparkVersion % Test classifier "tests",
  "org.apache.spark" %% "spark-sql" % sparkVersion % Test classifier
  "test-sources",
  "org.scalatest" %% "scalatest" % "3.2.2" % Test,
  "org.scalamock" %% "scalamock-scalatest-support" % "3.4.2" % Test,
  "com.holdenkarau" %% "spark-testing-base" % "3.0.1_1.0.0" % Test
)
```

The special syntax Test informs the build process to load the following libraries only during evaluation of our tests. This way, we don't ship the additional libraries along with our application since we won't be using them again after the test execution.

Spark Testing Base   I want to call out the spark-testing-base library by Holden Karau. Holden is a PCM member of the Apache Spark project. Her dedication to the foundations of Spark and to this testing library mean that you can spend your time focusing on writing great applications.

Let's look next at writing your first unit test.

Exercise 7-4: Writing Your First Spark Test Suite

Open the test directory of the application, which is located under the application root.

```
app/src/test/scala/
  com.coffeeco.data/
```

You'll see the following files located in the test directory.

```
com.coffee.data/
  SharedSparkSql.scala
  SparkEventExtractorSpec.scala
  TestHelper.scala
```

Let's begin by looking at the SparkEventExtractorSpec. Open the file. You'll notice right off the bat that the class constructor extends AnyFlatSpec, which is provided by the scalatest framework, and mixes in the scalatest Matchers as well as the SharedSparkSql trait.

```
class SparkEventExtractorSpec extends AnyFlatSpec
  with Matchers with SharedSparkSql
```

The SharedSparkSql trait enables you to run an embedded, configurable version of Apache Spark inside your unit tests. This can be easily mixed into any of your unit tests rather than attempting to mock out everything that you are doing. The SharedSparkSql trait is shown in Listing 7-13.

Listing 7-13. The SharedSparkSql Trait Mixes in the SparkContextProvider from com.holdenkarau.spark.testing

```
trait SharedSparkSql extends BeforeAndAfterAll with SparkContextProvider {
  self: Suite =>

  @transient var _sparkSql: SparkSession = _
  @transient private var _sc: SparkContext = _

  override val sc: SparkContext = _sc

  def conf: SparkConf
  val sparkSql: SparkSession = _sparkSql
```

```scala
  override def beforeAll() {
    _sparkSql = SparkSession.builder()
      .config(conf).getOrCreate()

    _sc = _sparkSql.sparkContext
    setup(_sc)
    super.beforeAll()
  }

  override def afterAll() {
    try {
      _sparkSql.close()
      _sparkSql = null
      LocalSparkContext.stop(_sc)
      _sc = null
    } finally {
      super.afterAll()
    }
  }
}
```

The SharedSparkSql trait in Listing 7-13 requires each test suite to provide its own SparkConf. This is important since each test suite, for example the SparkEventExtractorSpec, may require a certain set of configurations that may not be necessary, or accurate, in other test suites. This guarantees that there will be exactly one SparkContext created for each test suite, ensuring that tests can run quickly without having to wait to spin up and down a local Spark runtime for every test. Now back to the SparkEventExtractorSpec.

Configure Spark in the SparkEventExtractorSpec

You'll see that this test suite begins by implementing the conf method of the SharedSparkSql trait, which is shown in Listing 7-14.

Listing 7-14. Setting a Custom SparkConf for Use Within Your Spark Application Unit Tests

```
override def conf: SparkConf = {
  val sparkWarehouseDir = fullPath("src/test/resources/spark-warehouse")
  val testConfigPath = fullPath("src/test/resources/application-test.conf")

  // override the location of the config to our testing config
  sys.props += ( ("config.file", testConfigPath ))
  // return the SparkConf object
  SparkEventExtractorApp.sparkConf
    .setMaster("local[*]")
    .set("spark.app.id", appID)
    .set("spark.sql.warehouse.dir", sparkWarehouseDir)
}
```

Something interesting is unfolding within the conf method in Listing 7-14. This is the first concrete example of overriding the base application.conf. Given there is only one SparkContext created per SparkSession and given that the BeforeAndAfterAll trait enforces this behavior, that means you only have one opportunity to provide these overrides. Luckily, the SparkApplication trait you created earlier *lazily initializes* both the SparkConf and the SparkSession as two separate functions, effectively delaying initialization until each object is required by the application.

When the following block is executed (from Listing 7-14).

```
SparkEventExtractorApp.sparkConf
  .setMaster("local[*]")
  .set("spark.app.id", appID)
  .set("spark.sql.warehouse.dir", sparkWarehouseDir)
```

This is being done at the time of SparkSession initialization and this is called once for each separate test in the test suite. You can see how this is called in the beforeAll method of the SharedSparkSql trait.

```
_sparkSql = SparkSession.builder()
  .config(conf).getOrCreate()
```

An equally important tidbit is how this works. Given the `SparkContext` is created exactly once, and given that the `SparkConf` is immutable, that means when the `SparkSession` builder is called a second time, only the initial configuration will stick. This is perfect for our testing use case.

Testing for Success

The test case shown in Listing 7-15 is an example of testing the behavior of the `SparkEventExtractor` transform method.

Listing 7-15. Testing the SparkEventExtractor Application Delegate

```
"SparkEventExtractor" should
  "join customerRating events with customers" in {
  val testSession = SparkEventExtractorApp.sparkSession
  import testSession.implicits._

  val ratingEvents = TestHelper
    .customerRatingsDataFrame(testSession)

  SparkEventExtractor(testSession)
  .transform(ratingEvents)
  .select(
    $"firstName",
    $"lastName",
    $"rating").head shouldEqual Row("Milo", "Haines", 4)
}
```

The test case in Listing 7-15 begins by fetching the `SparkSession` of the `SparkEventExtractorApp`. Because the `SparkSession` is a lazily initialized singleton object, we can guarantee that that all tests will reference the same session (object). This is the `getOrCreate` conditional fork seen in the `SparkSession` builder shown in Listing 7-16.

Listing 7-16. SparkSession Lazy Initalization from the SparkApplication Trait

```
lazy implicit val sparkSession: SparkSession = {
  SparkSession.builder()
    .config(sparkConf)
    .enableHiveSupport()
    .getOrCreate()
}
```

Afterward, we create some data for the test using a method on the TestHelper that returns a DataFrame representing a small number of customer ratings. This data enables us to test the behavior of the SparkEventExtractor transform method without having to wire up a full end-to-end test of the application. To run a simple equality test on the output of our transformation, we can do a simple select and then see if the resulting row is what we are expecting.

This technique is good for asserting expectations on both success and failure scenarios. Given that the transform method on the SparkEventExtractor requires the customers table to exist on the SparkSession, we can simply drop that table from the current active SparkSession and ensure that we throw the appropriate exception.

Testing for Failures

Testing for failure is as important as testing for success in your Spark applications, if not more so. The code block in Listing 7-17 shows you how to test for the not-so-happy path!

Listing 7-17. Testing for Failures

```
"SparkEventExtractor" should
  "fails to join customerRating events when customer table is missing" in {
  val testSession = SparkEventExtractorApp.sparkSession
  testSession.sql("drop table customers")

  val ratingEvents = TestHelper.customerRatingsDataFrame(testSession)

  assertThrows[RuntimeException](
    SparkEventExtractor(testSession)
      .transform(ratingEvents))
}
```

This simple unit test ensures that there are no shenanigans at play in the application logic. Say you end up changing the behavior of the transform method later on down the road. This will catch your future self, or other future engineers, from having to debug to determine why things broke down the line.

Fully Testing Your End-to-End Spark Application

No application is fully complete without a full, hands-off, end-to-end test. This means that you are testing the full expectations of the application. Listing 7-18 shows a full end-to-end test using the main application as opposed to the application delegate, which was used in the previous two examples.

Listing 7-18. Testing the End-to-End Behavior of the Spark Application

```
"SparkEventExtractor" should
  "handle end to end processing" in {
  val testSession = SparkEventExtractorApp.sparkSession

  TestHelper.customersTable(testSession)
  TestHelper.customerRatingsTable(testSession)

  val destinationTable = "silver.customerRatings"

  testSession.conf
    .set(SourceTableName, "bronze.customerRatings")

  testSession.conf
    .set(DestinationTableName, destinationTable)

  testSession.conf
    .set(SaveModeName, "Overwrite")

  testSession.sql(s"drop table if exists $destinationTable")

  SparkEventExtractorApp.run()

  testSession.catalog
    .tableExists("silver", "customerRatings") shouldBe true

  testSession.table(destinationTable).count shouldEqual 1
}
```

The full end-to-end test ensures that the `bronze.customerRatings` and `customers` tables are both initialized in the SparkSQL catalog, since this is a requirement for the application to run. It also ensures we drop the `destination` table, if it exists, prior to calling the `run` method on the `SparkEventExtractorApp`. This way we have to have done something in order for this table to exist. We go one step further and test that the table does in fact exist, and that the one rating we have provided shows up.

Exercise 7-4: Summary

Running unit tests is critical for any of your Spark applications. It is especially important for larger, more complicated Spark applications. The beauty of writing unit tests that lean on Apache Spark in order to test Apache Spark applications is that you can test all aspects of your application, from corrupt input data, to missing tables, and everything under the sun. As a bonus, rich application testing can also help you become better at your Apache Spark skills. In the same way you test things using Apache Zeppelin, you can also write your unit tests in a test-driven way, building your applications piece by piece, while ensuring that your tests capture the end-to-end intentions of your applications.

Summary

We covered a ton of ground in this chapter. You learned about the different flavors of Spark application, from the dynamic interactive applications (`spark-shell` and Apache Zeppelin), to the humble batch-mode style apps, which consisted of stateless and stateful batch processing. We finished the tour with an introduction to the two common modes of structured streaming that handle all of your stateful stream processing needs—micro-batch and continuous.

The journey then shifted gears and you turned your attention away from Spark applications and segued into thinking about how to solve common data modeling and requirement modeling use cases. Through this process, you were introduced to CoffeeCo, you learned about the mission statement of this fake company, and you walked through some thought exercises focused on the data modeling process. Given that software engineering has established rules and design patterns and given that data engineering is a subset of software engineering, then by association, we can follow the same patterns that have resulted in large, reliable, tested, and fault-tolerant applications.

Using the lens of traditional software engineering, you walked through a complete end-to-end use case that started with an idea and resulted in a fully functioning, fairly well tested Spark application. The important lessons learned here were that you can easily test your Spark applications in the same way you've grown accustomed to testing any traditional software, and it is only up to your imagination what you can build next.

This simple Spark application is ready to be whisked away to run as a data pipeline job. That is really great since that is the focus of the next chapter. Are you ready? It is time to shift gears and learn how to drive your data pipelines using the popular workflow orchestration framework, Airflow.

PART II

The Streaming Pipeline Ecosystem

## CHAPTER 8

Workflow Orchestration with Apache Airflow

Generally speaking, there are two kinds of problems you'll find yourself running into more often than not as a data engineer. The first stems from broken promises, aka bad upstream data sources, and the more general realm of the unknown unknowns with respect to data movement through your data pipelines. The second problem you'll find yourself up against is *time*. This is not the part in the book where I start to talk to you about life, death, and decision making, but rather time as a boundary or a threshold. Time exists between the physical runtime of jobs, as well as a very real line in the sand when it relates to data service level agreements (SLAs). These data contracts revolve around expectations in terms of the data format (aka *schemas*) as well as the agreed upon time when data should be expected to become available. Another way in which time gets the best of us is at the intersection of both of these common problems, e.g., upstream problems married happily with stale data, or missed SLAs.

If you think about these two kinds of problems (bad data and time constraints), you may start to understand the need for workflow orchestration, as there is this butterfly effect that stems from missing or stale data problems. Bad upstream data in the pipeline leads to pipeline traffic jams, which propagate in turn throughout the data pipeline as a whole. This can be caused by downtime in any APIs and services that feed directly into your data pipeline, as well as broken or misconfigured jobs at any point in the data pipeline that cause SLAs to be broken.

I tend to think about data pipelines like the paths, roads, waypoints, and destinations on a map. *Jobs* can be thought of as waypoints as well as final destinations, and the data flow itself can be likened to an entity in transit across the map, be it bicycles, trains, cars, and anything in between. When things go wrong, things tend to back up (and in less extreme cases fall through the cracks). You've likely seen this yourself at specific times of day while commuting, for example during rush hour traffic, and also at random points in

© Scott Haines 2022
S. Haines, *Modern Data Engineering with Apache Spark*, https://doi.org/10.1007/978-1-4842-7452-1_8

time when you find yourself at the whims of unexpected road construction or when you find yourself with an unexpected flat tire. The solution to fix these problems is to remove the obstacles from the road, and with time, the normal flow of traffic returns. When things are running smoothly again, you can sit back and relax (breathe) and just enjoy the ride.

Workflow orchestration isn't only used to get unstuck. It is also used to build blueprints for reliably accomplishing a set of tasks, such as moving and transforming data through a data pipeline. Let's talk briefly about workflow orchestration and then take a look at how Airflow steps up to assist with these all-too-common problems.

Workflow Orchestration

A *workflow* is a common set of actionable tasks that need to be executed in a specific order to achieve a desired effect. In other words, a workflow looks at the big picture. It's results oriented, aka it isn't burdened with the nitty gritty regarding how each task is completed as long as it is completed on time and doesn't interfere with the final results.

With respect to data, workflows are traditionally encapsulated by pipeline jobs, which in turn work as tasks and grant responsibilities to individual operators, which work together in a continuous, distributed data-processing network. Figure 8-1 shows a zoomed in view of pipeline job (A) to the left of a complete workflow (jobs A-E).

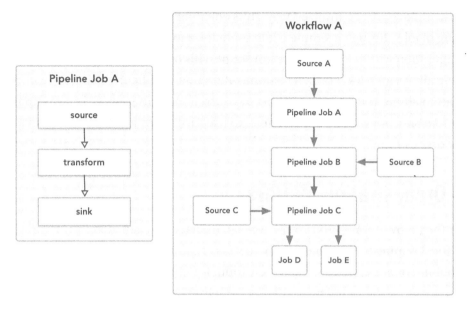

Figure 8-1. *From pipelines to workflows. Workflows encapsulate the chain of inputs and outputs that makes up the distributed network of jobs being run across a data pipeline network*

Another way to think about workflow orchestration is to associate it with the typical hierarchical structure seen at the center of most companies and organizations. A director works with one or more managers, who in turn manage a group of employees, who complete the work required of them. How the work is completed according to schedule, as a cohesive unit, falls under the orchestration side of workflow orchestration. You can think of this component as the deadlines and demands passed down across the network from director to manager and onto the teams completing the work.

When you put this all together, you have a unit that acts together, accomplishing a distributed goal, according to a schedule that optimizes across the total number of resources available. Now keep this concept in mind as we turn our attention to Apache Airflow and move from the theoretical to the actionable.

Apache Airflow

Apache Airflow is the most popular open-source workflow orchestration platform (at the time of this writing). The project started out as an internal project at Airbnb late in 2014. It materialized from the general need to solve the common problem of increasingly complex workflows in a centralized way, thus enabling programmatic authoring and scheduling of jobs through a simple UI. Essentially, a job defines its own schedule and configuration within a simple Python script. This code-driven approach to job scheduling allows for a level of rich customization that is difficult to achieve with static configuration alone. You may be thinking to yourself that this is great for a company the size of Airbnb, but why would you need to use Airflow, or any kind of automated workflow orchestrator?

When Orchestration Matters

Consider the following. You and your team can manage an ever-increasing workload up until a critical tipping point, where each new task requires another task to be abandoned and left behind or automated to make new room for more tasks. If your team is small, you can manage the task load manually for a while at least. For example, kicking off daily batch jobs or running manual reports for different teams in the company. However, automation almost always plays a key role in the success of the team long-term, and by removing the work that can be easily automated, the team can turn to new tasks, accomplish new goals, reap new rewards, and extended their collective reach and surface area, without needlessly scaling the team linearly with the workload.

How do you know where to start? If you find yourself or others on your team spending more time on redundant work and toiling mindlessly on repetitive tasks, that is your catalyst for automation. Make a plan of action to automate the unfulfilling, or repetitive work, to free up development cycles and make room for whatever comes next. Funny enough at a certain point you'll find that even these automated tasks need to be managed through automation, and that is the point where Apache Airflow and other orchestration frameworks come into play.

Working Together

Looking back to last chapter, where you learned the difference between stateless and stateful batch jobs, Airflow can be used as an intersection of these two modes of operation. It allows you to define robust workflows that sprinkle in more dynamic configuration powered by Python helper libraries, additional Airflow plugins, and our friend, Apache Spark. Let's pick up from last chapter and work on automating the run of our Spark application using Airflow.

Exercise 8-1: Getting Airflow Up and Running

Getting up and running with Apache Airflow can be a simple process when you can manage things using Docker Compose. Head over to your IDE and open the exercise contents for Chapter 8.

Exercise Materials   You can find the materials for this chapter in the `ch-08/` folder of the book's GitHub resources at `https://github.com/newfront/ spark-moderndataengineering`.

We will start by simply spinning up Airflow. Open your terminal and change your working directory to the `ch-08` directory to get started.

Installing Airflow

From the command line, change your working directory to the `airflow` directory. Run `cd ch-08/airflow`. From the `airflow` directory, there are a few housekeeping chores you need to do to have a smooth experience.

Add the Directories

The following Airflow directories will be volume-mounted into the Airflow container runtime, thus enabling you to retain your job definition, and runtime logs all from a simple local location. Run the command in Listing 8-1 from the `airflow` directory.

Listing 8-1. Adding the Dags, Logs, and Plugins Directories Under the airflow Directory

```
mkdir ./dags ./logs ./plugins
```

Now on to the next order of business.

Add Environment Variables for Docker Compose

Did you know you can export environment variables into an .env file that will be used in the Docker Compose process? The .env file can be used to plug configurable paths, users, Docker images, passwords, and more into your Docker Compose. This allows you to reuse `docker-compose.yaml` definitions while not needing to hard-code usernames, passwords, or anything that could end up being an eventual security exception.

For example, the .env file from the chapter exercises (located at /ch-08/airflow/.env) is pre-filled. If you wanted to update the user and group ID associated with Airflow at runtime, or even change the Airflow UI port, you can edit the file directly or append new variables using the append operator >>, which is shown in Listing 8-2.

Listing 8-2. Using the Shell (Command Line), Echo, and the Append Operators to Modify the Airflow .env File

```
echo -e "AIRFLOW_UID=$(id -u)\nAIRFLOW_GID=0\n _AIRFLOW_WEB_SERVER_
PORT=8088" >> .env
```

Port Conflicts   Airflow will bind to port 8080 by default, which is the same port as your internal Zeppelin host. The update shown in Listing 8-2 allows the two services to run together by switching the Airflow port to 8088. If you don't want to remember the ports where the services run, you could run `nginx` and route based on hostnames.

Initialize Airflow

It is time to download, set up, and initialize Airflow. From the ch-08/airflow working directory, execute the command in Listing 8-3 from the command line. This will prepare the environment and will bootstrap database tables, but you won't start the Airflow containers just yet.

Listing 8-3. Using Docker Compose to Initialize Airflow

```
docker compose up airflow-init
```

The Docker Compose initialization command in Listing 8-3 may take some time downloading all the dependent Docker containers. Next the `airflow-init` command will activate and bootstrap the Airflow metadata database, which stores all your Airflow job configurations and then sets up the Redis database. That database is used with the job scheduler. Redis is an incredibly fast, performant, and popular key/value and data structure store used by Airflow to manage scheduling and caching. You will learn to use Redis to power your first structured streaming application in Chapter 10.

You will notice a lot of logging during the initialization process. This process will run the database migrations, bootstrapping and all other initialization scripts and procedures. You can scroll through the output if you are interested (or to get ideas for writing your own initialization routines); otherwise, you are ready to run Airflow.

Running Airflow

With the initialization complete, you can execute the command in Listing 8-4.

Listing 8-4. Running the docker compose up Command from the Airflow Working Directory to Start Your Local Airflow Cluster

```
docker compose up
```

The Docker Compose process will spin up a total of seven Docker containers. This is what is needed to have a fully functioning Airflow cluster up and running on your local environment. One thing to point out is that—unlike in the prior chapters where we used the `run.sh` script to start or stop the Zeppelin environment—the `docker-compose up` command will only run for the life of the terminal window. If you want to run things as a long-lived process, you'll learn to run Airflow in detached mode next.

Running Airflow in Detached Mode

Running in detached mode allows you to run each container as a background process, which continues running until you explicitly stop the container process (or shut down the machine). There are two things to think about when running in detached mode. First is the process used to start up and tear down your containers. For example, do you prefer to use command-line aliases or script files? The second thing to consider is how you want to manage the composite runtime environments created using multi-service Docker Compose definitions, and where you want your shared volumes and directories to be collocated and referenced. Let's look at running Airflow in detached mode and then provide a solution for creating a more streamlined local Docker environment.

Start Things Up

The Docker Compose process accepts an explicit file pointer `-f`, which can be used to pass an absolute path, as well as the `-d` flag, which changes the process to run in the background rather than in the foreground. The difference between these two modes is that in the foreground you will see all the logs (across all containers) controlled by your `docker-compose.yaml` definition. In the background, you must reference the actual container identifier, `docker logs -f container_ID`, if you want to reattach to the output of the container and view the output. Listing 8-5 shows the longer `docker compose` command. You can run this command from the Airflow working directory.

Listing 8-5. Passing a File Reference to docker compose and Running in Detached Mode

```
docker compose \
  -f docker-compose.yaml \
  up -d
```

Tear Things Down

When you are ready to shut down your Airflow cluster, you can use `docker compose` again to stop all containers. Execute the command in Listing 8-6 to shut your Airflow cluster back down.

Listing 8-6. Using docker compose to Tear Down All Services Directly Defined in the docker-compose.yaml Definition

```
docker compose \
  -f docker-compose.yaml \
  down --remove-orphans
```

The process of starting up and tearing down Airflow (or any Docker Compose-controlled services) from within a local directory (such as `/ch-08/airflow/`) is simple enough. However, the process is harder than it needs to be in the long term unless you work to organize your local environments. For example, if you have many environments controlled by many `docker-compose.yaml` definitions that support many different local data engineering use cases, container naming collisions or other environment issues can creep up. It can be much easier in the long run to create a service hierarchy, nestled neatly under a root directory. We'll look at optimizing your local environment now, to reduce complexity as we move forward.

Optimizing Your Local Data Engineering Environment

As you add more and more technologies to your local data engineering environment, you may find it useful to create a single location (parent directory) that can be used to create an organized local data platform, broken down by services, organized along with specific service volume mounts, `docker-compose.yaml` definitions, and .env specifications to create a consistent environment for Docker to run a given technology.

Let's optimize your environment now. Let's copy all the hard work you've done so far in this chapter into the `dataengineering` parent directory. Execute the commands in Listing 8-7 to add the `dataengineering` directory to your home directory.

Listing 8-7. Creating the dataengineering Root Directory and Copying the airflow Directory and Contents from the Chapter Exercises

```
mkdir ~/dataengineering &&
  cp -R /path/to/ch-08/airflow ~/dataengineering
```

With Airflow nestled nicely in its new home, the next step is to add a simple alias to start and stop Airflow. Aliases can be added into your local Bash or shell environment settings (`.bashprofile`, `.bashrc`, or `.zshrc`), which may differ depending on the command line you run. Listing 8-8 shows how to add the `airflow2_start` and `airflow2_stop` aliases for UNIX-style systems.

Listing 8-8. Using Bash Aliases to Simplify Starting and Stopping Airflow

```
### Helpers for MDE for Apache Spark
export DATA_ENGINEERING_BASEDIR="~/dataengineering"
alias airflow2_start="docker compose -f ${DATA_ENGINEERING_BASEDIR}/
airflow/docker-compose.yaml up -d"
alias airflow2_stop="docker compose -f ${DATA_ENGINEERING_BASEDIR}/airflow/
docker-compose.yaml down --remove-orphans"
```

After making the changes to your profile, make sure you run `source` to load the changes into your current session. For example, executing the command `source` `~/.zshrc` adds the two Airflow aliases to my current terminal session. Now, whenever you need to start the local Airflow cluster, you only need to remember `airflow2_start`, and likewise when you are finished or want to gain some system resources back, it is as easy as calling `airflow2_stop`. You've made it easier to start and stop Airflow. You are also probably dying to check things out. Let's get back to it.

Sanity Check: Is Airflow Running?

Presuming you have started Airflow using the `airflow2_start` alias, or with either command in Listing 8-4 or Listing 8-5, then the simplest way to see if things are running is to open your favorite browser and head on over to `http://localhost:8088`. You will be greeted by a login screen, which is shown in Figure 8-2.

Figure 8-2. *Airflow login screen*

The login screen is confirmation that at least the Airflow webserver process is up and running. You can log in with the username (`airflow`) and the password (`airflow`). This should take you to Airflow's main UI (see Figure 8-3).

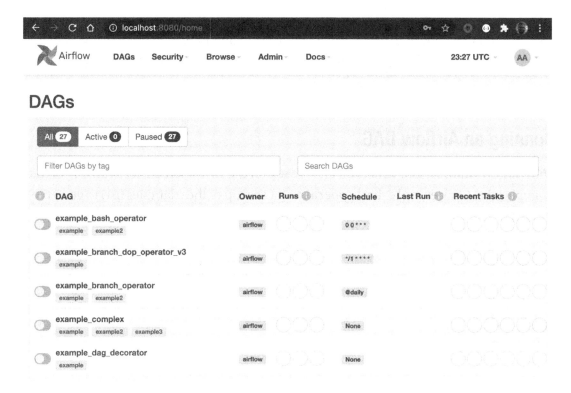

Figure 8-3. *Airflow's default UI*

Caution   Your browser may tell you that the username and password have been compromised. This is because the username:password combination `airflow:airflow` is the Airflow default and should never be used for anything outside of local testing.

From the Airflow main screen, you'll see all your job definitions, job metadata, and statistics about the prior runs (if any) across each job. This is also a branching-off point for general security and common administration tasks. We will spend a little time going through the user/role controls and look at how to add secure configuration such as passwords in this chapter.

For starters, your job definitions are officially called *DAGs*. This is because each job is defined as a directed acyclic graph (DAG) of one or more operations, and essentially each DAG is in and of itself a pipeline operation or an entire complex workflow.

Note   You can choose to keep or delete these default jobs. I recommend keeping them around so you can look through the examples and get an idea of what is possible with Airflow.

Running an Airflow DAG

With Airflow running, and while you are already on the DAGs screen, let's run one of the example jobs. Click example_bash_operator (at the top of the examples list) to go to the DAG overview screen (see Figure 8-4).

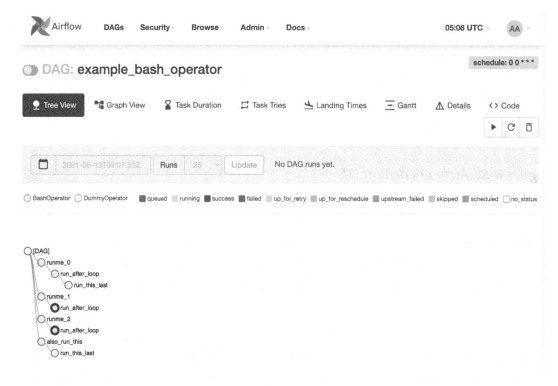

Figure 8-4.   *Airflow's DAG details UI is where you will do much of your work*

The DAG UI in Figure 8-4 gives you a general overview of a multi-step pipeline job. In order to run this example job, you need to do the following two things:

1.   Toggle the switch icon (top left of the word DAG in Figure 8-4) to the enabled position. DAGs by default are disabled to save resources in the cluster. You opt in to running each schedule or one-off execution.

2. Click the Play icon (right side, next to the refresh and delete buttons) to start running the pipeline.

The job will kick off (after confirming the optional JSON configuration) when you press the Trigger button, and then you will see the job transition into the Running state. The UI in Figure 8-5 shows the live Airflow job view.

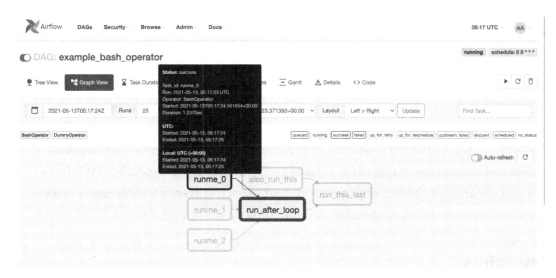

Figure 8-5.   *The runtime job view of a live Airflow job*

It is worth mentioning that the job will complete almost immediately; however a nice feature of the updated Airflow 2 UI is the auto-refresh option (shown on the right side of Figure 8-5). If you turn this option on, you can watch the progress of your pipeline as it is actively switching between stages of operation, which can help you monitor/observe progress, or faults, as tasks complete or fail in the DAG.

Exercise 8-1: Summary

Apache Airflow is available to you now whenever you want to spin it up. However, there is still more that you'll need to accomplish to kick off Spark batch jobs from Airflow. The next section teaches you about the components of Airflow, and then we will install the Spark operator.

The Core Components of Apache Airflow

Now that you've run your first Airflow DAG, it is time to discuss the components that allowed the DAG to run in the first place. Mainly Airflow's tasks, operators, schedulers, and executors.

Tasks

Within each DAG, there exists a series of one or more *tasks* that are executed to fulfill the obligations set forth by a given DAG. Tasks are the basic unit of execution in Airflow, which is also the case for Spark applications. The overlap here is that with Spark, you have jobs, and each job is comprised of stages that distribute tasks. With Airflow, your DAGs are composed of tasks, which run in a specific pre-defined order, and work is completed within the context of the Airflow executor managing a specific DAG. Tasks can execute arbitrary code using operators, and so each DAG can handle many tasks. Each task can execute local or remotely depending on the operator in use.

Operators

Airflow operators are templated tasks. Think of this as a blueprint that can be reused to wrap generic use cases, enabling common Python code to be reused to fulfill a given operation. In the example DAG (example_bash_operator), you were introduced to two of the core Airflow operators—BashOperator and DummyOperator.

The BashOperator does exactly what you think it does; it will run a bash (or command-line) operation. The DummyOperator is used to ensure the DAG has a final common node in a graph of tasks. The DummyOperator can also be used as a placeholder while you are building more complicated DAGs. Listing 8-9 shows the Python code for the example_bash_operator DAG.

Listing 8-9. Creating a DAG of Synchronous and Asynchronous Tasks

```
from datetime import datetime, timedelta

from airflow import DAG
from airflow.operators.bash import BashOperator
from airflow.operators.dummy import DummyOperator
```

```python
with DAG(
    dag_id='example_bash_operator',
    schedule_interval='0 0 * * *',
    start_date=datetime(2021, 12, 28),
    dagrun_timeout=timedelta(minutes=60),
    tags=['example', 'example2'],
    params={"example_key": "example_value"},
) as dag:
    run_this_last = DummyOperator(task_id='run_this_last',)
    run_this = BashOperator(
        task_id='run_after_loop',
        bash_command='echo 1',
    )
    run_this >> run_this_last

    for i in range(3):
        task = BashOperator(
            task_id='runme_' + str(i),
            bash_command='echo "{{ task_instance_key_str }}" && sleep   1',
        )
        task >> run_this

    also_run_this = BashOperator(
        task_id='also_run_this',
        bash_command='echo "run_id={{ run_id }} | dag_run={{ dag_run }}"',
    )
    also_run_this >> run_this_last

    this_will_skip = BashOperator(
        task_id='this_will_skip',
        bash_command='echo "hello world"; exit 99;',
        dag=dag,
    )
    this_will_skip >> run_this_last

    if __name__ == "__main__":
        dag.cli()
```

Each task is defined in Listing 8-9 as an instance of either the DummyOperator or BashOperator. Each task you define acts as a node in the graph and as the engineer designing the pipeline you establish the processing order by appending tasks (nodes) as children or parents of each other.

Viewing your DAG in Graph mode (shown in Figure 8-6) can help you as you are planning more complex workflows. Essentially each task can be complex or simple. The SparkOperator is an example of a complex task that can be added to a workflow.

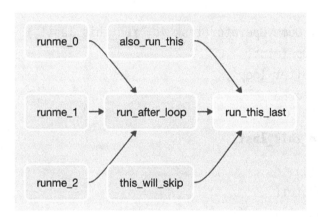

Figure 8-6.   *The graphical view of interconnected and independent tasks making up the simple example_bash_operator DAG*

We will look at the specialized SparkSubmitOperator and the SparkSQLOperator in more detail later in this chapter. To round out your working experience with all that Airflow can do, my recommendation is to have fun and play around in your new local sandbox environment. Next, we look at the process behind how Airflow runs each task within a DAG.

Schedulers and Executors

The executor is the process in Airflow responsible for kicking off the tasks in your DAG based on a schedule. For example, say you have a job that needs to be run every day at the end of the day. The properties for scheduling this DAG are shown in Listing 8-10.

Listing 8-10.   An Example DAG Configuring a Daily Cron Job

```
DAG(
  dag_id="daily_active_users_reporting_job",
```

```
  start_date=days_ago(2),
  default_args=args,
  tags=["coffee_co","core"],
  schedule_interval="@daily",
)
```

DAG definitions are parsed by Airflow initially on startup, and periodically to support adding DAGs at runtime. When a new `dag_id` is encountered, the DAG metadata is written to the RDBMS and the scheduler executor process will trigger the DAG run (if the job is not disabled) as soon as the threshold of the `start_date` is crossed.

The schedule configured in Listing 8-10 sets the `start_date` to `days_ago(2)` with a `schedule_interval` of `@daily`. This is an example of a backfill job. It will look at today's date (on the Airflow server) and rewind so that the prior two days run before it begins running on a daily schedule, which runs at the start of each day (or midnight). For more information you can look at the Cron Presets section under DAG Runs in the official Airflow documentation.

The Airflow executor waits for the scheduler to notify it that a DAG is ready to run. In fact, the executor itself runs within the scheduler process. Due to this relationship, you can only assign one mode of execution for a given Airflow cluster, given the tight coupling between scheduling and execution.

Airflow ships with several useful executors, offering local or remote styles of execution.

Local Execution

The `SequentialExecutor` or `LocalExecutor` can be used if you are running a small cluster. Local execution means that task execution will be collocated within the executor process, and the executor process itself runs inside the scheduler process. This process inception can be good for testing things out, but this pattern won't scale to run multiple DAGs in parallel.

Remote Execution

For production environments where you have critical workflows that must run, you have no choice but to use remote execution. This enables the workloads of multiple DAGS to be executed in parallel across a network of Airflow worker nodes. To view what execution mode your Airflow cluster is running in, execute the command shown in Listing 8-11.

Listing 8-11. Viewing the Airflow Executor Config

```
docker exec \
  -it airflow_airflow-webserver_1 \
  airflow config get-value core executor
```

The output will show you the `CeleryExecutor`, which is one of the more popular remote executor options available in Airflow. The `CeleryExecutor` uses Redis for scheduling and distributes its work across the Airflow Worker processes. Let's turn our attention now to integrating Apache Spark with Airflow.

Scheduling Spark Batch Jobs with Airflow

Airflow is a robust external scheduler you can rely on to run your mission-critical Apache Spark batch jobs. Furthermore, workflow orchestration is becoming an essential data platform component used to automate increasingly complex data pipelines and meet the needs of many internal and external data customers. As you learned in Exercise 8-1, running a DAG on Airflow is a piece of cake (once you've learned the basics). To get started with Airflow and Spark, all that is needed is the *Spark Airflow Provider,* which enables you to use a few different Spark operators.

Exercise 8-2: Installing the Spark Airflow Provider and Running a Spark DAG

This exercise will focus on teaching you to install the Apache Spark provider. You will learn how to install things manually, and then we will turn our attention to using a pre-built Airflow container that has the things we'll need neatly packaged. Then you'll learn to configure some runtime configuration in Airflow using the Admin UI and finish off by running the batch application from the last chapter using Airflow.

Let's begin by using the version of Airflow you set up in Exercise 8-1. You'll be executing the `pip install` commands directly on the webserver and worker Airflow containers because the Spark Python modules must be available locally.

Locating the Airflow Containers

To run the `install` command, you need to list the full name of the Airflow containers. Execute the command in Listing 8-12 to view the container names.

Listing 8-12. Using Docker to Find Your Airflow Container Metadata

```
docker ps | grep airflow_airflow-w*
```

Once you have the container names, you can use the Docker CLI to run the install script (using pip/Python).

Manually Installing the Spark Provider

The following process shows how to authenticate as the `airflow` user and install the Spark provider. Use Listings 8-13 and 8-14 as a guide to install the provider.

Note   The example shows how to install the provider on the webserver. You need to repeat the process on the worker(s) as well, or read on to use the pre-built container.

Listing 8-13. Opening a Bash Session as the Airflow (Admin) User to Gain the Correct Permissions to Install a New Provider

```
docker exec \
    --user airflow \
    -it airflow_airflow-webserver_1 \
    bash
```

From the Docker container, execute the `pip install` command shown in Listing 8-14.

Listing 8-14. Install the Spark Provider Using Pip

```
pip install \
  --no-cache-dir \
  --user \
  apache-airflow-providers-apache-spark
```

You've completed the work on the webserver. Unfortunately it is time to start the process all over again for each worker instance. This will ensure the providers are installed consistently across the Airflow cluster. Consistency is critical since Airflow operates as a distributed service and you can't always guarantee where a DAG will be scheduled.

Note   You can use named queues with the `CeleryExecutor` to assist in distributing work across your workers, and you can also distribute tasks across pre-configured processing pools, but queues and pools are out of the scope of this book.

When the DAG is triggered, for each independent DAG run, all tasks encapsulating the DAG will run until they have all succeeded or a critical task fails, thereby blocking the rest of the tasks from completing. In the real-world use, you'll see worker nodes easily spanning tens to hundreds of instances, and at an even larger scale within enterprise clusters. Now imagine that you were tasked at ensuring that this full system was in sync, keeping in mind that you'd have to repeat this step for each deployment to ensure consistency. Would you want to be tasked with keeping everything in sync?

Of course not. This is clearly an anti-pattern, especially given we are running things using containers in the first place. Sure, you could add some scripts to run when the container is starting up, but that just means you are subject to potential runtime errors due to failed dependencies. You've probably seen such issues when centralized package repositories (like Pypi or Maven) have outages, or when package locations change (as is often the case with archived artifact versions).

Using Containers for Runtime Consistency

A runtime environment based on consistent containers removes the environmental guesswork when troubleshooting, so you can narrow your focus and get things running again quickly. For us this means we don't have to mess with manually installing anything required for our DAG's operations.

Note   The section titled "Advanced Operations" in the `ch-08/airflow/` `README.md` includes directions for building a custom Airflow docker image with the Apache Spark provider and Java 11 (JRE) preinstalled.

In order to use the pre-built Docker container (`newfrontdocker/apache-airflow-spark:2.1.0`), you have to make a quick adjustment to the environment file located at `~/dataengineering/airflow/.env`. Then for the changes to take effect you also need to restart your Airflow cluster. As a first step, use the `airflow_stop` alias to shut down your existing cluster.

Note   If you have yet to set up your local environment, go back to Listing 8-7 for directions.

Now, edit the `.env` file (using an editor or the command line) to uncomment the single line shown in Listing 8-15.

Listing 8-15. Adding the Custom Airflow Image to the Docker Environment File

```
AIRFLOW_IMAGE_NAME=newfrontdocker/apache-airflow-spark:2.1.0
```

When you are finished, save the change and start the cluster back up using your `airflow_start` alias. You won't need to reinitialize anything since you are only switching out the runtime container (and not switching to a different RDBMS or executor backend).

Now we can move on to running a Spark DAG.

Running Spark Jobs with Apache Airflow

You've taken care of all the required dependencies, and Airflow is now ready to run your Spark batch jobs. In this next section, you'll be using the Airflow UI to add some additional configurations and helpers for running your Spark jobs.

Add Airflow Variables

Airflow variables are similar to environment variables, but they are configured in the metadata database and accessible through `airflow.models` Python package. Variables can be managed directly in Airflow using the Admin UI (Admin ➤ Variables), or by using the Airflow command line directly. This service allows you to configure key/value pairs that can be reused in your DAGs.

Listing 8-16 adds the SPARK_HOME variable to the set of runtime variables accessible through your Airflow DAGs. Variables can augment your DAGs, and more importantly, variables are controlled by administrators and require role-based user privileges to add or remove.

Listing 8-16. Airflow Variables Enable You to Inject Configuration Directly Into Your DAGs

```
docker exec \
  -it airflow_airflow-webserver_1 \
  airflow variables set SPARK_HOME /opt/spark
```

Now open the Airflow UI (http://localhost:8088) and click the Variables control (in the Admin menu) to see the new variable in the list (shown in Figure 8-7).

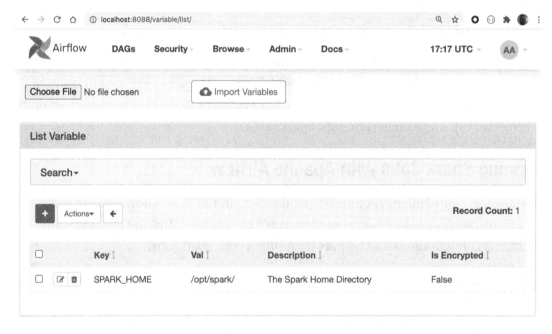

Figure 8-7. *The SPARK_HOME variable is seen in the list*

You can also edit the metadata for any of the variables you add to Airflow to add descriptions. This simple action can help to document the usage of a given variable and clearly relay your intentions.

Note   The `admin` menu is only available to users with the Admin role, so these variables can't be edited by just anyone. The "Continued Explorations" section of this chapter shows you how to create a regular user and introduces the notion of Airflow roles (role-based access control).

Add Airflow Connections

Connections enable Airflow to communicate with external services, such as Apache Spark, and new connections can be set up directly using the Admin UI (Admin ➤ Connections) or with the command line (`airflow connections list`).

Given the Spark Airflow provider is a module extending Airflow, it would make sense that there is a simple means to connect to a given Spark cluster (or process). Open the Airflow UI and go to the Admin ➤ Connections page. Click the + button (on the left side) to create a new connection for Spark. This action will take you to a new page with an empty form.

Use the form to create `local_spark_connection` using the Spark type from the Conn Type menu. The completed form is shown in Figure 8-8.

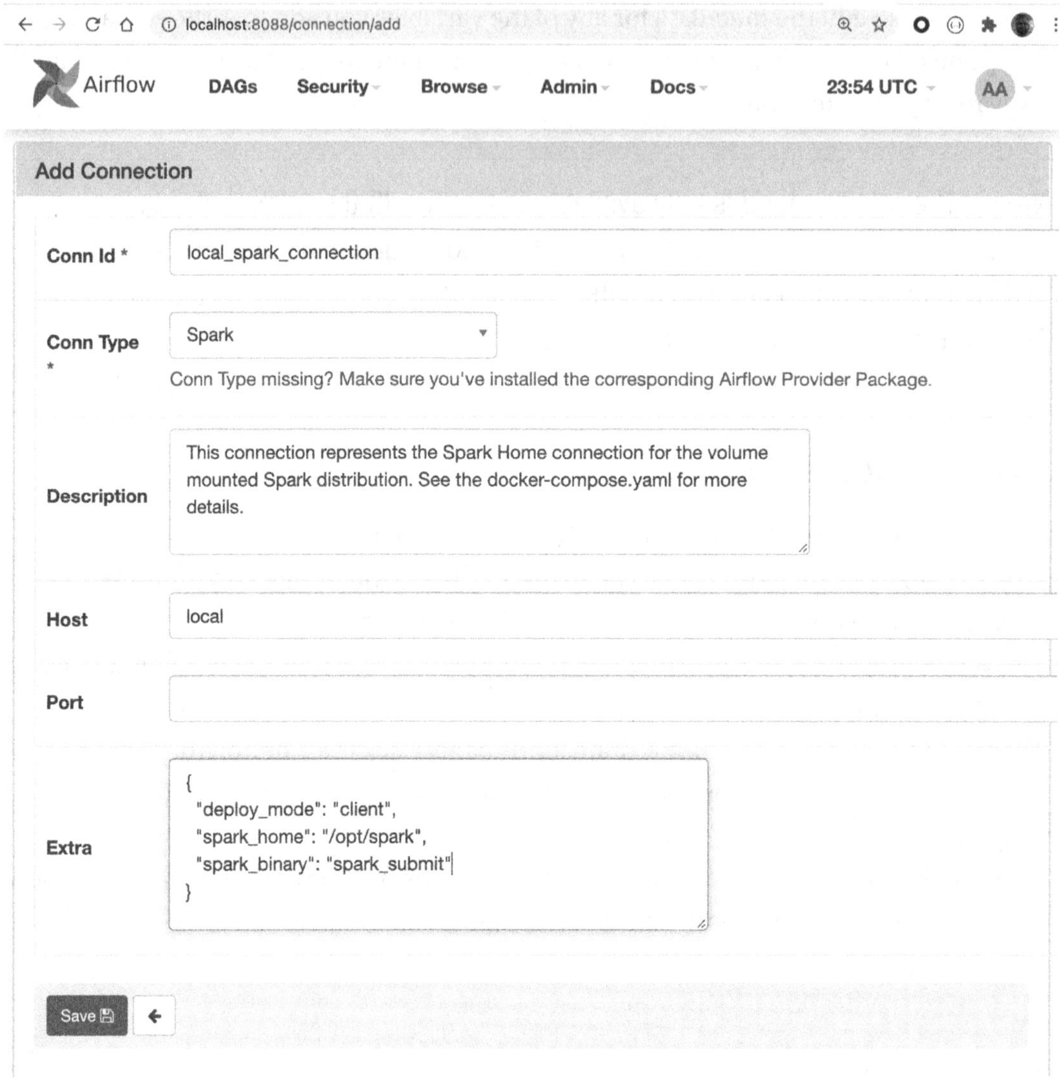

Figure 8-8. *Using the Airflow UI to add a Apache Spark connection*

With the connection created, let's move on to writing your first Spark DAG. You will be using the SparkSubmitOperator, which is included in the Spark provider you installed in your Airflow cluster.

Writing Your First Spark DAG

Spark ships with a few example applications. One of these is called SparkPI. We'll use this example Spark application to get familiar with writing a new DAG. Let's roll with the idea of having some "daily pi."

To get started, create a file named daily_spark_pi.py inside the Airflow DAGs directory (~/dataengineering/airflow/dags/daily_spark_pi.py). Then type (or copy) the code in Listing 8-17 and save your changes.

Listing 8-17. Using the SparkSubmitOperator to Get Your Daily Dose of PI

```
from airflow.providers.apache.spark.operators.spark_submit import
SparkSubmitOperator
from airflow.models import DAG, Variable
from airflow.utils.dates import days_ago

args = {
  'owner': 'airflow',
}
spark_home = Variable.get("SPARK_HOME")

with DAG(
  dag_id='daily_spark_pi',
  default_args=args,
  schedule_interval='@daily',
  start_date=days_ago(1),
  tags=['coffeeco', 'core'],
) as dag:
  spark_pi_job = SparkSubmitOperator(
    application=f'{spark_home}/examples/jars/spark-
    examples_2.12-3.1.1.jar',
    conn_id="local_spark_connection",
    java_class="org.apache.spark.examples.SparkPi",
    task_id="spark_pi_job"
  )
```

The DAG in Listing 8-17 starts by importing the `SparkSubmitOperator` and some helper libraries that enable you to access the Airflow variables, along with some additional utilities used for scheduling. We then create a dictionary that currently stores the ownership metadata for the DAG. Next, we fetch the `spark_home` variable for use in creating the `spark_pi_job`. Then we specify the DAG metadata and schedule.

```
with DAG(
  dag_id='daily_spark_pi',
  default_args=args,
  schedule_interval='@daily',
  start_date=days_ago(1),
  tags=['coffeeco', 'core'],
) as dag:
```

And we add a single task to the DAG block using the `SparkSubmitOperator`.

```
spark_pi_job = SparkSubmitOperator(
  application=f'{spark_home}/examples/jars/spark-examples_2.12-3.1.1.jar',
  conn_id="local_spark_connection",
  java_class="org.apache.spark.examples.SparkPi",
  task_id="spark_pi_job"
)
```

Now that you have written the DAG, let's run it.

Running the Spark DAG

All new DAGs start off in the paused position. This can be configured differently, but in general it is a good idea since this behavior is used to protect your Airflow cluster resources from being overwhelmed by many DAGs competing to start up at the same time. Let's unpause this DAG using the command line.

```
docker exec \
  -it airflow_airflow-webserver_1 \
  airflow dags unpause daily_spark_pi

Dag: daily_spark_pi, paused: False
```

With the DAG unpaused, you can manually trigger it.

```
docker exec \
  -it airflow_airflow-webserver_1 \
  airflow dags trigger daily_spark_pi
```

Assuming everything worked as expected, you should be able to list the successful runs of this DAG.

```
docker exec \
  -it airflow_airflow-webserver_1 \
  airflow dags list-runs \
  -d daily_spark_pi \
  --state success
```

Using the Airflow UI, you can also look at the logs that were generated during the run of the DAG. To do this, follow these steps:

1. Go to the Airflow DAGs home screen and filter by active jobs using the coffeeco tag (/home?status=active&tags=coffeeco).

2. Click the daily_spark_pi DAG and switch to the Graph View (/graph?dag_id=daily_spark_pi).

3. Click the spark_pi_job. You will see a button at the top named Log; click it to view the logs from this run of the task. See Figure 8-9.

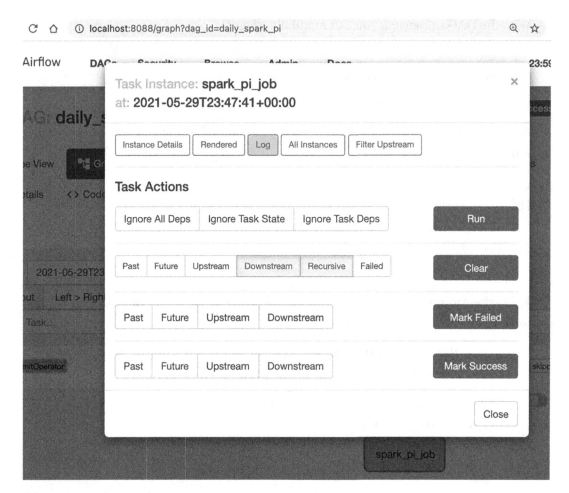

Figure 8-9. *Viewing the logs from the run of your Spark PI DAG*

The logs will show you the output of the job, including the results of running the Spark DAG. If you scroll down the page, you will see the following:

```
{spark_submit.py:526} INFO - Pi is roughly 3.1411757058785295
```

This now confirms that you are able to run your Spark workflows from Apache Airflow. Although this is exciting, what you are going to do next is set up a DAG to run the Spark application you created in Chapter 7.

Exercise 8-2: Summary

This exercise taught you to install the Apache Spark provider, configure variables and connections, and write and run a simple Spark DAG using the `SparkSubmitOperator`. In the final exercise of the chapter, you learn to run the `SparkEventExtractorApp` from Chapter 7 using the skills you've acquired in the first two exercises.

Running the SparkEventExtractorApp using Airflow

In the last chapter, you wrote and compiled your first Spark application. You learned how to create a configuration-driven application that simply joins data from two tables backed by the local data warehouse. This simple ETL-like process combined data from the `customers` table and joined it against the `customerRatings` table.

For the rest of the chapter, you will be setting up an Airflow DAG to configure and run this application with Airflow.

Starting with a Working Spark Submit

Given our goal is to set up a DAG to run our `SparkEventExtractorApp`, the first place to start is looking at what is needed to run the application. To recap, we had to provide a good amount of configuration, the JAR location, and the driver app configuration to launch the app.

```
$SPARK_HOME/bin/spark-submit \
  --class "com.coffeeco.data.SparkEventExtractorApp" \
  --packages=org.mariadb.jdbc:mariadb-java-client:2.7.2 \
  --conf "spark.event.extractor.source.table=..." \
  --conf "spark.event.extractor.destination.table=... " \
  --conf "spark.event.extractor.save.mode=..." \
  --conf "spark.sql.warehouse.dir=..." \
  --driver-java-options "-Dconfig.file=conf/local.conf" \
  target/scala-2.12/spark-event-extractor...jar
```

When we use the `SparkSubmitOperator` in Airflow, it is essentially generating a similar `spark-submit` command within the Airflow cluster context. This requires us to modify the Spark submit code slightly in order to run this as a DAG. Let's start backward and look at the final DAG, and then move the required configurations and JARs into place so things can work.

Exercise 8-3: Writing and Running the Customer Ratings Airflow DAG

Create a new file in the ~/dataengineering/airflow/dags directory named customer_ratings_etl_dag.py. Then copy (or paste) the following Python code fragments to assemble the final DAG.

```
from airflow.providers.apache.spark.operators.spark_submit import
SparkSubmitOperator
from airflow.models import DAG, Variable
from airflow.utils.dates import days_ago
```

Next, we'll add our user attribution, and we'll reuse the Spark home variable from our last DAG.

```
args = {
    'owner': 'scotteng',
}
spark_home = Variable.get("SPARK_HOME")
```

So far so good. Next, we'll create a new dictionary to store the full Spark configuration for our job.

```
customer_ratings_conf = {
  "spark.sql.warehouse.dir": "s3a://com.coffeeco.data/warehouse",
  "spark.hadoop.fs.s3a.impl": "org.apache.hadoop.fs.s3a.S3AFileSystem",
  "spark.hadoop.fs.s3a.endpoint": "http://minio:9000",
  "spark.hadoop.fs.s3a.access.key": "minio",
  "spark.hadoop.fs.s3a.secret.key": "minio_admin",
  "spark.hadoop.fs.s3a.path.style.access": "true",
  "spark.sql.catalogImplementation": "hive",
  "spark.sql.hive.metastore.version": "2.3.7",
  "spark.sql.hive.metastore.jars": "builtin",
  "spark.sql.hive.metastore.sharedPrefixes": "org.mariadb.jdbc,com.mysql.
  cj.jdbc",
```

```
    "spark.sql.hive.metastore.schema.verification": "true",
    "spark.sql.hive.metastore.schema.verification.record.version": "true",
    "spark.sql.parquet.compression.codec": "snappy",
    "spark.sql.parquet.mergeSchema": "false",
    "spark.sql.parquet.filterPushdown": "true",
    "spark.hadoop.parquet.enable.summary-metadata": "false",
    "spark.hadoop.mapreduce.fileoutputcommitter.algorithm.version": "2",
    "spark.sql.hive.javax.jdo.option.ConnectionUserName": "dataeng",
    "spark.event.extractor.source.table": "bronze.customerRatings",
    "spark.event.extractor.destination.table": "silver.customerRatings",
    "spark.event.extractor.save.mode": "ErrorIfExists"
}
```

This dictionary can be used directly in SparkSubmitOperator, reducing the chore of scanning the configuration inside of the DAG itself. Before you write the actual DAG, you'll need to add the following variables (app_jars and driver_class_path) to increase the readability of the final DAG.

app_jars=f'{spark_home}/user_jars/hadoop-aws-3.2.0.jar,{spark_home}/user_jars/hadoop-cloud-storage-3.2.0.jar,{spark_home}/user_jars/mariadb-java-client-2.7.2.jar,{spark_home}/user_jars/mysql-connector-java-8.0.23.jar,{spark_home}/user_jars/aws-java-sdk-bundle-1.11.375.jar'

driver_class_path=f'{spark_home}/user_jars/mariadb-java-client-2.7.2.jar:{spark_home}/user_jars/mysql-connector-java-8.0.23.jar:{spark_home}/user_jars/hadoop-aws-3.2.0.jar:{spark_home}/user_jars/hadoop-cloud-storage-3.2.0.jar:{spark_home}/user_jars/aws-java-sdk-bundle-1.11.375.jar'

Finally, we can now use the variables and properties to create our ETL DAG.

```
with DAG(
    dag_id='customer_ratings_etl_dag',
    default_args=args,
    schedule_interval='@daily',
    start_date=days_ago(1),
    tags=['coffeeco', 'core'],
) as dag:
```

```
customer_ratings_etl_job = SparkSubmitOperator(
  application=f'{spark_home}/user_jars/spark-event-extractor.jar',
  jars=app_jars,
  driver_class_path=driver_class_path,
  conf=customer_ratings_conf,
  conn_id="local_spark_connection",
  name='daily-customer_ratings',
  verbose=True,
  java_class="com.coffeeco.data.SparkEventExtractorApp",
  status_poll_interval='20',
  task_id="customer_ratings_etl_job"
)
```

This is it. Save the file and give yourself a quick pat on the back.

The lion's share of the setup work is all configurations. You probably noticed many new Spark settings being used to wire up the job. Let's look first at the Spark SQL warehouse configurations.

The Spark Configuration

The job configurations declare the settings for our Spark SQL warehouse.

```
"spark.sql.warehouse.dir": "s3a://com.coffeeco.data/warehouse",
"spark.hadoop.fs.s3a.impl": "org.apache.hadoop.fs.s3a.S3AFileSystem",
"spark.hadoop.fs.s3a.endpoint": "http://minio:9000",
"spark.hadoop.fs.s3a.access.key": "minio",
"spark.hadoop.fs.s3a.secret.key": "minio_admin",
"spark.hadoop.fs.s3a.path.style.access": "true"
```

The spark.sql.warehouse.dir property was used in the last chapter. As a refresher, this is the single immutable warehouse path that is provided for the SparkContext when your application is starting up. We've configured this to point to Amazon S3 using another service called MinIO. MinIO works like Amazon S3 and can run locally or in the cloud as a drop-in replacement. We'll cover how this works in more detail after this section is complete. However, by moving from a locally mounted directory to an S3 clone, we are simplifying future work for ourselves (in the case where we will be using S3) by introducing a local service that is fully interoperable. This makes it painless to switch between our local environment and other environments (including production) without needing to change our mental models too much.

Local Job Configuration

We are also providing the three required configurations to inform our Spark batch job where to look to find the data we are intending to join. These configurations were introduced in Chapter 7 and allow you to modify the database tables and names.

```
"spark.event.extractor.source.table": "bronze.customerRatings",
"spark.event.extractor.destination.table": "silver.customerRatings",
"spark.event.extractor.save.mode": "ErrorIfExists"
```

Spark Submit Properties

The properties for the SparkSubmitOperator enable us to configure how we launch a Spark application, with almost the same conventions as the actual spark-submit command.

```
SparkSubmitOperator(
  application=f'{spark_home}/user_jars/spark-event-extractor.jar',
  jars=app_jars,
  driver_class_path=driver_class_path,
  conf=customer_ratings_conf,
  conn_id="local_spark_connection",
  name='daily-customer_ratings',
  verbose=True,
  java_class="com.coffeeco.data.SparkEventExtractorApp",
  status_poll_interval='20',
  task_id="customer_ratings_etl_job"
)
```

In the case of running this Spark application, we need to provide some additional JARs to the runtime. We add the jars and driver_class_path configuration to tell Spark to load some driver-specific JARs (primarily the hive JDBC JARs), as well as the Amazon S3 JARs to enable us to communicate with the MinIO S3 client.

We also flipped on verbose logging (verbose=True) to monitor the job from the Airflow side of things. Lastly, we set the job completion polling to every 20 seconds (status_poll_interval='20'). This allows us to continue to fetch updates from the job and get fresh log updates every 20 seconds, rather than seeing nothing while the job is waiting to fail or complete.

Docker Inception   Because Airflow is running inside of Docker, we must ensure that the Spark JARs and configuration are all available to the Airflow workers. As a recap, this is because we are running the CeleryExecutor, and this means we are running remotely on the Airflow workers. This process can be simplified by providing a common file system to your Docker instances, or even by using a Cloud Object store (like Amazon S3). Then your dependencies (JARs) and additional application configuration can be fetched on-demand by Airflow.

Copy the Spark Application JAR into the Shared Spark JARs Location

Since you are reusing the Spark application you created in Chapter 7, you'll need to copy the JAR into the spark/jars directory referenced in ~/dataengineering/airflow/docker-compose.yaml.

```
cp /path/to/ch-07/app/target/scala-2.12/spark-event-extractor-assembly-0.1-
SNAPSHOT.jar ~/dataengineering/airflow/spark/jars/spark-event-extractor.jar
```

Now, let's run the DAG.

Running the Customer Ratings ETL DAG

To run the DAG, you need to go through the extended environment setup. The directions are available in the main README file located in the chapter materials. The setup process will give you all the things you need to run your new DAG. Keep in mind that to run the job more than once, you have to toggle the spark.event.extractor.save.mode to Overwrite. Alternatively, you can just delete the silver.customerratings directory from the Spark SQL data warehouse location.

Using the Airflow UI, go to the main DAGs page and click the customer_ratings_etl_dag DAG. From the UI, click the Trigger Now button. You should see a success status like the example shown in Figure 8-10.

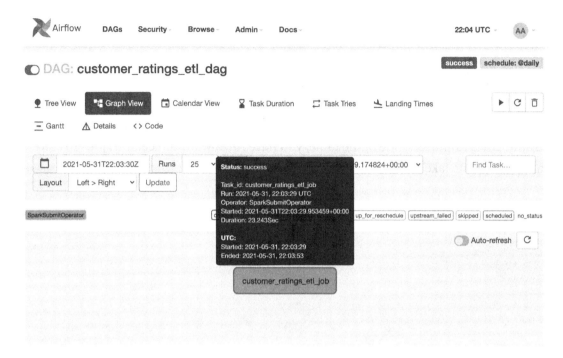

Figure 8-10. *Successfully running the SparkEventExtractorApp from Airflow*

You can look at the Spark logs and see how all the moving pieces came together, but essentially, what you have accomplished this chapter is worth a round of applause.

Exercise 8-3: Summary

The last exercise of the chapter showed you how to launch a more complex Spark batch job using an Airflow DAG. There was a good amount of work that went into getting things set up to use the final `docker-compose-chapter-end.yaml`, in place of the `docker-compose.yaml` that was used in Exercise 8-2. Before we close the chapter, let's look at the underlying architecture and components required to make running the DAG possible.

Looking at the Hybrid Architecture

There are now quite a few moving pieces to keep in mind when you run this DAG. For starters, you have to manage your local MySQL backed (Hive Metastore), as well as the Airflow Docker containers, and somehow provide additional shared volumes for your Spark SQL data warehouse as well. All these central data platform components come together to form the critical pieces of your infrastructure and enable Airflow to run the Spark DAG.

Figure 8-11 is an example of the simplified current state of your local data platform. You have everything running to make Airflow work. You have the MySQL backed Hive Metastore, which is responsible for storing the metadata for your distributed data tables, and you also have your new MinIO S3 data lake. You can now reuse shared configuration from your local Spark installation and reuse the configs and JARs between Zeppelin and Airflow. This provides you with a Spark UI for interacting with the data you produce from Airflow. On the opposite side of the table, you have Airflow, which can run jobs to store data in your data lake or in MySQL as a JDBC table.

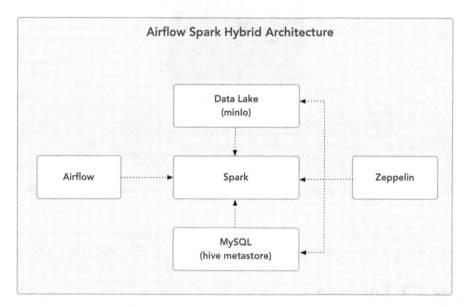

Figure 8-11. *The component architecture behind the Customer Ratings ETL DAG*

Setting Up the Extended Environment

The main README in the chapter materials provides you with step-by-step instructions for getting your extended local environment up and running. There are some additional points worth going over, with respect to both MinIO as well as migrating your Hive Metastore.

The extended environment moves all of the key pieces of this hands-on environment into a single location on your laptop. This will enable you to reuse all of these common architectural components from here on out, without needing to remember which chapter exercise we added something. The path is set to your home directory in a folder named dataengineering.

```
~/dataengineering
  -/airflow
  -/minio
  -/mysql
  -/spark
  -/zeppelin
```

Migrating to MinIO from the Local File System

To help simplify things moving forward, there are instructions for using the MinIO distributed file system located inside the chapter materials. In a nutshell, MinIO is an Amazon S3 clone and is fully compatible with the S3 APIs. This will reduce some of the complexity moving forward in future chapters. One point to keep in mind is that when first running MinIO, you have to set some permissions on the file system. Luckily, you can do this via the UI. See Figure 8-12.

Figure 8-12. *Switch the file system from the default Read Only to Read and Write*

You'll need to open the MinIO UI (`http://localhost:9001`). You can also add a record in your `/etc/hosts` folder to make this easier to find. Mine is just `http://minio:9001`. From the home screen, you will be prompted to authenticate (the user and password is stored in the `docker-compose.yaml` file for MinIO). Or if you haven't changed anything, it is username: minio, password: minio_admin. This will allow you to apply your prefix permissions.

Apply the Bucket Prefix Permissions

On the left side of the screen (shown in Figure 8-13), you will see com.coffeeco.data. This is the default bucket name, and you just need to hover over the name and click the little dots that show up to open up the Permissions menu.

Figure 8-13. *Editing inside the MinIO browser*

Just add the prefix /warehouse/* and set it to Read and Write, and you will be golden.

The next thing you may want to do is migrate your Hive Metastore tables to a better common location.

Bootstrapping MySQL in a Common Location

Your Hive Metastore is most likely located in the ch-06 directory. You can either export the SQL databases and tables from your current metastore and move the existing contents into ~/dataengineering/mysql, or you can start from scratch and bootstrap things in the ~/dataengineering/mysql directory. The mysql directory in the chapter materials includes a README.md that provides step-by-step directions to start over with a clean hive installation.

From here on out, we will be reusing this common setup as we traverse the data waters.

Continued Explorations

This chapter was a complete look at using Airflow to schedule Apache Spark applications. As you know, there is always much more that you can learn (but not everything fits into the book). As a continued exploration, I suggest the following:

- You can add DAG override variables when you go to run your DAGs. This enables you to provide additional changes to the underlying DAG without needing to go back and change the underlying Python code. Try adding a variable to change the following variable:

  `spark.event.extractor.save.mode`

The `available` options are `ErrorIfExists`, `Overwrite`, `Ignore`, and Append.

- Create another Spark application using the common traits from Chapter 7 and run it using Airflow.

- Try generating a new dataset or thinking about a new problem you can solve with the data in your local data warehouse.

In addition to the continued explorations, the following section goes over the process of creating individual users and assigning roles.

Creating a User

Creating a new user is simple using the Airflow command line. Essentially, all of the user commands are nestled within the `users` sub-command.

```
docker exec \
  -it airflow_airflow_webserver-1 \
  airflow users [sub-command]
```

Say, for example, you needed to create a new user and grant them limited access within Airflow. You can use the `User` role to assign normal permissions. Execute the command in Listing 8-18 to generate the `scotteng` user, for example.

Listing 8-18. Generating a New User with the Airflow Command Line

```
docker exec \
  -it airflow_airflow_webserver-1 \
  airflow users create \
  --username scotteng \
  --firstname Scott \
  --lastname Haines \
  --role User \
  --email scott@coffeeco.com
```

This command will ask you for a password, so just follow the prompt to finish creating the new user. When you are finished, log out of Airflow, and then log back in as the user scotteng. You will notice that the Admin or Security menus are now gone, but you still have access to run DAGs (see Figure 8-14).

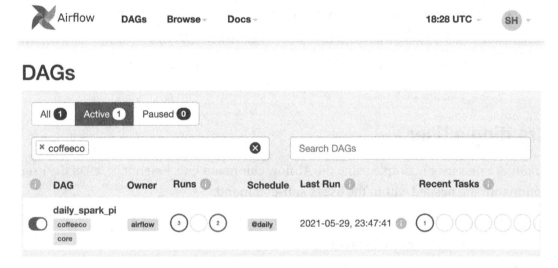

Figure 8-14. *Viewing the Airflow UI as a normal user*

That is all for now.

Summary

This chapter covered at a high level the concept of workflow automation and introduced you to Apache Airflow. While we could easily spend the rest of the book discussing how to create more and more complex Airflow DAGs, and dive deeper into the additional nuances of Airflow, that exercise will have to be left to you. While we will touch on Airflow again as we continue our journey, our next stop will take us to the exciting topic of writing Structured Streaming applications with Apache Spark. Buckle up, as things are about to start moving quickly!

CHAPTER 9

A Gentle Introduction to Stream Processing

Learning to tackle and optimize data engineering problems can be challenging due to the many dimensions each problem can take on. At the outset of each new problem, you must think about data discovery, wrangling, ingestion, transformation, and data accountability, which is an umbrella relating to data contracts (strictly defined data definitions), as well as the need to optimize the data ingestion footprint (since data at scale can easily eat into operation costs). There are additional concerns relating to data access, lineage, and governance that need to be back of mind as well. Understanding how to use your collective knowledge to create quick plans of data attack is a skill that will get you far as a modern data engineer.

Think about the various ways you have learned to work with data over the course of this book, whether it was starting off small using the `spark-shell` to explore initial ideas, pursuing those ideas further using notebooks (Apache Zeppelin), and finally taking the lessons you learned and rolling them up into Spark applications. Through this process, you learned to create reusable, tested, Spark applications, and you also learned the important skill of problem-solving through exploration. You learned how to work through the stages of problem identification and decomposition—understanding the problem, creating a plan of attack, testing and validating your results, and then wrapping up the solution as a finished application. What is even more important is that you learned, by doing, the steps to break these end-to-end data questions up by writing transformative queries with Spark SQL. You learned how to test the behavior of these micro-applications (using queries) and even how to schedule their periodic running (using Apache Airflow).

These steps have all been leading up to now. As promised at the end of the last chapter, things are about to start moving a lot faster. Why is that? Because right now we are starting the journey from batch processing to stream processing, or the processing of unbounded datasets.

© Scott Haines 2022
S. Haines, *Modern Data Engineering with Apache Spark*, https://doi.org/10.1007/978-1-4842-7452-1_9

For many use cases, processing data in-stream, or as it becomes available, can help reduce a really enormous data problem (due to the scale of the event data) into one that is much more manageable. Simply by processing a smaller number of data points more often, you can divide and conquer a data problem that may otherwise be cost-prohibitive. However, how you transition from a batch mindset to a streaming mindset can also be tricky.

For example, say you are tasked with creating an application that must process around 1 billion events (1,000,000,000) a day. While this might feel far-fetched due to the sheer size of the data, it often helps to step back and think about the intention of the application/process. If the data can be broken down (partitioned) and processed in parallel as a streaming operation (aka, in-stream), you only have to create an application that can ingest, and process, a mere 11.5 thousand (k) events a second (or around 695k events a minute if the event stream is constant).

Although these numbers may seem out of reach, this is where distributed stream processing can really shine. Essentially, you are reducing the perspective, or scope, of the problem to accomplish a goal over time, in a more distributed fashion over a partitioned dataset. While not all problems can be handled in-stream, a surprising number of problems do lend themselves to this processing pattern.

This chapter is a gentle introduction to stream processing, making room for you to jump directly into building your own end-to-end Structured Streaming application in Chapter 10, without having to backtrack and discuss a lot of the theory behind the decision-making process.

By the end of the chapter, you should understand the following (at a high level):

- How to think about streaming data and streaming data problems

- Why data accountability starts with the producer

- Methods for handling data quality at the API edge

- Why binary serializable structured data is essential

Stream Processing

Streaming data is *not stationary*. In fact, you can think of it as being alive (if even for a short while). This is because streaming data is data that encapsulates the *now—it records events and actions as they occur in flight*. Let's look at a practical, albeit theoretical, example that begins with a simple event stream of sensor data. Fix into your mind's eye the last parking lot (or parking garage) you visited.

Use Case: Real-Time Parking Availability

Imagine you just found a parking spot all thanks to some helpful signs that pointed you to an open space. Now let's say that this was all because of the data being emitted from a connected network of local parking sensors. Sensors that operate with the sole purpose of being used to identify the number of available parking spaces at that precise moment in time.

This is a real-time data problem where real-time accuracy is measurable and physically noticeable by a user of the parking structure. Enabling these capabilities all began with the declaration of the system scenario.

> *"We would like to create a system that keeps track of all available parking spaces, identifies when a car parks, and knows how long the car remains in a given spot. This process should be automated as much as possible."*

Optimizing a system like this can begin with a simple sensor located in each parking spot (associated with a `sensor.id`/`spot.id` reference). This sensor would be responsible for emitting data in the form of a spot identifier, timestamp, and simple bit (0 or 1), to denote if a spot is empty or occupied. This data can be encoded into a compact message format, as shown in Listing 9-1, which would be efficiently encoded and sent periodically from each device.

Listing 9-1. An Example Sensor Event (Encapsulated in the Google Protocol Buffer Message Format)

```
message ParkingSensorStatus {
  uint32 sensor_id = 1;
  uint32 space_id  = 2;
  uint64 timestamp = 3;
  bool   available = 4;
}
```

During the normal flow of traffic throughout the day, the state of each of these sensors would flip on or off (binary states) in unpredictable ways due to the dynamic schedules of the individual drivers. This sensor data can inform drivers, in real-time, that there are now X total number of available spots in the garage. This data can help to automate the human decision-making process and could even be made available online, through a simple web service, for real-time status tracking. Additionally, this data can be used to track when each sensor last checked in, which can be used to diagnosis faulty sensors, and even track how often sensors go offline or fail. Nowadays, more technologically advanced garages even go so far as to direct the driver (via directional signs and cues) to the available spots in the structure. This acts to reduce inter-garage traffic and congestion, which in turn raises customer satisfaction, all by simply capturing a live stream of sensor data and processing it in near-real-time.

Given the temporal information gathered from these streams of sensor events, a savvy garage operation could use prior trends to decrease or increase the daily or hourly prices, based on the demand for parking spots, with respect to current availability in real-time. By optimizing the pricing (within realistic limits) an operator could find the perfect threshold where the price per hour/price per day leads to a full garage. In other words, "at what price will most people park and spots won't go unused?"

This is an example of an optimization problem that stems from the collection of real-time sensor data. It is becoming more common for organizations to look at how they reuse data to solve multiple problems at the same time. The Internet of Things (IoT) use cases are just one of the numerous possible streams of data you could be working with when writing streaming applications. Earlier in the book, we discussed creating a system that could take information about Coffee store occupancy that would inform folks what shop nearest to them has seating for a party of their size. At that point in the story, we simply created a synthetic table that could be joined to showcase this example. However, this is another problem that can be solved with sensors, or something as simple as a check-in system that emits relevant event data that can be passed reliably downstream via streaming data pipelines. Both these examples employ basic analytics and could benefit from simple machine learning to uncovering new patterns of behavior that could lead to more optimal operations. Before we get too far ahead of ourselves, let's take a short break to dive deeper into the capabilities that streaming data networks provide.

Time Series Data and Event Streams

Moving from a stationary data mindset to one that interprets data as it flows over time, in terms of streams of unbounded data across many views and moments in time, is an exercise in perspective but also one that can be challenging to adopt. Often when you think about streaming systems, the notion of streams of continuous events bubble to the surface. This is one of the more common use cases and can be used as a gentle introduction to the concept of *streaming data*. Take for example the abstract time series shown in Figure 9-1.

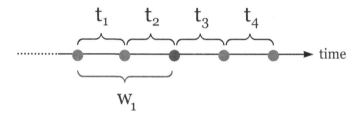

Figure 9-1. *Events occur at precise moments of time and can be collected and processed individually (t1->t4) or can be aggregated across windows of time (w1)*

As you can see, data itself exists across various states depending on the perspective applied by a given system (or application). Each event (T1->T4) individually understands only *what has occurred* in its narrow pane of reference, or to put that differently, events capture a limited (relative) perspective of *time*. When a series of events are processed together in a bounded collection (window), you have a series of data points (events) that encapsulate either fully realized or partially realized ideas. When you zoom out and look at the entire timeline, you can paint a more accurate story of what happened from first event to last. Let's take this idea one step further.

Do Events Stand Alone?

Consider this simple truth. Your event data exists as a complete idea, or as partial ideas or thoughts. I have found that thinking of data as a story over time helps give life to these bytes of data. Each data point is therefore responsible for helping to compose a complete story, as a series of interwoven ideas and thoughts that assemble or materialize over time.

This data composition concept can be used as a lens as you work on adopting a distributed data view of things. I also find it lends itself well while building up and defining new distributed data models, as well as while working on real-world data networks at scale. Viewed as a composition, these events come together to tell a specific story, whose event-based breadcrumbs can inform of the order in which something came to be and is greatly enhanced with the timestamp of each occurrence. Events without time paint a flat view of how something occurred while the addition of time grants you the notion of momentum or speed, or a slowing down and stretching of the time between events or for a full series of data points. Understanding the behavior of the data flowing through the many pipelines and data channels is essential to data operations and requires reliable monitoring to keep data flowing at optimal speeds. Let's look at a use case where the dimension of time helps paint a better story of a real-world scenario.

Use Case: Tracking Customer Satisfaction

What if I told you two customers came into the coffee shop, ordered drinks, and left the store with their drinks. You might ask me why I bothered to tell you that since that is what happens in coffee shops. What if I told you that the two coffee orders were made around the same and that the first customer in the story was in and out of the coffee shop in under five minutes. What if I told you that it was a weekday, and this story took place during morning rush hour? What if I told you that the second customer, who happened to be next in line, was in the coffee shop for 30 minutes? You might ask if the customer stayed to read the paper or use the facilities. Both are valid questions.

If I told you that the customer was waiting around because of an error that the occurred between Steps 3 and 4 of a four-step coffee pipeline, then we'd have a better understanding of how to streamline the customer's experience in the future. The four steps are:

1. Customer orders: `{customer.order:initialized}`

2. Payment made: `{customer.order:payment:processed}`

3. Order queued: `{customer.order:queued}`

4. Order fulfilled: `{customer.order:fulfilled}`

Whether the error was in the automation, or because of a breakdown in the real-world system (printer jam, barista missed an order, or any other reason), the result here is that the customer needed to step in and inform the operation (the coffee pipeline) that it appears that someone forgot their drink.

At this point, the discussion could turn toward how to handle the customer's emotional response, which could swing widely across both positive and negative reactions: from happy to help (1), to mild frustration (4), all the way to outright anger (10) at the delay and breakdown of the coffee pipeline. But by walking through a hypothetical use case, we are all now more familiar with how the art of capturing good data can be leveraged for all kinds of things.

The Event Time, Order of Events Captured, and the Delay Between Events All Tell a Story

Without the knowledge of how much time elapsed from the first event (`customer.order:initialized`) to the terminal event (`customer.order:fulfilled`), or how long each step typically takes to accomplish, we'd have no way to score the experience or really understand what happened, essentially creating a blind spot to abnormal delays or faults in the system. It pays to know the statistics (average, median, and 99th percentiles) of the time a customer typically waits for a variable sized order, as these historic data points can be used via automation to fix a problem preemptively when, for example, an order is taking longer than expected. It can literally mean the difference between an annoyed customer and a lifetime customer.

This is one of the big reasons that companies solicit feedback from their customers, be it a thumbs up/thumbs down on an experience or rewarding application-based participation (spend your points on free goods and services). This data—collected and captured through real-world interactions, encoded as events, and processed for your benefit—is helpful if it positively affects the operations and reputation of the company. Just be sure to follow data privacy rules and regulations and ultimately don't creep out your customers.

This little thought experiment was intended to shed light on the fact that the details captured in your event data (as well as the lineage of the data story over time) can be a game changer. Time is the dimension that gives these journeys momentum. There is just one problem with time.

The Trouble with Time

While events occur at precise moments in time, the trouble with time is that it is also subject to the problems of time and space (location). Einstein used his theory of relativity to explain this problem on a cosmic scale, but this is also a problem on a more localized scale as well. For example, I have family living in different parts of the United States. It can be difficult to coordinate time where everyone's schedule syncs up. This happens for simple events like catching up with everyone over video or meeting up in the real-world for reunions. Even when everything is coordinated, people have a habit of just running a little bit late.

Zooming out from the perspective of my family, or people in general, with respect to central coordination of events, you will start to see that the problem isn't just an issue relating to synchronization across time zones (east/central/west coast). If you look closer you can see that time, relative to our local/physical space, is subject to some amount of temporal drift or clock skew.

Take the modern digital clock. It runs as a process on your smart phone, watch, or any number of many "smart" connected devices. What remains constant is that time stays noticeably in sync (even if the drift is on the order of milliseconds). Many people still have analog, non-digital, clocks. These devices run the full spectrum from incredibly accurate, in the case of high-end watches ("timepieces") to cheap clocks that sometimes need to be reset every few days.

The bottom line here is that it is rare that two systems agree on the precise time in the same way that people also have trouble coordinating within both time and space. Therefore, a central reference (or point of view) must be used to synchronize the time with respect to systems running across many time zones.

Correcting Time   Servers running in any modern cloud infrastructures utilize a process called Network Time Protocol (NTP) to correct the problem of time drift. The NTP process is charged with synchronizing the local server clock using a reliable central time server. This process corrects the local time to within a few milliseconds of the Universal Coordinated Time (UTC). This is an important concept to keep in mind since an application running in a large network and producing event data will be responsible for creating timestamps, and these timestamps need to be precise in order for distributed events to line up. There is also the sneaky problem of daylight savings, so coordinating data from systems across time zones as well as across local datetime semantics requires time to be viewed from this central, synchronized, perspective.

We've looked at time as it theoretically relates to event-based data but to round out the background we should also look at time as it relates to the priority in which data needs to be captured and processed in a system (streaming or otherwise).

Priority Ordered Event Processing Patterns

You may be familiar with this quote. "Time is of the essence." This is a way of saying something is important and a top priority. The speed to resolution matters. This sense of priority can be used as an instrument, or defining metric, to make the case for real-time, near-real-time, batch or eventual (on-demand) processing when processing critical data. These four processing patterns handle time in a different way by creating a certain focus on the data problem at hand. The scope here is based on the speed in which a process must complete, which in turn limits the complexity of the job as a factor of time. Think of these styles of processing as being deadline-driven. There is only a certain amount of time in which to complete an action.

Real-Time Processing

The expectations of real-time systems are that end-to-end latency from the time an upstream system emits an event, until the time that event is processed and available to be used for analytics and insights, occurs in the milliseconds to a few seconds. These events are emitted directly to an event stream processing service, such as Apache Kafka, which under normal circumstances enables listeners/consumers to immediately use that event once it is written. There are many typical use cases for true real-time systems, including logistics (like the parking space example as well as finding a table at a coffee shop). Processes impacting a business on a whole new level, such as fraud detection, active network intrusion detection, or other bad actor detection where a longer mean time to detection (average milliseconds/seconds to detection) can lead to devastating consequences in terms of reputation or financial losses.

For other systems, it is more than acceptable to run in near real-time. Given that answering tough problems requires time, real-time decision-making requires a performant, pre-computed or low-latency answer to the questions it will ask. This is pure in-memory stream processing.

Near Real-Time Processing

Near real-time is what most people think of when they consider real-time. A similar set of time based constraints as real-time, the only difference is that the expectations of end-to-end latency are relaxed from a number of seconds to a handful of minutes. For most systems, there is no real reason to react immediately to every event as it arrives, so while time is still of the essence, the priority of the SLA for data availability is extended.

Operational dashboards and metric systems that are kept up to date (refreshing graphs and checking monitors every 30s to five minutes) are usually fast enough to catch problems and give a close representation of the world. For all other data systems, you have the notion of batch or on-demand.

Batch Processing

We covered batch processing and reoccurring scheduling in the last two chapters, but for clarity, having periodic jobs that push data from a reliable source of truth (data lake or database) into other connected systems has been, and continues to be, how much of the world's data is processed in the real world. The reason for this is cost. This factors down to both the cost of operations and the human cost for maintaining large streaming systems.

Streaming systems demand full-time access to a variable number of resources from CPUs and GPUs to Network IO and RAM, with an expectation that these resources won't be scarce, since delays (blockage) in stream processing can pile up quick. Batch on the other hand can be easier to maintain in the long run, assuming the consumers of the data understand that there will always be a gap from the time data is first emitted upstream, until the data becomes available for use downstream.

The last consideration to keep in mind is on-demand processing (or just-in-time processing).

On-Demand or Just-In-Time Processing

Let's face it. Some questions (aka queries) are asked very rarely or in a way that is just not suitable to any predefined pattern.

For example, custom reporting jobs and exploratory data analysis are two styles of data access that lend themselves nicely to these paradigms. Most of the time, the backing data to answer these queries is loaded directly from the data lake and then processed using shared compute resources or isolated compute clusters. The data that is made available by these queries may be the by-product of other real-time or near-real-time systems that were processed and stored for batch or historic analysis.

Using this pattern, data can be defrosted and loaded on-demand by importing records from slower commodity object storage, like Amazon S3 into memory, or across fast-access solid state drives (SSDs). Depending on the size, format, and layout of the data, it can be queried directly from the cloud object store. This pattern can be easily delegated to Apache Spark using Spark SQL. This enables ad hoc analysis via tools like Apache Zeppelin, or directly in-app through JDBC bindings using the Apache Spark thrift-server and the Apache Hive Metastore.

The differentiator between these four flavors of processing is time.

Circling back to the notion of views and perspective, each approach or pattern has its *time and place*. Stream processing deals with events captured at specific moments in time and, as we've discussed during the first half of this chapter, how we associate time and how we capture and measure a series of events (as data) all comes together to paint a picture of what is happening now or what has happened in the past. It's important to also talk about the foundations of stream processing. In this next section, we'll walk through some of the common problems and solutions for dealing with continuous, unbounded streams of data. It would only make sense to therefore discuss data as a central pillar and expand outward from there.

Foundations of Stream Processing

Foundations are the unshakable, unbreakable base upon which structures are placed. When it comes to building a successful data architecture, the data is the core central tenant of the entire system and the principal component of that foundation. The most popular way that data now makes its way onto the data platforms is through stream processing platforms like Apache Kafka. This means the data contracts surrounding data's schema, availability, and validity become the critical underpinnings of that data foundation (given the decentralized, distributed stream-processing capabilities presented by Apache Kafka).

Building Reliable Streaming Data Systems

As data engineers, building reliable data systems is literally our job, and this means data downtime should be measured like any other component of the business. You've probably heard of the terms SLAs, SLOs, and SLIs at one point or another. In a nutshell, these acronyms are associated to the contracts, promises, and measures in which you grade a system, and they ultimately hold the service owners accountable for its successes and failures. For example, Service Level Agreements (SLAs) between your team or organization and your customers are used to create a binding contract with respect to the service you are providing. For data teams, this means identifying and capturing metrics (KPMs: key performance metrics) based on your Service Level Objectives (SLOs). The SLOs are the promises you intend to keep based on your SLAs, and they can be anything from a promise of near perfect (99.999%) service uptime or something as simple as a promise of 90-day data retention for a particular dataset. Lastly, your Service Level Indicators (SLIs) are the proof that you are operating in accordance with the service level contracts and are typically presented in the form of operational analytics (dashboards) or reports.

Knowing where we want to go can help establish the plan to get there. This journey begins at the inset (or ingest point), and with the data. Specifically, with the formal structure and identity of each data point. Considering the observation that "more and more data is making its way into the data platform through stream processing platforms like Apache Kafka" it helps to have compile-time guarantees, backward compatibility, and fast binary serialization of the data being emitted into these data streams. Data accountability can be a challenge in and of itself. Let's look at why.

Managing Streaming Data Accountability

Streaming systems operate 24 hours a day, 7 days a week, and 365 days a year. This can complicate things if the upfront effort isn't applied to the problem, and one of the problems that tends to rear its head from time to time is corrupt data, aka data problems in flight.

Dealing with Data Problems in Flight

There are two common ways to reduce data problems in flight. First, you can introduce gatekeepers at the edge of your data network that negotiate and validate data using traditional Application Programming Interfaces (APIs), or as a second option, you can create and compile helper libraries, or Software Development Kits (SDKs), to enforce the data protocols and enable distributed writers into your streaming data infrastructure. You can even use both strategies in tandem.

Data Gatekeepers

The benefit of adding gateway APIs at the edge of your data network is that you can enforce *authentication* (can this system access this API?), *authorization* (can this system publish data to a specific data stream?), and *validation* (is this data acceptable or valid?) at the point of data production. Figure 9-2 shows the flow of the data gateway.

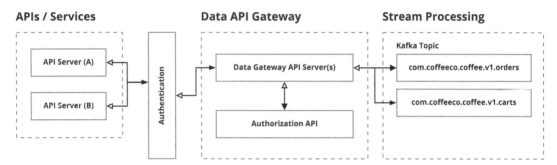

Figure 9-2. *Data gateway services can act as the gatekeepers to your data network. They provide mechanisms for authenticating API access, authorize which upstream services (APIs/services) are allowed to publish data, and can also be used to validate data from the publishers before corrupt or bad data makes it into the data network (stream processing)*

This means that the upstream system producing data can *fail fast* when producing data. This stops corrupt data from entering the streaming or stationary data pipelines at the edge of the data network. It's a means of establishing a conversation with the producers regarding exactly why and how things went wrong in a more automatic way via error codes and helpful messaging.

```
{
  "error": {
    "code": 400,
    "message": "The event data is missing the userId, and the timestamp is
    invalid (expected a string with ISO8601 formatting). Please view the
    docs at http://coffeeco.com/docs/apis/customer/order#required-fields to
    adjust the payload."
  }
}
```

This approach has its pros and cons. The pros are that most programming languages work out of box with HTTP (or HTTP/2) transport protocols (or with the addition of a tiny library), and JSON data is about as universal a data exchange format that you can get these days. On the flip side (cons), one can argue that there is yet another service to manage, and without some form of API automation or adherence to an open specification like OpenAPI, each new API route could begin to take more time than necessary. The API becomes a central point of failure, and more changes don't necessarily mean better functionality! To flip this problem on its head, this is why unit and functional testing are so important. Ideally, the data itself (schema/format) could dictate the rules of its own data-level contract by enabling field-level validation (predicates), producing helpful error messages, and acting in its own self-interest. Hey, with a little route- or data-level metadata and some creative thinking, the API could automatically generate self-defining routes and behavior.

Lastly, gateway APIs can be seen as centralized troublemakers, as each failure by an upstream system to emit valid data (e.g., blocked by the gatekeeper) causes valuable information (event data or metrics) to be dropped. The problem of blame here also tends to go both ways, as a bad deployment of the gatekeeper can blind an upstream system that isn't set up to handle retries in the event of gateway downtime (if even for a few seconds).

Putting aside all the pros and cons, using a gateway API to stop the propagation of corrupt data before it enters the data platform means that there are fewer places to look to understand what happened. This sure beats debugging a distributed network of data pipelines, services, and the myriad final data destinations. What about SDKs?

Software Development Kits

SDKs are libraries (or micro-frameworks) that are imported into a codebase to streamline an action, activity, or an otherwise complex operation. They are also known by another name, *clients*. Take the example from earlier about using good error messages and error codes. This process is necessary in order to inform a client that their prior action was invalid. However, it can be advantageous to add appropriate guardrails directly into an SDK to reduce the surface area of any potential problems. For example, let's say we have an API set up to track customer's coffee-related behavior through event tracking. A client SDK could theoretically include all the tools necessary to manage the interactions with the API server, including authentication, authorization, and validation.

In fact, if the SDK does its job, the validation issues would go out the door. Listing 9-2 shows an example of an SDK that could be used to reliably track customer events.

Listing 9-2. SDKs Enable Applications to Reliably Emit Event Data in a Consistent Way

```
import com.coffeeco.data.sdks.client._
import com.coffeeco.data.sdks.client.protocol._

Customer.fromToken(token)
  .track(
    eventType=Events.Customer.Order,
    status=Status.Order.Initalized,
    data=Order.toByteArray
  )
```

With some additional work (aka the client SDK), the problem of data validation or event corruption can just about go away. Additional problems can be managed in the SDK itself, such as how to retry sending a request when the server is offline. Rather than having all requests retry immediately, or in some loop that floods a gateway load balancer indefinitely, the SDK can take smarter actions like employing exponential backoff.

The Thundering Herd Problem   Let's say you have a single gateway API server. You've written a fantastic API and many teams across the company are sending event data to this API. Things are going well until one day, a new internal team starts to send invalid data to the server. (Instead of respecting your HTTP status codes, they treat all non-200 HTTP codes as a reason to retry. They forgot to add any kind of retry heuristics like exponential backoff, so all requests just retry indefinitely—across an ever-increasing retry queue.) Mind you, before this new team came on board, there was never a reason to run more than one instance of the API server, and there was never a need to use any sort of service-level rate limiter either, because everything was running smoothly within the agreed-upon SLAs.

Well, that was before today. Now your service is offline. Data is backing up, upstream services are filling their queues, and people are upset because their services are starting to run into issues because of your single point of failure.

These problems all stem from a form of resource starvation coined "the thundering herd problem." This problem occurs when many processes are awaiting an event, like system resources being available, or in this example, the API server coming back online. Now there is a scramble as all of the processes compete to attempt to gain resources, and in many cases the load on the single process (the API server) is enough to take the service back offline again. Unfortunately, this starts the cycle of resource starvation over again. This is of course unless you can calm the herd or distribute the load over a larger number of working processes, which would decrease the load across the network to the point where the resources have room to breathe again.

While this initial example is more of an unintentional distributed denial of service attack (DDoS), these kinds of problems can be solved at the client (with exponential backoff or self-throttling) and at the API edge via load balancing and rate limiting.

Ultimately, without the right set of eyes and ears enabled by operational metrics, monitors, and system level (SLA/SLO) alerting, data can play the disappearing act, and this can be a challenge to resolve.

Whether you decide to add a data gateway API to the edge of your data network, employ a custom SDK for upstream consistency and accountability, or decide to take an alternative approach when it comes to dealing with getting data into your data platform, it is good to know what your options are. Regardless of the path in which data is emitted into your data streams, this introduction to streaming data wouldn't be complete without a proper discussion of data formats, protocols, and the topic of binary serializable data. Who knows, you may just uncover a better approach to handling your data accountability problem!

Selecting the Right Data Protocol for the Job

When you think of structured data, the first thing to come to mind might be JSON data. JSON data has structure, is a standard web-based data protocol, and if nothing else it is super easy to work with. These are all benefits in terms of getting started quickly, but over time, and without the appropriate safeguards in place, you could face problems when it comes to standardizing on JSON for your streaming systems. The first problem is that JSON data is mutable. This means as a data structure, it is flexible and therefore fragile. Data must be consistent to be accountable, and in the case of transferring data across a network (on-the-wire) the serialized format (binary representation) should be highly compactable. With JSON data, you must send the keys (for all fields) for each object represented across the payload. Inevitably this means that you'll typically be sending a large amount of additional weight for each additional record (after the first) in a series of objects.

Luckily, this is not a new problem, and it just so happens that there are best practices for these kinds of things and multiple schools of thought regarding what is the best strategy for optimally serializing data. This is not to say that JSON doesn't have its merits. Just when it comes to laying a solid data foundation, the more structure the better and the higher level of compaction the better as long as it doesn't burn up a lot of CPU cycles.

Serializable Structured Data

When it comes to efficiently encoding and transferring binary data, two serialization frameworks tend to come up: Apache Avro and Google Protocol Buffers (protobuf). Both libraries provide CPU efficient techniques for serializing row-based data structures, and both technologies also provide their own *remote procedure call* (RPC) frameworks and capabilities. Let's look at avro, then protobuf, and we will wrap up looking at remote procedure calls.

Avro Message Format

With avro, you define declarative schemas for your structured data using the concept of records. These records are simply JSON-formatted data definitions files (schemas) stored with the file type avsc. Listing 9-3 shows an example of the coffee schema.

Listing 9-3. coffee.avsc: Defining a Schema for Coffee Can Be Done Fairly Easily

```
{
  "namespace": "com.coffeeco.data",
  "type": "record",
  "name": "Coffee",
  "fields": [
    ("name": "id", "type: "string"},
    {"name": "name", "type": "string"},
    {"name": "boldness", "type": "int", "doc": "from light to bold. 1
    to 10"},
    {"name": "available", "type": "boolean"}
  ]
}
```

Working with avro data can take two paths that diverge, related to how you want to work at runtime. You can take the compile-time approach, or the figure things out on-demand at runtime. This enables a flexibility that can enhance an interactive data discovery session. For example, avro was originally created as an efficient data-serialization protocol for storing large collections of data as partitioned files, long-term within the Hadoop file system. Given data was typically read from one location and written to another within HDFS, avro could store the schema (used at write time) once per file.

Avro Binary Format

When you write a collection of avro records to disk the process encodes the schema of the avro data directly into the file itself (once). There is a similar process when it comes to Parquet file encoding, where the schema is compressed and written as a binary file footer. We saw this process firsthand, at the end of Chapter 4, when we went through the process of adding StructField-level documentation to our StructType. This schema was used to encode our DataFrame, and when we wrote to disk it preserved our inline documentation on the next read.

Enable Backward Compatibility and Preventing Data Corruption

In the case of reading multiple files as a single collection, problems can arise in the case of schema changes between records. Avro encodes binary records as byte arrays and applies a schema to the data at the time of deserialization.

This means you have to take the extra precaution to preserve backward compatibility, or you'll find yourself running into issues with `ArrayIndexOutOfBounds` exceptions.

This can happen to the schema in subtle ways. For example, say you need to change an integer value to a long value for a specific field in your schema. Don't. This will break backward compatibility due to the increase in byte size from an `int` to a `long`. This is due to the use of the schema definition for defining the starting and ending position in the byte array for each field of a record. To maintain backward compatibility, you'll need to deprecate the use of the integer field moving forward (while preserving it in your avro definition) and add (append) a new field to the schema to use moving forward.

Best Practices for Streaming Avro Data

Moving from static avro files, with their useful embedded schemas, to an unbounded stream of well binary data, the main differentiator is that you need to bring your own schema to the party. This means that you'll need to support backward compatibility (in the case that you need to rewind and reprocess data before and after a schema change), as well as forward compatibility, in the case that you have existing readers already consuming from a stream.

The challenge here is support both forms of compatibility given that avro doesn't have the ability to ignore unknown fields, which is a requirement for supporting forward compatibility. In order to support these challenges with avro, the folks at Confluence open-sourced their schema registry (for use with Kafka), which enables schema versioning at the Kafka topic (data stream) level.

When supporting avro without a schema registry, you'll have to ensure you've updated any active readers (Spark applications or otherwise) to use the new version of the schema prior to updating the schema library version on your writers. The moment you flip the switch, you could find yourself at the start of an incident.

Protobuf Message Format

With protobuf, you define your structured data definitions using the concept of messages. Messages are written in a format that feels more like defining a struct in C. These message files are written into files with the .proto filename extension. Protocol buffers have the advantage of using *imports*. This means you can define common message types and enumerations that can be used within a large project, or even imported into external projects, enabling wide-scale reuse. A simple example of creating the Coffee record (message type) using protobuf is shown in Listing 9-4.

Listing 9-4. coffee.proto: Defining the Coffee Datatype Using Protobuf

```
syntax = "proto3";
option java_package="com.coffeeco.protocol";
option java_outer_classname="Common";

message Coffee {
  string id        = 1;
  string name      = 2;
  uint32 boldness  = 3;
  bool available   = 4;
}
```

With protobuf, you define your messages once, and then compile down for your programming language of choice. For example, we can generate code for Scala using the coffee.proto file from the standalone compiler in the ScalaPB project (created and maintained by Nadav Samet).

Code Generation

Compiling protobuf enables simple code generation. The following example is taken from the /ch-09/data/protobuf directory. The directions in the chapter's README cover how to install ScalaPB and include the steps to set the correct environment variables to execute the command in Listing 9-5.

Listing 9-5. Compiling Protobuf Messages for Scala

```
$SCALAPBC/bin/scalapbc -v3.11.1 \
  --scala_out=/Users/`whoami`/Desktop/coffee_protos \
```

```
--proto_path=$SPARK_MDE_HOME/ch-09/data/protobuf/ \
coffee.proto
```

This process saves time in the long run by freeing you up from having to write additional code to serialize and deserialize your data objects (across language boundaries or within different codebases).

Protobuf Binary Format

The serialized (binary wire format) is encoded using the concept of binary field level separators. These separators are used as markers that identify the data types encapsulated within a serialized protobuf message. In the example, `coffee.proto`, you probably noticed that there was an indexed marker next to each field type (`string id = 1;`). This is used to assist with encoding/decoding of messages on/off the wire. This means there is a little additional overhead compared to the avro binary, but if you read over the encoding specification, you'll see that other efficiencies more than make up for any additional bytes (such as bit packing, efficient handling of numeric data types, and special encoding of the first 15 indices for each message). With respect to using protobuf as your binary protocol of choice for streaming data, the pros far outweigh the cons in the grand scheme of things. One of the ways that it more than makes up for itself is with support for both backward and forward compatibility.

Enable Backward Compatibility and Prevent Data Corruption

There are similar rules to keep in mind when it comes to modifying your protobuf schemas like we discussed with avro. As a rule of thumb, you can change the name of a field, but you never change the type or change the position (index) unless you want to break backward compatibility. These rules can be overlooked when it comes to supporting any kind of data in the long term and can be especially difficult as teams become more proficient with their use of protobuf. There is this need to rearrange, and optimize, that can come back to bite you if you are not careful. (See "Maintaining Data Quality Over Time" for more context.)

Best Practices for Streaming Protobuf Data

Protobuf supports both backward and forward compatibility, which means that you can deploy new writers without having to worry about updating your readers first, and the same is true of your readers. You can update them with newer versions of your protobuf definitions without worrying about a complex deploy of all your writers. Protobuf supports forward compatibility using the notion of unknown fields. This is an additional concept that doesn't exist in the avro specification, and it is used to track the indexes and associated bytes it was unable to parse due to the divergence between the local version of the protobuf and the version it is currently reading. The beneficial point here is that you can also opt in, at any point, to newer changes in the protobuf definitions.

For example, say you have two streaming applications (a) and (b). Application (a) is processing streaming data from an upstream Kafka topic (x), enhancing each record with additional information, and then writing it out to a new Kafka topic (y). Application (b) reads from (y) and does its thing. Say there is a newer version of the protobuf definition, and application (a) has yet to be updated to the newest version. The upstream Kafka topic (x) and application (b) are already updated and are expecting to use some new fields available from the upgrade. The amazing thing is that it is still possible to pass the unknown fields through application (a) and onto application (b) without even knowing they exist.

Tip   Maintaining Data Quality Over Time: When working with either avro or protobuf, you should treat the schemas no different than you would code you want to push to production. This means creating a project that can be committed to your company's GitHub (or whatever version control system you are using), and it also means you should write unit tests for your schemas. Not only does this provide living examples of how to use each message type, but the important reason for testing your data formats is to ensure that changes to the schema don't break backward compatibility. The icing on the cake is that in order to unit test the schemas, you'll need to first compile the (.avsc or .proto) files and use the respective library code generation. This makes it easier to create releasable library code, and you can also use release versioning (version 1.0.0) to catalog each change to the schemas.

One simple method to enable this process is to serialize and store a binary copy of each message across all schema changes as part of the project lifecycle. I have found success adding this step directly into the unit tests themselves, using the test suite to create, read, and write these records directly into the project test resources directory. This way, each binary version, across all schema changes, is available within the codebase.

With a little extra upfront effort, you can save yourself a lot of pain in the grand scheme of things, and rest easy at night knowing your data is safe (at least on the producing and consuming sides of the table).

You now know that there are benefits to using avro or protobuf when it comes to your long-term data accountability strategy. By using these language-agnostic, row-based, structured data formats, you reduce the problem of long-term language lock-in, leaving the door open to whatever popular programing language is used later down the line. It can be a thankless task to support legacy libraries and codebases. Additionally, the serialized formats help reduce the network bandwidth costs and congestion associated with sending and receiving large amounts of data. This helps reduce the storage overhead costs for retaining your data long-term as well.

Lastly, let's look at how these structured data protocols enable additional efficiencies when it comes to sending and receiving data across the network using remote procedure calls.

Remote Procedure Calls

RPC frameworks, in a nutshell, enable client applications to transparently call remote (server-side) methods via local function calls by passing serialized messages back and forth. The client and server-side implementations use the same public interface definition to define the functional RPC methods and services available. The Interface Definition Language (IDL) defines the protocol and message definitions and acts as a contract between the client and server-side. Let's see this in action by looking at the popular open-source RPC framework, gRPC.

gRPC

First conceptualized and created at Google, gRPC which stands for "generic" remote procedure call, is a robust open-source framework used for high-performance services ranging from distributed database coordination, as seen with CockroachDB, to real-time analytics, as seen with Microsoft's Azure Video Analytics.

Using protocol buffers for message definitions, serialization, as well as the declaration and definition of services, gRPC can simplify how you capture data and build services. For example, let's say we wanted to continue the exercise of creating a tracking

API for customer coffee orders. The API contract could be defined in a simple services file, and from there the server-side implementation and any number of client-side implementations could be built using the same service definition and message types.

Define a gRPC Service

You can define a service interface, with its request and response objects, as well as the message types that need to be passed between the client and server, as easily as the example shown in Listing 9-6.

Listing 9-6. services.proto: Defining a gRPC Service and Messages

```
syntax = "proto3";

service CustomerService {
    rpc TrackOrder (Order) returns (Response) {}
    rpc TrackOrderStatus (OrderStatusTracker) returns (Response) {}
}

message Order {
    uint64 timestamp    = 1;
    string orderId      = 2;

    string userId       = 3;
    Status status       = 4;
}

enum Status {
  unknown_status = 0;
  initalized    = 1;
  started       = 2;
  progress      = 3;
  completed     = 4;
  failed        = 5;
  canceled      = 6;
}

message OrderStatusTracker {
  uint64 timestamp = 1;
```

```
  Status status    = 2;
  string orderId   = 3;
}

message Response {
    uint32 statusCode = 1;
    string message    = 2;
}
```

With the addition of gRPC, it can be much easier to implement and maintain the server-side and client-side code used in your data infrastructure. Given that protobuf supports backward and forward compatibility, this means that older gRPC clients can still send valid messages to newer gRPC services without running into common problems and pain points.

Figure 9-3 shows an example of gRPC at work. The server-side code is written in C++ for speed, while clients written in both Ruby and Java can interoperate with the service using protobuf messages as their means of communicating.

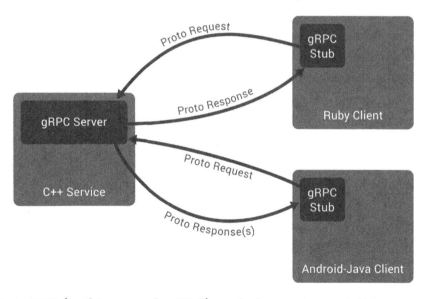

Figure 9-3. *RPC (in this example gRPC) works by passing serializing messages to and from a client and server. The client implements the same Interface Definition Language (IDL) interface and this acts as an API contract between the client and server. (Credit:* `https://grpc.io/docs/what-is-grpc/introduction/`*)*

gRPC Speaks HTTP/2

As a bonus, with respect to modern service stacks, gRPC can use HTTP/2 for its transport layer. This also means you can take advantage of modern data meshes (such as Envoy) for proxy support, routing, and service-level authentication, all while reducing the problems of TCP packet congestion seen with standard HTTP over TCP.

Mitigating data problems in flight and achieving success when it comes to data accountability starts with the data and fans outward from that central point. Putting processes in place when it comes to how data can enter your data network should be considered a prerequisite before diving into the torrent of streaming data.

Summary

The goal of this chapter was to lay out the moving parts, concepts, and nuances with respect to transitioning from a stationary batch-based mindset to one that considers the risks and rewards of working with real-time streaming data. Harnessing data in real-time can lead to fast, actionable insights, and opens the doors to state-of-the-art machine learning and artificial intelligence. However, distributed data management can also become a data crisis if the right steps aren't taken ahead of time. Remember that without a strong, solid data foundation, built on top of valid (trustworthy) data, the road to real-time will not be a simple endeavor.

In the next chapter, you build an end-to-end Structured Streaming application based on the concept of creating a system that processes real-time information about available occupancy in each of the coffee shops in order to enable people to find the shop nearest to their current location that has seating for a party of their size. You'll learn to use Kafka, as well as Redis (Redis Streams) along the way. There is a nice overlap between these two streaming technologies, as well as differences, which lend themselves nicely when it comes to Spark Structured Streaming.

## CHAPTER 10

Patterns for Writing Structured Streaming Applications

In the last chapter, we built up a streaming frame of reference for working with data in-flight. By introducing common problems and pitfalls that tend to occlude the path to streaming success, you discovered the nuances between local and global coordinated time, aka the trouble with time. You also learned about the complex problems that arise when corrupt or invalid data makes its way into your data network, in data problems in flight. In a perfect world, each event or data point would magically appear in the correct time-aligned (synchronous) order, but in a world devoid of perfection, there are tools you can use to lend a helping hand.

In this chapter, you'll discover effective design patterns used to solve streaming data problems when working with Spark Structured Streaming. For starters, you'll be introduced to the core programming model and concepts that encapsulate Spark Structured Streaming. Starting small and building up, you'll be writing a series of Structured Streaming applications that tackle the necessary core semantics for reading and writing unbounded data streams.

This chapter teaches you to:

- Process structured steaming data using *Redis Streams*. This reinforces the lessons introduced in the last chapter regarding the necessity of strict schemas for architecting accountable data systems.

- Learn what checkpoints are and how they work to power stateful structured streaming applications.

© Scott Haines 2022
S. Haines, *Modern Data Engineering with Apache Spark*, https://doi.org/10.1007/978-1-4842-7452-1_10

- Control your applications with rate limiting through the marriage of triggers and Spark configuration.

- Understand how to design Structured Streaming applications.

Let's begin building the foundation by learning the core tenants of Spark Structured Streaming.

What Is Apache Spark Structured Streaming?

In a nutshell, Apache Spark Structured Streaming enables you to tap into the power of real-time streaming data in an approachable way. It enables you, the engineer, to treat streaming data as a point-in-time (view) across an unbounded series of micro-batches (sets of data), or as a continuous stream of individual events (data points), using the foundation provided by the structured Spark APIs.

At the core of the Spark Structured Streaming programming model sits our friend SparkSQL. As a unifying process, SparkSQL enables you to apply the same mental model when writing applications that process streaming data, data at rest (batch data), or a mix of both. By blurring the lines defining *tables* and *views*, Structured Streaming effectively removes the cognitive burden associated with processing data across these two data access models. In the next section, you learn about how this processing paradigm works, beginning with the concept of *unbounded tables*.

Unbounded Tables and the Spark Structured Streaming Programming Model

Unbounded tables are a construct that makes it easier to conceptualize the processing patterns used to transform and query streaming data in the context of your application. If you were to store a truly unbounded table, eventually the application would collapse under the weight (number of data points) and pressure (memory and CPU) of all that data. In most real-world Structured Streaming applications, the data streams being ingested into your streaming application have a processing deadline, or a data time to live (TTL), before each data point becomes unnecessary, uninteresting, or useless for real-time. It is probably safer to say that when you write a Structured Streaming application, you are in fact creating an application that processes and stores a bounded set of data in memory, across an infinite or unbounded stream of eventual data over time. Figure 10-1 presents the concept of the unbounded table.

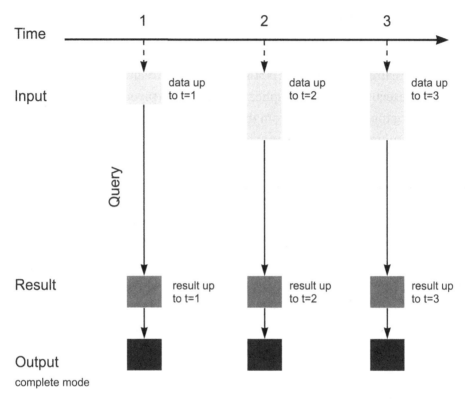

Figure 10-1. *Spark Structured Streaming programming model*

Figure 10-1 is a high-level sequence illustrating the processes at work that enable the reliable ingestion of structured unbounded streaming data over time within Spark. In the example, as new data is ingested into the application, a deterministic cycle begins with each of these ticks, aka micro-batches, of the system. Within each of these batches, Spark executes a directed series of steps (input --> query --> result --> output) before moving on and repeating the processing for the next set of data. This deterministic series of steps (called a DAG) represents the four essential processing paradigms of Spark Structured Streaming:

- Execution/processing modes (micro-batch or continuous modes)
- Triggers (effective processing windows/boundaries)
- Incremental queries (progressive data processing across time)
- Streaming output modes (complete, update, and append)

You'll explore these paradigms next and understand the roles they play in Spark Structured Streaming.

Processing Modes

In the beginning, micro-batch processing was introduced as the only mode of execution for Structured Streaming and today it is still the default. Streaming micro-batch scheduling and execution enabled application authors to pivot to a reliable, "structured" stream processing model and away from the do-it-yourself approach introduced by Spark's first streaming programming model.

Apache Spark Streaming with *Discretized Streams (DStreams)*

With DStreams, engineers had to be intimately aware of the RDD API (*core data structure and programming model of Apache Spark*). This meant engineers had to learn to work effectively with the lower level RDD APIs. This meant understanding how to recover from failure, implement receivers, efficiently serialize/deserialize their code and data, and scale out and performance-tune their applications. These are all things that you still need to learn, just at a higher level, using the abstractions made available by the DataFrames and Datasets APIs and SparkSQL. You can consider all of this as more of a natural stepping stone along the way to Structured Streaming, one that showcased the good, the bad, and the ugly with respect to design patterns adopted by the community to overcome the hurdles.

Fortunately, this book does not cover Apache Spark Streaming and instead presents the more approachable, performant, and self-optimizing *Apache Spark Structured Streaming*. I recommend taking a stroll down Spark memory lane to understand where things were and to appreciate how they have evolved. Now back to the Structured Streaming processing modes.

Continuous execution mode was introduced for true real-time processing with extremely low end-to-end latency. Both modes of execution enable applications to manage the complexities of stream processing, while changing only the semantics of how data is processed in the application. Figure 10-2 shows how the general flow of data remains consistent between the input sources and output sinks while the main differentiator resides in triggering.

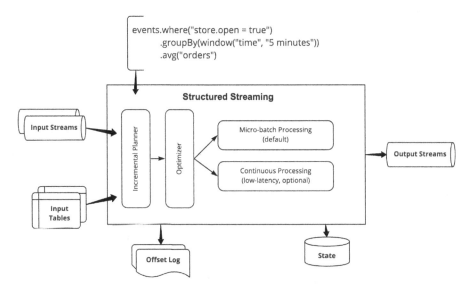

Figure 10-2. *The processing flow of Structured Streaming*

Understanding when to use each processing mode is a critical decision you will have to make before writing any initial code. Let's see why.

Micro-Batch Processing

As stated, micro-batch processing was introduced at the inception of Structured Streaming. This processing paradigm enables application authors to write streaming applications with the structured Spark APIs (DataFrames/Datasets). The mechanics of micro-batch execution relies on the concept of periodic polling for new data. This allows the Spark Structured Streaming engine to then generate a new batch of data for processing. It is worth pointing out that these applications are not just structured in terms of the data, which is bound to a very real structured schema (StructType), but also in the order of planning and execution surrounding each micro-batch processed.

Batches begin with a plan of execution, which is the result of a chain of pre-processing steps including: planning, analysis, and optimization by the Spark engine, which can be seen in Figure 10-3.

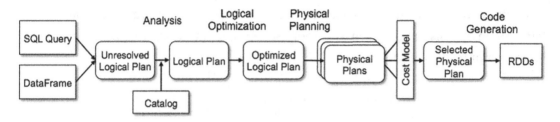

Figure 10-3. *The optimization process for each batch (or micro-batch) of data processed through the Spark*

After generating a physical plan (the process shown in Figure 10-3 and described in more detail in Chapter 6), the input data source(s) and their relevant metadata is written ahead (write-ahead logging) to create a reference point, called a checkpoint, for each micro-batch. This enables an application to fail mid-process and still rewind and recover from the last complete save point, in a deterministic fashion.

On the other side of each micro-batch, the output writer acknowledges the success of its actions, in what is known as a *commit*. It is also worth pointing out that each micro-batch is processed to successful completion before a new micro-batch can begin processing. This acknowledgment is handled by fully committing all work done within a micro-batch, or aborting all further execution when a terminal, unrecoverable, failure occurs.

By combining these tenants, you have an application that processes structured data from reliable data source(s), as a series of deterministic micro-batches, providing a clear and consistent pattern across all input and outputs, for the lifecycle of the application, for a seemingly infinite stream of data. We will see this process in action during the chapter exercises in the next section.

Continuous Processing

Shortly after the release of Structured Streaming, the lower-latency (sub-second) continuous processing execution framework was introduced. Rather than rely on periodic polling of data sources, the Spark execution framework instead relies on long-running processes to handle the continuous end-to-end processing of data, from the data source all the way to the data sink. This means data is processed as fast as possible, across each event individually, without the overhead required by waiting for all event data across each batch to be fully committed. A high-level diagram is shown to assist with the mental model in Figure 10-4.

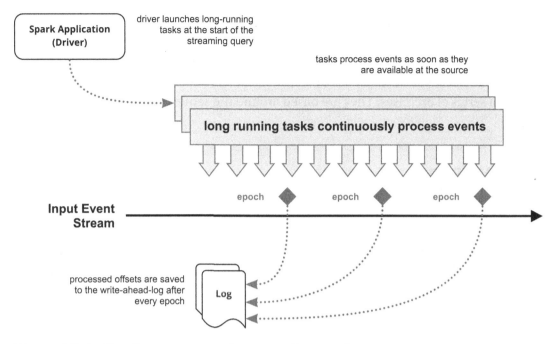

Figure 10-4. *Continuous processing execution mode*

While this execution mode can be used to achieve extremely low latency (~100ms end-to-end), it is best suited for the processing of streams of individual events rather than micro-batches of events that can be grouped, analyzed, or processed holistically. For example, if you were tasked with building a system that notified an individual that their coffee order is ready for pickup (like Apple's push notification service), you could technically distribute this load across your Spark cluster to send each message exactly once. You wouldn't want to annoy your customers and have them opt-out of being notified.

You'll be learning to use checkpoints and triggers in the third and fourth exercises. This will also be an opportunity to experiment with continuous and micro-batch processing.

Exercise Materials   The exercise materials for Chapter 10 are located at https://github.com/newfront/spark-moderndataengineering/tree/main/ch-10. The README files located within the material will provide you with more context regarding any additional installation or gotchas. These exercises depend on the extended ~/dataengineering/ setup from Chapter 8, which provides the common MySQL (Hive Metastore) and MinIO (S3).

Exercise Overview

The following exercises are aimed at teaching you to use the core features of Apache Spark Structured Streaming. You'll be using different approach patterns as you progress through a series of exercises. They are based around the same core challenge and extend from it.

The Challenge

You are asked to create a system that ingests real-time event data across all coffee shops owned by CoffeeCo. This system will be used to track customer usage patterns across the network, starting with order tracking. For each order, an accountable upstream gateway (like the examples described in Chapter 9) can be expected to produce reliable events that you will ingest into your Spark application.

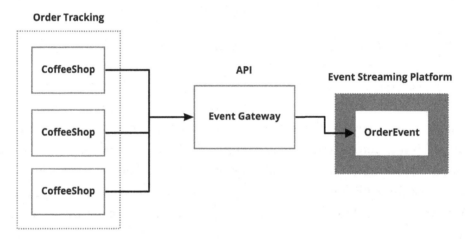

Figure 10-5. *The order tracking ingress path. For each order, an event is emitted through the event Gateway API and enqueued into an event streaming platform*

Figure 10-5 shows the high-level system at work providing the necessary data for our streaming application. For the sake of simplicity, this chapter exercises focus on the data from the event streaming platform onward. With that in mind, we'll need an event format.

The OrderEvent Format

Although it isn't impossible to write an event-based application without having an agreed-upon data format, it is harder and complicates things in the long term. In this case, let's say that the format of each event will need to provide the IDs for the order, the customer associated with the order, and the store where the transaction took place, as well as the number of items ordered and the total price of the transaction. This will be encapsulated in what we will call the OrderEvent. We know what we'll eventually need to ingest; we just need to formulize the actual event.

Note   In the real world, it's not uncommon for data engineers to be tasked with the job of creating the structured data formats and even writing the upstream systems producing the data fed into their own data domains. This way they own the data contract for their own domains and therefore understand the data lineage of other upstream data providers/producers in their network.

As mentioned in the last chapter, events should be consistent and reliable; otherwise your streaming application can abruptly stop functioning and you'll have a mess (and outage) on your hands. This doesn't mean that you need to spend a lot of time getting your upstream data ducks in a row before testing an idea. In fact, it is often the right approach to start off with the rough concept/idea, and then harden things in a phase-based approach.

Let's take that approach now. We will start by crafting the data we will be working with inside of the Spark engine.

Leaning on the Spark Struct

Given that the structured data transformations you make with Spark SQL rely on the StructType, you can begin the process of conceptualizing applications with simple DataFrames. This enables you to move quickly before formalizing the data contracts for a given application. Remember to always start small and build.

```
new StructType()
  .add(StructField("timestamp", LongType, false))
  .add(StructField("orderId", StringType, false))
  .add(StructField("storeId", StringType, false))
```

```
.add(StructField("customerId", StringType, false))
.add(StructField("numItems", IntegerType, false))
.add(StructField("price", FloatType, nullable false))
```

Now that we have a general representation of the structure of the data we will be using in our application, we can start the first exercise.

Exercise 10-1: Using Redis Streams to Drive Spark Structured Streaming

Redis Streams can be used to quickly build and test full end-to-end streaming applications. As you'll see over the course of this exercise, that can help with the discovery process associated with constructing applications.

Redis Streams runs with little overhead, thanks to it being a native data structure of Redis, which is a highly optimized data structure store that is very well known and well liked in the community. If you have never used Redis before, you can still learn from this exercise.

Exercise 10-1   The exercise materials.

```
/ch-10
    /applications/redis-streams
    /exercises/01_redis-streams
```

Spinning Up the Local Environment

Use the provided docker-compose file located in exercises/01_redis-streams to get started. You'll start the environment again using Docker compose, which means you will need to have Docker started as well.

```
docker compose \
  -f docker-compose.yaml \
  up -d --remove-orphans
```

This will spin up the external dependencies of the exercise, which are Redis, MinIO, and MySQL. Then it is time to connect to the new Redis container.

Connecting to Redis

Open the terminal or create a new session. You will use the terminal attached to the Redis Docker container and provide a command-line connection to Redis through the redis-cli. This connection will be used to push new event data into your event stream. This is also a convenient way to read, write, and manage the full lifecycle of the data for your event streams (or other Redis needs). As a bonus, you don't have to write any application code to get started.

Connect to the Redis CLI

With the assistance of the Redis container, you can now connect to the redis-cli by executing the command in Listing 10-1.

Listing 10-1. Connecting to Redis Container and Using the redis-cli (Command Line)

```
docker exec \
  -it redis redis-cli
```

Execute ping in the prompt. You will see PONG returned to you, meaning things are working.

```
127.0.0.1:6379> ping
PONG
```

Next, you'll learn to monitor the flow of commands into Redis. This is a great way to see what is happening between the different points in the system.

Enable the Redis Command Monitor

Open another terminal window and execute the command in Listing 10-2. You should be greeted with an OK once you are connected.

Listing 10-2. Using the redis-cli from the Redis Container to Monitor Commands Sent to Redis (Locally or Remotely)

```
docker exec \
  -it redis redis-cli monitor
```

The monitor process is a useful view into what commands are being sent to Redis (in real-time). If you wait around for a few seconds, you'll see a ping message coming through the monitor from the Redis container's health checks.

```
1626138124.541908 [0 127.0.0.1:59704] "ping"
```

Container Health Checks   The healthcheck block provided in the docker-compose file (ch-10/exercises/01_redis-streams/) enables a service to monitor itself using simple commands. In the case of Redis, the health check simply sends a ping through the redis-cli and makes sure it doesn't time out. This means that the container is operating as expected.

```
healthcheck:
    test: ["CMD", "redis-cli", "ping"]
    interval: 5s
    timeout: 30s
    retries: 50
```

Now it is time to create your first stream with Redis.

Creating the Redis Stream

A *Redis Stream* is generated automatically and becomes instantly available as you add data. Since Redis is an in-memory database, there is minimal process overhead for stream generation given there are no immediate files to create, new disk space allocations, or expensive locking associated with stream generation. You simply publish the first event, and the stream magically comes into existence. Behind the scenes Redis persists the entire database periodically to the RDB, which is a point-in-time snapshot of your entire Redis database. If you are familiar with Redis, the stream type is conceptually a mix of the list, sorted set, and the hashmap data structures.

Stream Guarantees

Redis *guarantees the order* of the fields in each event in your stream, as well as the consistency of the insert order. Each event identifier will always be greater than any event inserted before it.

Why Field Level Ordering Matters

Field-level ordering is important for more than peace of mind. Each encoded payload stored as an event encapsulated by the Redis Stream is a pointer to a hashmap. By design, hashmaps can be randomly accessed by their keys, but they don't retain any specific order out of the box. For example, there is a big difference between a HashMap and a LinkedHashMap in Java. The linked variant is used when you want to maintain the order of the keys within the EntrySet. For reliability and consistency within an Redis Stream, Redis takes care of the field-level ordering, but you must still provide a reliable event. We will look at how this is handled later in the exercise.

Event Identifiers

You add new events using the XADD command. This command includes the option of using an auto-generated ID rather than providing your own identifier. The valid event ID format is <number>-<number>. In the case of autogenerated IDs, Redis will use the concatenation of the UNIX epoch milliseconds (timestamp) when the event was ingested into Redis and the monotonically increasing ID for events received at the same millisecond.

```
1625766684875-0
```

This pattern ensures that each event identity is greater than any event that came before it. If you decide to bring your own identifiers to the party, you'll have to keep this in mind. Otherwise, you'll get see ERR Invalid stream ID specified as stream command argument due to the broken expectations of the event ID.

Redis Streams Are Append-Only

It is important to understand that streams in Redis are an append-only streaming data structure. Append only means you can't change the insert order. This forces upstream systems that want to reorganize events (sorting, adding missing data) to read the stream, reorganize it, and either push events into a new stream, or push these results into a different reliable datastore. When you have an immutable, append-only streaming source, you maintain consistency. This reduces the complexity of using this data and provides you with a reliable data source that is ideal for streaming.

Let's publish an event and generate your first stream.

Event Publishing to a Redis Stream

You create a Redis Stream by publishing an event. Copy the command in Listing 10-3 from xadd onward and paste it into the `redis-cli`.

Listing 10-3. Publishing an Event to a Redis Stream

```
127.0.0.1:6379> xadd com:coffeeco:coffee:v1:orders  * timestamp
1625766472513 orderId ord126 storeId st1 customerId ca183 numItems 6
price 48.00
```

Redis will acknowledge your command by returning the event ID for the event you published. For example, "1625766684875-0".

Consume Events from a Redis Stream

Now that you have an event published and a stream generated. You can read (consume) all available published events using the `redis-cli`. To do this, you'll use the xrange or xrevrange commands, which allow you to read the items in the stream. Execute the command in Listing 10-4 using the `redis-cli`.

Listing 10-4. Using xrange to Read Events from the Redis Stream

```
127.0.0.1:6379> xrange com:coffeeco:coffee:v1:orders - + COUNT 1
```

This will fetch the first item added to the stream in ascending insert order, which is based on the tail of the stream (think about how linked lists work with head or tail pointers). Conceptually this process is the SQL equivalent to an order by asc limit 1.

```
select * from v1.orders
order by timestamp asc limit 1
```

The reverse of this is the xrevrange option. It uses descending insert order, reading from the head of the stream (newest items first). So now that you have a general idea how Redis Streams operates, it is time to get into the Spark application itself.

The SparkRedisStreamsApp

Open the SparkRedisStreamsApp, which is located in the applications directory in the chapter materials at /ch-10/applications/redis-streams. This application is a throwback to Chapter 7, when you created your first compiled Spark application in Scala. The general structure of the application is the same, with the addition of a Dockerfile and one unmanaged library item (*lib/spark-redis). *At the time of writing, the spark-redis library was not generally available for Spark 3x.

redis-streams/
  conf/
  lib/
  project/
  src/
  Dockerfile
  build.sbt

This project provides all the necessary building blocks to create your end-to-end application. Rather than look through the code first, let's flip this strategy on its head and compile first, then ask questions later.

Let's start by compiling the application using sbt.

Compile the Application

From the project root directory (redis-streams), compile the application.

```
sbt clean assembly
```

Note   Compiling the app requires Redis to be running in Docker. Return to the start of the exercise and execute the docker compose command in the exercise directory.

Once the application is compiled, your JAR will be located in target/scala-2.12/ spark-redis-streams-assembly-0.1-SNAPSHOT.jar. We'll use this resource when we wrap the application in its own Docker container next.

Build the Docker Container

With the application compiled, you can now build your Docker container. Open the Dockerfile located in the project root directory (shown in Listing 10-5).

Listing 10-5. The Dockerfile Used to Create Our Consistent Redis Streams Application

```
FROM newfrontdocker/apache-spark-base:spark-3.1.2-jre-11-scala-2.12

COPY target/scala-2.12/spark-redis-streams-assembly-0.1-SNAPSHOT.jar /opt/
spark/app/spark-redis-streams.jar

EXPOSE 4040
```

The container definition file (Dockerfile) simply adds a new layer on top of the apache-spark-base base container, which copies your application JAR to the /opt/spark/app directory (created by the spark-base container). It also exposes port 4040, which is the port the Spark UI runs on. From the project root directory, execute the docker build command in Listing 10-6.

Listing 10-6. Building the spark-redis-streams Docker Container

```
docker build . \
  -t `whoami`/spark-redis-streams:latest
```

You'll see that the container takes any time at all to build (assuming the apache-spark-base container is cached in your local Docker).

```
[+] Building 0.4s (7/7) FINISHED
 => [internal] load build definition from Dockerfile
0.0s
 => => transferring dockerfile: 37B
0.0s
 => [internal] load .dockerignore
0.0s
 => => transferring context: 34B
0.0s
```

```
 => [internal] load metadata for docker.io/newfrontdocker/apache-spark-
base:spark-3.1.2-jre-11-scala-2.12
0.0s
 => CACHED [1/2] FROM docker.io/newfrontdocker/apache-spark-
base:spark-3.1.2-jre-11-scala-2.12
0.0s
 => [internal] load build context
0.2s
 => => transferring context: 11.90MB
0.2s
 => [2/2] COPY target/scala-2.12/spark-redis-streams-assembly-0.1-SNAPSHOT.
jar /opt/spark/app/spark-redis-streams.jar
0.1s
 => exporting to image
0.1s
 => => exporting layers
0.1s
 => => writing image sha256:2e542656e6e8c5e165fac6e308c68252f2396152a0078
5c7094c15f55a702939
0.0s
 => => naming to .../spark-redis-streams:latest
0.0s
```

The build completes quickly because essentially all you are doing is adding one JAR (which you already compiled locally) and defining one port to be exposed.

Tip   Think about how you can reuse containers to create stackable environments. In the example in Listing 10-5, the stack starts with `apache-spark-base`, and you add to the stack with your application JAR.

You are now ready to run the application and consume the Redis Stream.

Running the Application

The application acts like a simple echo server. When you add a new event to the Redis Stream, the application will (almost) immediately process the event and output the formatted DataFrame and the batch identifier (micro-batch). This application represents a first pass, ensuring things are wired up correctly. Remember to always start small and build.

This is the point where the `redis-cli` monitor (from Listing 10-2) becomes interesting. If you have the screen real estate, bring up the `redis-cli` and the `redis-cli` monitor windows (in Listings 10-1 and 10-2) and then open a third terminal window where you'll run the application.

Execute the command in Listing 10-7 to start the container and trigger the Spark application to run.

Listing 10-7. The docker run Command to Launch the Spark Redis Streams Application (in Container) Running in Your Local Data Platform

```
docker run \
  -p 4040:4040 \
  --hostname spark-redis-test \
  --network mde \
  -v ~/dataengineering/spark/conf:/opt/spark/conf \
  -v ~/dataengineering/spark/jars:/opt/spark/user_jars \
  -it `whoami`/spark-redis-streams:latest \
  /opt/spark/bin/spark-submit \
  --master "local[*]" \
  --class "com.coffeeco.data.SparkRedisStreamsApp" \
  --deploy-mode "client" \
  --jars /opt/spark/user_jars/mariadb-java-client-2.7.2.jar \
  --driver-class-path /opt/spark/user_jars/mariadb-java-client-2.7.2.jar \
  --conf "spark.sql.warehouse.dir=s3a://com.coffeeco.data/warehouse" \
  --conf "spark.app.source.stream=com:coffeeco:coffee:v1:orders" \
  /opt/spark/app/spark-redis-streams.jar
```

The Docker container will start up using the SparkRedisStreamApp JAR you compiled earlier and wrapped into the container. Inspecting the logs, you will notice a lot of info-level logs from Spark. You can always reduce this chatter using log4j (changing the root-level logging to WARN or ERROR or opting into application only logs). For now, the logs can teach you more about how things work.

Redis Streams Consumer Groups and Offsets

When the application starts up, it will connect to the Redis Stream identifier provided in the spark.app.source.stream configuration value. This process can be seen in the output logs. Just look for the following output.

```
INFO RedisSource: Getting batch...
  start: None
  end: {"offsets":{"com:coffeeco:coffee:v1:orders":{"groupName":
  "spark-source","offset":"1626552490734-0"}}}
```

During this process, there is a quick conversation happening between Spark and Redis, which you can observe using the redis-cli monitor. You'll see the following output in the monitor.

```
"XGROUP" "create" "com:coffeeco:coffee:v1:orders" "spark-source" "$"
"mkstream"

"XINFO" "STREAM" "com:coffeeco:coffee:v1:orders"
"XREADGROUP" "group" "spark-source" "consumer-1" "count" "100" "block"
"500" "noack" "streams" "com:coffeeco:coffee:v1:orders" ">"

"XINFO" "STREAM" "com:coffeeco:coffee:v1:orders"
```

These commands are part of the handshake process when the application starts up with no prior state. (You'll learn to store the application state with checkpoints in Exercise 10-3.) This information-gathering flow queries the stream and creates new consumer groups for the application.

Consumer Groups

To keep track of the position of the stream and to associate the Spark application with a specific group, the XGROUP command is used. Given that the application has never been run before, a new group is first created, and then the stream is queried (XINFO) to get all the metadata related to the stream so Spark can begin to ingest new events.

You can run the `xinfo` command in the `redis-cli` to see what data is returned.

```
xinfo STREAM com:coffeeco:coffee:v1:orders
```

Then, using XREADGROUP, the application creates an initial position to block and await new events.

Now that the application understands its initial state, it can move on and begin to do work. This is the critical first step in any Structured Streaming application. In fact, you'll see that this process also creates a first batch.

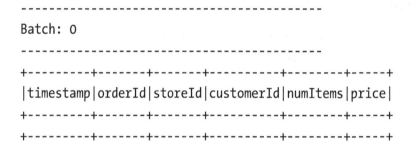

```
-------------------------------------------
Batch: 0
-------------------------------------------
+---------+-------+-------+----------+--------+-----+
|timestamp|orderId|storeId|customerId|numItems|price|
+---------+-------+-------+----------+--------+-----+
+---------+-------+-------+----------+--------+-----+
```

This process will now only repeat when new data arrives from the Redis Stream. Let's add another event.

Creating the Next Batch

You can copy the XADD command from Step 3 and paste it into the `redis-cli` to trigger a new batch.

```
xadd com:coffeeco:coffee:v1:orders  * timestamp 1625766472513 orderId
ord126 storeId st1 customerId ca183 numItems 6 price 48.00
```

From the Spark application container, you'll see the application react and process this new event. From the logs, the important piece of output is the acknowledgment of the batch and the formatted output of the batch.

```
--------------------------------------------
Batch: 1
--------------------------------------------
+-------------+-------+-------+----------+--------+-----+
|    timestamp|orderId|storeId|customerId|numItems|price|
+-------------+-------+-------+----------+--------+-----+
|1625766472513| ord126|    st1|     ca183|       6| 48.0|
+-------------+-------+-------+----------+--------+-----+
```

Try pasting the following events into the `redis-cli`. What do you notice about the behavior of the application?

```
> xadd com:coffeeco:coffee:v1:orders  * timestamp 1625766472503 orderId
ord124 storeId st1 customerId ca153 numItems 1 price 10.00
> xadd com:coffeeco:coffee:v1:orders  * timestamp 1625766472513 orderId
ord125 storeId st1 customerId ca173 numItems 2 price 7.00
> xadd com:coffeeco:coffee:v1:orders  * timestamp 1625766472513 orderId
ord126 storeId st1 customerId ca183 numItems 6 price 48.00
> xadd com:coffeeco:coffee:v1:orders  * timestamp 1625766472613 orderId
ord127 storeId st1 customerId ca184 numItems 1 price 3.80
```

You'll notice that the application will process each event in its own batch. Well, why is that, you might be asking? That doesn't seem to be in line with what we might expect from micro-batch, which feels more continuous. This is true and this behavior can be controlled by adding triggers into the control path of the application.

For now, you can stop the docker process (^+C) or just exit the window where it is running.

We'll be diving into the application code itself in Exercise 10-2 to see the pieces that make this application tick. You'll learn some tricks that help make writing applications simple along the way.

Exercise 10-1: Summary

This first exercise taught you how Redis Streams work from a high level. You learned to use the `redis-cli` and what the `redis-cli` monitor is used for as you figured out how to navigate these new streaming waters. You finished up the exercise with a preview of the Spark Redis Streams application, the contents of which you'll be looking at next.

Exercise 10-2: Breaking Down the Redis Streams Application

The application you ran in Exercise 10-1 is a partial Structured Streaming application. This is due to the fact that it doesn't yet keep track of its own state, making it stateless, and it also doesn't do anything productive with the events it consumes. This is all by design though; the current state of the application is shown in Figure 10-6.

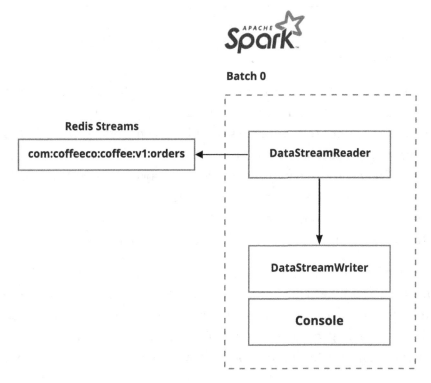

Figure 10-6. *The starting point of Structured Streaming. It connects to Redis and continuously polls for new events*

Figure 10-6 represents the application you just ran in Exercise 10-1. Using this as a mental model, open the redis-streams project in your favorite IDE and navigate to SparkRedisStreamsApp.scala located at /ch-10/applications/redis-streams/ SparkRedisStreamsApp.scala.

You'll notice that we are again using the SparkApplication trait that we introduced in Chapter 7.

```
object SparkRedisStreamsApp extends SparkApplication
```

As a refresher, this trait provides the application with a `SparkSession`, a handle to the application name, and the immutable `SparkConf` that ultimately reduces a lot of the boilerplate code when setting up each new application.

DataStreamReader

Using the diagram for reference, let's begin with the *DataStreamReader*. This is one of the core Structured Streaming interfaces and it offers a common means to load streaming datasets into Spark. In the application, you'll find the following block.

```
lazy val inputStream: DataStreamReader = {
  sparkSession.readStream
    .format("redis")
    .option("stream.keys", inputStreamName)
    .schema(streamStruct)
}
```

Just like with our batch applications, we use the `SparkSession` as an interface to configure our data source and generate a new reader, or ingress point. The only difference between what we've done in the book up until now is that we are specifying the data source as a streaming data source by using `readStream` instead of `read`.

Extending Spark   This may also be the first time you are seeing a custom data source format. Spark is a pluggable platform, and it understands that `format("redis")` will be provided by the `spark-redis` library through the use of the DataSourceRegistry and StreamSourceProvider interfaces. You can create your own source and sink formats to add new capabilities to Spark without having to reinvent the wheel.

Then we fill in the required configuration for the data source. In this case, just the name of the stream (key name of the stream) and the expected schema from this source.

Is the Schema Required?

The *schema* is required for most streaming data sources. The JSON data source can infer the schema, but it does that using sampling, which can cause inconsistencies. This means you won't have a deterministic schema in the case of sparse event data (e.g., optional fields). For your own sanity, it is better to provide the schema based on your upstream data contracts. That way, you are writing your applications with solid expectations.

You were introduced to the order event format in the exercise's introduction. The following struct is the formal structured data contract for this application.

The OrderEvent Schema

```
lazy val streamStruct = new StructType()
  .add(StructField("timestamp", LongType, false))
  .add(StructField("orderId", StringType, false))
  .add(StructField("storeId", StringType, false))
  .add(StructField("customerId", StringType, false))
  .add(StructField("numItems", IntegerType, false))
  .add(StructField("price", FloatType, false))
```

It is worth mentioning that the schema builder method of the DataStreamReader takes either a fully formatted StructType (as seen above) or you can also provide a DDL string. Spark will parse the DDL string and assemble things for you.

Generating a StructType from a DDL String. This Is Equivelent to Calling StructType. fromDDL(streamStruct.toDDL)

```
StructType.fromDDL("""
  `timestamp` BIGINT,
  `orderId` STRING,
  `storeId` STRING,
  `customerId` STRING,
  `numItems` INT,
  `price` FLOAT""")
```

This flexibility enables you to write rich streaming applications that can be config-driven while still adhering to a strict data contract.

That covers how the application knows where to connect and what schema to expect from an upstream. Given the simplicity of this first streaming application, we'll be moving right along to the DataStreamWriter.

DataStreamWriter

The DataStreamWriter is the central interface used to write streaming datasets to external storage in Spark. The interface provides a smaller subset of methods than its batch mode equivalent, the DataFrameWriter. We'll cover using outputMode, partitionBy, and trigger in Exercise 10-3. Looking at the run method will show you this in action.

The Streaming Implemenetation of the run Method from the SparkApplication Trait

```
override def run(): Unit = {
  ...
  val writer: DataStreamWriter[Row] = SparkRedisStreams(sparkSession)
  .transform(inputStream.load())
  .writeStream
  .queryName("orders")
  .format("console")

    startAndAwaitApp(writer.start())
  }
```

The run method takes advantage of some simple application design patterns which might not be immediately noticeable. We'll cover the subtle use of lazy invocation and dependency injection next. These techniques work alongside the Spark delegation pattern introduced in Chapter 7.

Lazy Invocation

The run method leans on Scala companion class construction and lazy dependency injection to reduce the complexity required for Spark to read the intentions of this application and generate an execution plan.

```
SparkRedisStreams(sparkSession)
  .transform(...)
```

The DataFrame that's passed to the transform method is the result of calling load on our DataStreamReader.

```
.transform(inputStream.load())
```

This inputStream is a lazy singleton that acts as a promise to generate a DataFrame at a future point in time. This is all Spark needs to know at this point.

347

Depending on Dependency Injection

As an aside, this pattern also enables you to use dependency injection to provide the `transform` method with a DataFrame. This makes testing easier and provides flexibility with regard to upstream data sources. At this moment the application is connected to Redis, but you can just as easily plug Kafka into place and the application doesn't change aside from upstream configuration. This pattern also enables you to reuse the application for either batch or streaming data sources given the interface is bound to the DataFrame and not a DataStreamReader.

This leaves us with the final piece of the puzzle. To have a Structured Streaming application, you need to have both a data source as well as a data sink. We've covered the upstream data source, which is Redis. As a quick recap, this streaming data source is generated using the `readSteam` method on the Spark Session, which results in a DataStreamReader. Now to fulfill the last requirement, all we need is a sink.

Writing Streaming Data

The `writeStream` method is a member of the `Dataset` class. A DataFrame is simply a `Dataset[Row]` and the application's `transform` method returns a DataFrame. Now when we call `writeStream` we are transforming our DataFrame to a DataStreamWriter.

In the application, the output format used is `"console"`. This is a special sink that can be used to observe the output of your application as it doesn't write any data to a reliable data store.

```
.writeStream
.queryName("orders")
.format("console")
```

With the full application wired up (source --> transform --> sink), all that is left to do is start the app. The `start` method on the DataStreamWriter creates a `StreamingQuery`.

```
startAndAwaitApp(writer.start())
```

Streaming Query

The StreamingQuery interface encapsulates the end-to-end structured stream. It provides methods for introspecting the query while it is running in your Spark application. It is also thread-safe, so you can observe the stream from anywhere on the driver. You'll be using this interface directly as this application is extended and hardened with tests in Chapter 11.

For now, you need to only concern yourself with one method: awaitTermination.

```
def startAndAwaitApp(query: StreamingQuery): Unit = {
  streamingQuery.awaitTermination()
}
```

When you call awaitTermination on the streaming query, the application will continue running until there is an unrecoverable exception thrown, or until the process is exited, such as in the case of stopping the Docker container or sending a termination signal.

This end-to-end process is all kicked off by calling run() from within the application entry point.

Application Entry Point

The application entry point is the class providing the main method. This application extends the SparkApplication trait, which extends the App trait from Scala. So, when you ran this application earlier, you were essentially calling the main method of SparkRedisStreamsApp, which calls run().

```
/opt/spark/bin/spark-submit \
  --master "local[*]" \
  --class "com.coffeeco.data.SparkRedisStreamsApp" \
  ...
  /opt/spark/app/spark-redis-streams.jar
```

Now that you understand how the initial application works, we can turn things up a notch and take a second pass over the application. In Exercise 10-3, you'll be learning to make the application stateful using *checkpoints*.

Exercise 10-2: Summary

In Exercise 10-2, you learned how the application introduced in Exercise 10-1 works. You walked through the steps to connect a streaming data source to a streaming data writer and learned how the events received (across each batch) flow from source to sink. This simple exercise introduced you to the core Spark interfaces for Structured Streaming: the `DataStreamReader`, `DataStreamWriter`, and `StreamingQuery` interfaces. Lastly, you learned how lazy invocation and dependency injection can be layered to create easy-to-follow and simple-to-maintain Spark applications.

In this next exercise, you learn to add checkpoints to this application.

Exercise 10-3: Reliable Stateful Stream Processing with Checkpoints

In Exercise 10-1, when the application started there was a handshake that occurred between Spark and Redis. This process generated the initial state for our stateful application. Unfortunately, when you killed the process, all this temporary state was essentially wiped out. *This* means that the application skips processing any data added to the Redis Stream while it is offline.

You probably know what I'm going to say next, and yes this is not an ideal situation, and no it doesn't help to build confidence in the application.

Why Checkpoints Matter

Checkpoints give you a save point. Like in any classic video game, this provides you with a safe place to return to if you die. Now with our application, if it dies, we need to be able to start over exactly where we left off.

Reliable Checkpoints

Reliable checkpoints begin with the concept of consistency and durability. This should have you thinking about fault tolerance, and it isn't too difficult a leap to connect this all to atomicity. Apache Spark enables reliable checkpointing by leaning on the foundations of the distributed file system (aka HDFS) to store application checkpoints (state) off box, so that an application can be restarted in the event that the machine it was running on shuts down.

Enabling Reliable Checkpoints

Open the StatefulSparkRedisStreamsApp. This application changes very little to introduce reliable checkpointing. In fact, there are only two noticeable changes.

The first change is introduced in the run method.

Adding the checkpointLocation to the DataStreamWriter

```
SparkRedisStreams(sparkSession)
  .transform(inputStream.load())
  .writeStream
  .queryName("orders")
  .option("checkpointLocation", appCheckpointLocation)
  .format("console")
```

The value of the variable passed into this parameter is passed to the application through the SparkConf.

```
lazy val appCheckpointLocation: String = sparkConf.get("spark.app.
checkpoint.location")
```

Now let's run the application.

Running the Stateful Application

You can test drive this application using a slight modification to the Docker run script you used in Exercise 10-1.

```
docker run \
  ...
  --class "com.coffeeco.data.StatefulSparkRedisStreamsApp" \
  --conf "spark.sql.warehouse.dir=s3a://com.coffeeco.data/warehouse" \
  --conf "spark.app.source.stream=com:coffeeco:coffee:v1:orders" \
  --conf "spark.app.checkpoint.location=s3a://com.coffeeco.data/apps/spark-
  redis-streams-app/1.0.0/simple" \
  /opt/spark/app/spark-redis-streams.jar
```

The two highlighted changes correspond to the application entry point and the additional configuration needed to provide the application checkpoint location.

Observing the Stateful Behavior

With the application running, open your redis-cli (if you've closed it) and paste in one of the example OrderEvents from Exercise 10-1.

```
> xadd com:coffeeco:coffee:v1:orders  * timestamp 1625766472503 orderId
ord124 storeId st1 customerId ca153 numItems 1 price 10.00
```

This will result in the batch being processed. It's the same behavior as before. So now for the exciting part. Stop the Spark application (press Control+C). Before you start the application back up, add three more events to the Redis Stream.

```
> xadd com:coffeeco:coffee:v1:orders  * timestamp 1625766472503 orderId
ord124 storeId st1 customerId ca153 numItems 1 price 10.00
> xadd com:coffeeco:coffee:v1:orders  * timestamp 1625766472613 orderId
ord127 storeId st1 customerId ca184 numItems 1 price 3.80
> xadd com:coffeeco:coffee:v1:orders  * timestamp 1625766472513 orderId
ord125 storeId st1 customerId ca173 numItems 2 price 7.00
```

Start the Spark application back up. The first observation you may make (if you scan the logs) is that the Redis Source output now has a start parameter.

```
INFO RedisSource: Getting batch...
  start: Some({"offsets":{"com:coffeeco:coffee:v1:orders":{"groupName":
  "spark-source","offset":"1626658307940-0"}}})
  end: {"offsets":{"com:coffeeco:coffee:v1:orders":{"groupName":
  "spark-source","offset":"1626658362223-0"}}}
```

You'll also see that the application picks right back up, as if it were never offline. Instead of starting with batch 0 again, you'll see the next batch is non-zero. For example:

```
-------------------------------------------
Batch: 2
-------------------------------------------
+------------+-------+-------+----------+--------+-----+
|   timestamp|orderId|storeId|customerId|numItems|price|
+------------+-------+-------+----------+--------+-----+
|1625766472503| ord124|    st1|     ca153|       1| 10.0|
|1625766472613| ord127|    st1|     ca184|       1|  3.8|
|1625766472513| ord125|    st1|     ca173|       2|  7.0|
+------------+-------+-------+----------+--------+-----+
```

This means you can now survive application downtime without missing a beat.

How Checkpoints Work

For each batch, the streaming data source will write the explicit offsets used for a batch to reliable storage, aka the distributed file system. This is called write-ahead logging. After each successful batch, there is an acknowledgment written to reliable storage in the form of a commit. You can see the process in Figure 10-7.

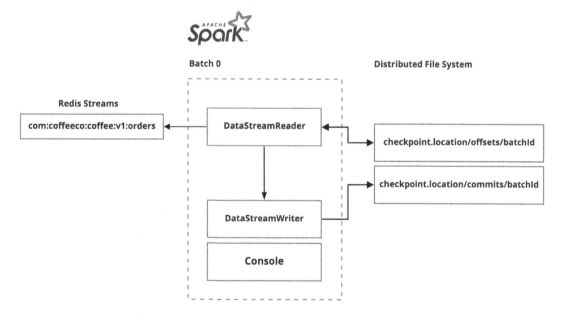

Figure 10-7. *The checkpointing process*

This process is repeated for each batch. This means that a failure in a batch won't move the cursor so to speak, and your application will restart at its last successful save point.

You can inspect the application checkpoints using the MinIO browser at `http://127.0.0.1:9000/minio/com.coffeeco.data/apps/spark-redis-streams-app/1.0.0/simple/` or on the file system under `~/dataengineering/minio/`.

Deleting Checkpoints

You only need to delete checkpoints when an application is end of life, or when an application is being updated and the underlying DAG has changed. The DAG changes when you change data sources or sinks, or when you add more data sources to the table, such as with streaming joins from multiple upstreams.

Exercise 10-3: Summary

This third exercise introduced you to the process that makes Structured Streaming reliable—the checkpoint. You learned that each batch is controlled using a write-ahead log (WAL) and each successful batch is marked complete using commits. This simple process keeps track of your application state even in the case of a terminal failure of the physical server or machine where your application is running.

In the final exercise of this chapter, you learn how to control the runtime behavior of the application by adding processing triggers.

Exercise 10-4: Using Triggers to Control Application Runtime Behavior

Triggers can be used to control the behavior of the Structured Streaming application. During this exercise, you learn how to apply three different flavors of trigger while also enhancing the application. The changes are shown in Figure 10-8.

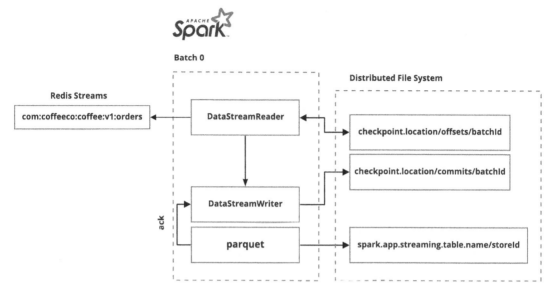

Figure 10-8. *The final updates being made to the Redis Streams application. This change will write the contents of the data read through Redis Streams and then write those events out into a durable distributed table*

Open the file named TriggeredStatefulSparkRedisStreamsApp. Here you'll find a few new methods and configurations and an update to the run method. Skip down to the outputStream method, since this is where we'll be starting. This method is responsible for building a partial DataStreamWriter.

```
lazy val outputStream: DataStreamWriter[Row] = {
  SparkRedisStreams(sparkSession)
    .transform(inputStream.load())
    .writeStream
    .queryName("orders")
    .format("parquet")
    .partitionBy("storeId")
    .option("path", streamingSinkLocation)
    .outputMode(OutputMode.Append())
}
```

The output format of the streaming sink is *parquet*. This means the application is converting the data read from the Redis Stream into parquet and then writing these events out, partitioned by `storeid` into the path defined by `streamingSinkLocation`. When you write out using `partitionBy`, you are essentially creating key/value named directories.

```
/tables/coffee.order.events/storeId={storeId}/
```

Tip   As a rule of thumb, you should partition your tables across lower cardinality columns. It is common to use the date (date=YYYY-mm-dd) and the parts of the year (year=YYYY/month=mm/day=dd). To mix it up, you can include columns that are specific to the queries you'll be making. In this exercise, that column is the `storeId`.

Lastly, we changed the output format to be append only: `OutputMode.Append()`. This will only let new rows in the streaming DataFrame/dataset be written to the sink. When we look at using aggregations in the next chapter, the output modes will make more sense, as they can help reduce the amount of output for stateful aggregations and other analytical processing. Let's look at the triggers now.

The Updated Run Method with Triggers

The `run` method now further customizes `DataStreamWriter` from the `outputStream` method using the provided trigger configuration. Let's look at the two config values.

```
lazy val triggerEnabled: Boolean = sparkConf.getBoolean(
    "spark.app.stream.trigger.enabled", true)
```

Why would you want to turn triggering on or off? In the case of recovery or replay of an application, the triggers can get in the way of achieving maximum throughput in the application. For example, the `ProcessingTime` trigger enables the Spark application to execute based on an interval (30 seconds, 5 minutes, or 1 hour). Each time the application ticks, it will run for as long as it needs to, and then the timer starts counting down until the next time it can trigger the next batch. If you want the application to process quickly, it is nice to be able to flip the switch and short-circuit triggering for these kinds of special use cases.

```
lazy val triggerType: String = sparkConf.get(
    "spark.app.stream.trigger.type", "processing")
```

The trigger type is associated with the values processing, once, or continuous. These are the three types of triggers available in Spark Structured Streaming.

Let's see how these two configs play together to control the configurations used to generate the final StreamingQuery in the run method.

```
override def run(): Unit = {
  super.run()
  val writer: DataStreamWriter[Row] = triggerType match {
    case "continuous" if triggerEnabled =>
      val checkpointInterval = "30 seconds"
      outputStream
        .option("checkpointLocation", s"$appCheckpointLocation/trigger/
        continuously/")
        .trigger(Trigger.Continuous(checkpointInterval))

    case "once" if triggerEnabled =>
      outputStream
        .option("checkpointLocation",
          s"$appCheckpointLocation/trigger/once/")
        .trigger(Trigger.Once())

    case "processing" if triggerEnabled =>
      outputStream
        .option("checkpointLocation", s"$appCheckpointLocation/trigger/
        processing/")
        .trigger(Trigger.ProcessingTime("10 seconds"))

    case _ => outputStream
      .option("checkpointLocation",
        s"$appCheckpointLocation/trigger/none/")
  }

  startAndAwaitApp(writer.toTable(streamingTableName))
}
```

We already covered what the `DataStreamWriter` interface does. We are just using a more functional approach to generate the final writer using the `triggerType` match.

```
triggerType match {
  case "continuous" if triggerEnabled =>
  case "once" if triggerEnabled =>
  case "processing" if triggerEnabled =>
  case _ =>
}
```

Pattern matching makes it easier to check if the `triggerType` is equal to continuous, once, or processing. Otherwise, the application will fall through to the default non-trigger based micro-batch processing mode. Within each of the blocks, something interesting is at play. Let's break those down now.

Continuous Mode Processing

Earlier in the chapter we covered what continuous processing mode is. Essentially it will process each event from the data source all the way to completion and commit asynchronously without concerning itself with the status of any other events (since there is no such thing as a batch).

```
Trigger.Continuous(checkpointInterval)
```

Unfortunately, we can only use this mode of execution with Kafka as a data source, so we have to hold off until the next chapter. (You can use the environment from Chapter 11 to test things out as a continued exploration.)

Periodic Processing

Sometimes you'll find yourself needing to control the frequency in which your application triggers each micro-batch. Say you need to run a job once a minute and you don't want to use Airflow to do that.

You can use this trigger to take the place of the Airflow scheduler and run a new batch every five minutes:

```
Trigger.ProcessingTime("5 minute").
```

```
case "processing" if triggerEnabled =>
    outputStream
```

```
.option("checkpointLocation", s"$appCheckpointLocation/trigger/
processing/")
.trigger(Trigger.ProcessingTime("10 seconds"))
```

In the application this interval has been clamped to ten seconds. You could easily add a configuration parameter to control the trigger in a more dynamic way.

Stateful Batch Jobs

Back in Chapter 7 you were introduced to stateful batch jobs. Well, here we are now, and you are looking at the controller used to change the application from a continuous stateful Structured Streaming application to a stateful batch-based application. While this is technically still a Structured Streaming application, the fact that it doesn't continue after the first run gives us wiggle room to call it what it is.

```
case "once" if triggerEnabled =>
     outputStream
       .option("checkpointLocation",
         s"$appCheckpointLocation/trigger/once/")
       .trigger(Trigger.Once())
```

Streaming Table Writer

The last order of business before testing the application is a brief discussion about the change made to the end of the run method.

```
startAndAwaitApp(
  writer.toTable(streamingTableName)
)
```

The DataStreamWriter has a method called toTable. This method, like the start method you saw before, results in a StreamingQuery. The magic going on here is that the table metadata is now being written to the Hive Metastore (like you learned to do in Chapter 6), while the rest of the physical partitioned parquet is being written into the distributed file system.

The streamingTableName is stored in the sparkConf, under spark.app.streaming.table.name. For the example, it is called coffee_orders.

Running the Stateful Application with Triggers

There is nothing left to do but run the application. You'll need your redis-cli open, and you should also have your order events at hand to start feeding to the application once it starts.

Run with Trigger Once

The necessary configuration is shown next. You can take the full command directly from the README in the application directory or make the necessary adjustments to the original full docker run command shown in Exercise 10-1.

```
docker run \
  ...
  --class "com.coffeeco.data.TriggeredStatefulSparkRedisStreamsApp" \
  --conf "spark.sql.warehouse.dir=s3a://com.coffeeco.data/warehouse" \
  --conf "spark.app.source.stream=com:coffeeco:coffee:v1:orders" \
  --conf "spark.app.stream.trigger.enabled=true" \
  --conf "spark.app.stream.trigger.type=once" \
  --conf "spark.app.checkpoint.location=s3a://com.coffeeco.data/apps/
  spark-redis-streams-app/canary/" \
  --conf "spark.app.streaming.sink.path=s3a://com.coffeeco.data/tables/
  coffee.order.events/" \
  --conf "spark.app.streaming.table.name=coffee_orders" \
  /opt/spark/app/spark-redis-streams.jar
```

The application will start up. If there are no checkpoints, then it will create an initial batch 0, and then the application will complete. However, just like with the example of the stateful application in Exercise 10-3, you can add as many events as you want to the Redis stream and then fire the application back up. It will process all your events and then stop running again. The beauty here is that you can run this application from Airflow using the Docker operator and run a fully consistent, reliable stateful batch jobs. Neat.

Run with ProcessingTime

This will start the application, giving it fresh checkpoints. You are then free to add as many new events as you want and watch the behavior of the application. It will pause for ten seconds between runs.

```
docker run \
  ...
  --class "com.coffeeco.data.TriggeredStatefulSparkRedisStreamsApp" \
  --conf "spark.sql.warehouse.dir=s3a://com.coffeeco.data/warehouse" \
  --conf "spark.app.source.stream=com:coffeeco:coffee:v1:orders" \
  --conf "spark.app.stream.trigger.enabled=true" \
  --conf "spark.app.stream.trigger.type=processing" \
  --conf "spark.app.checkpoint.location=s3a://com.coffeeco.data/apps/
  spark-redis-streams-app/canary/" \
  --conf "spark.app.streaming.sink.path=s3a://com.coffeeco.data/tables/
  other.coffee.order.events/" \
  --conf "spark.app.streaming.table.name=other_coffee_orders" \
  /opt/spark/app/spark-redis-streams.jar
```

Running Continuously

For now, we can't run in continuous mode. We need to add Kafka to the party, which is coming in the next chapter.

Viewing the Results

You can inspect the values stored in the Hive Metastore by jumping into the Docker instance running MySQL and querying the Metastore directly.

```
docker exec -it mysql bash
mysql -u dataeng -p

use metastore;
select * from TABLE_PARAMS
  where TBL_ID =(
    select TBL_ID from TBLS
      where TBL_NAME = "coffee_orders"
  );
```

This output will be the metadata related to the coffee_orders table.

The Hive Metastore Record Generated Using the toTable Method on the DataStreamWriter

```
+---------------------------------------------------------------+
| TBL_ID | PARAM_KEY                             | PARAM_VALUE|
+--------+--------------------------------------+------------+
|     16 | EXTERNAL                             | TRUE       |
|     16 | spark.sql.create.version             | 3.1.2      |
|     16 | spark.sql.partitionProvider          | catalog    |
|     16 | spark.sql.sources.provider           | parquet    |
|     16 | spark.sql.sources.schema.numPartCols | 1          |
|     16 | spark.sql.sources.schema.numParts    | 1          |
|     16 | spark.sql.sources.schema.part.0      | {schema}   |
|     16 | spark.sql.sources.schema.partCol.0   | storeId    |
|     16 | transient_lastDdlTime                | 1626675149 |
+--------+--------------------------------------+------------+
```

You'll see the table is an external table. This means you can delete the table from the Spark catalog but still retain the physical table located in the distributed file system.

Exercise 10-4: Summary

This final exercise taught you how to wire up multiple modes of execution within your application and use config-driven feature flags to change the behavior of the application. While you could take these examples much further, this is a good introduction to using Apache Spark Structured Streaming.

Summary

We covered a lot in this chapter. You were introduced to the core processing tenants of Spark Structured Streaming.

- Execution/processing modes (micro-batch or continuous modes)

- Triggers (effective processing windows/boundaries)

- Incremental queries (progressive data processing across time)

- Streaming output modes (complete, update, and append)

You learned how to switch between micro-batch execution mode and continuous execution mode using triggers. While the correct components were not currently available (aka Kafka), you still have an application that is pluggable for that use case (which is right around the corner). You used the append-only output mode, which means only data that Spark hasn't seen before will be written through the sink, and you will learn to use the other modes during the next chapter. Lastly, you have a full suite of tools at your disposal, including new chops when it comes to containerizing your Apache Spark applications.

In the next chapter, we will be continuing our work with Structured Streaming with a focus on more advanced data processing, from streaming joins to aggregations and other analytical processing. You'll also be getting your hands on Kafka for use in your vast set of data engineering tools. Let's go.

## CHAPTER 11

Apache Kafka and Spark Structured Streaming

The last chapter was an introduction to using Apache Spark Structured Streaming. You learned how the popular Redis database can be used to create structured in-memory event streams and explored how to write stateful streaming applications.

This chapter expands on the skills acquired in the last chapter, which included an introduction to using the core Structured Streaming APIs—the `DataStreamReader` and the `DataStreamWriter`, how to utilize application checkpoints to create stateful streaming applications that can handle fault tolerance and restart, how to control the application processing frequency and operating style using triggers, and lastly how the flow of an application works in stateful stream and stateful batch. You'll continue to explore how Structured Streaming works by focusing on using Apache Kafka as a conduit between Spark and the rest of the data ecosystem.

Apache Kafka in a Nutshell

Apache Kafka was introduced in Chapter 1 and is the most widely adopted event and data stream processing framework in use today. The leading role Kafka plays in the data engineering ecosystem has to do with the stabilizing function it provides. It steps in to solve common problems related to communication between distributed systems and services.

Asynchronous Communication

APIs and other HTTP services require all dependent systems and services to be up and running for requests to be handled in a synchronous fashion. It is common for backend services to implement various levels of retry to back off from a non-responsive system

S. Haines, *Modern Data Engineering with Apache Spark*, https://doi.org/10.1007/978-1-4842-7452-1_11

or service and to retry each request multiple times before giving up and returning an error to the callee. Chapter 9 introduced the "thundering herd" problem and the trouble that resource scarcity introduces as processes scramble to make up for lost time. Kafka enables systems (that can handle asynchronous requests) to move away from the traditional request/response cycle to one where a request identifier can be queried for completion state or where the success or failure of the operation encapsulated by the API request notifies its final state through use of a webhook or callback.

Horizontal Scalability

Kafka efficiently handles the common problems introduced by increasing load on a system due to an increase in the number of consumers (readers), producers (writers), or both, of a topic (the durable storage mechanism for a collection of related events/types) by enabling the system to scale the number of servers in the cluster out (horizontally) as opposed to scaling the size (CPU/RAM/disk) of the servers themselves.

Unlike traditional databases that could become plagued by too many connections due to the restrictions and limitations of bounded resources on a single server, Kafka brokers can be scaled out, thereby increasing the throughput capabilities of the entire cluster holistically.

A broker is a server-side (cluster member) service responsible for handling a percentage of all distributed requests in the cluster. Each broker is assigned and responsible for a percentage of one or many distributed topic partitions (data files), with the assigned role of either a partition leader (the primary owner of a partition), or a replica (redundant backup of the current leader's partition). Brokers can be increased to distribute a higher load across the cluster with respect to reading, writing, or redundancy and recovery. This enables the platform to scale up to meet the data demands of a topic while also managing the quirks that come along with a distributed system's need to be available and consistent.

High Service Availability and Consistency

You were introduced to CAP theorem in Chapter 1, and given that Kafka is a distributed, network connected service, the issues with consistency, availability, and network partitioning tolerance are important. To maintain distributed consistency, the theorem states that a service can't handle network partitioning (blind spots in the cluster) and remains fully available. For this reason, Kafka uses partition replication, within topics, to achieve high availability and consistency. To ensure that there is more than one complete in-sync replica (ISR) available for redundancy if the broker (server) assigned as a leader for a partition goes offline (or the server is shut down/dies), each write is durably written to a write-ahead log and acknowledged to the writer (producer). The Kafka cluster can use the ISR convention to reliably select the next available ISR (redundant copy) of a lost partition and promote that broker as the new leader of the topic partition. This cycle repeats whenever a broker goes offline to restore topic consistency and availability with minimal interruptions to the consumers and producers of a given topic.

In the same way that Kafka distributes the contents of a topic across many partitions, the brokers themselves can be used to issue topic-level commands that propagate between brokers. For example, to increase the number of partitions of a topic with zero downtime any broker can be used to spread the word (gossip style). Topic-level configuration can also be adjusted in real-time, and zero-downtime, to handle increasing data demands due to increased size of the payload (events) being produced, or to handle an increase in the number of records being produced and consumed. Why should you scale out? This concept takes us back to the law of large numbers. All systems will slow down the larger the load they must handle within a single process, so splitting partitions of a topic (e.g., an increasing the number of partitions encapsulating the entire topic) is a way to handle the demands of big data systems at scale.

Disaster Recovery

By enabling applications to replay events (data) from an earlier point in time, systems and services can handle some downtime or service interruptions and recover from scenarios that would have otherwise resulted in data loss. This is especially important for data pipelines since a dropped event is data loss, but a degraded service with the option of recovery is still eventually consistent.

Event Streams and Data Pipelines Ecosystem

Apache Kafka solves a similar problem to that of the *data lake* (introduced in Chapter 1). Remember that the data lake provides a unified, elastically scalable source of truth for raw, unprocessed data from across the organization that acts as a centralized staging area for data to be reliably re-read, re-processed, and transformed into actionable data at a later point in time. This central data staging area solved a problem having to do with data pipelines and lossy transformation of data made too early in the lineage of a data pipeline. If you have no way to get back to the original source of truth for an event, then corrupt data simply flows downstream.

Apache Kafka, on the other hand, provides a unified event streaming solution that organizes event data across *topics*, which act in a similar way to the partitioned distributed file system underpinning the data lake. Topics enable the construction of data pipelines of encoded binary data that can be tapped into to produce rich data networks. The key difference between the data lake and Kafka has to do with time, specifically minimizing the time between when data is produced (written to a topic) and when it can be consumed (read) downstream.

Kafka enables connected clients to consume data (in near real-time) immediately after the group (quorum) of server-side brokers assigned to a topic commit and acknowledge receipt of published events (records) emitted from a connected data producer. The connected client simply polls a topic on a scheduled interval, fetching any new data in the form of a collection of records (events). Kafka intentionally minimizes the time (delta) between event creation and when an event can be processed, which is important for real-time (live) event streaming and for processing live data.

In this chapter, you learn to do the following:

- Configure and run a Kafka cluster with Docker

- Create and alter Kafka topics

- Connect and configure the Spark Kafka library

- Use the Kafka data source

- Transform binary data from Kafka into DataFrames and datasets

- Use the Kafka data sink

Exercise Materials   The exercise materials for Chapter 11 are located at `https://github.com/newfront/spark-moderndataengineering/tree/main/ch-11`. The README file located in each exercise will provide you with more context regarding any additional installation or gotchas.

Chapter Exercises

The following exercises will help as you continue to explore the features and capabilities of Spark's Structured Streaming framework. You'll start by getting Kafka running locally using `docker-compose`. This is also an opportunity to become acquainted with the various architectural components that make Kafka tick, before we dive deeper into patterns to take full advantage of Kafka's rich streaming ecosystem.

Exercise 11-1: Getting Up and Running with Apache Kafka

Open the chapter material and navigate to the directory named `ch-11` to start working through the exercises.

Exercise 11-1: Materials

The exercise materials are located under `ch-11`.

```
ch-11
├── docker
│   └── kafka
│       ├── README.md
│       └── docker-compose.yaml
└── exercises
    └── 01_kafka_up_and_running
```

Spinning Up Your Local Environment

Use the `docker-compose` file provided under `docker/kafka` to get started. Follow the steps in the `README.md` file to move Kafka into your data engineering toolbox. Now you can use the `docker compose` command to spin up a full Kafka cluster, running Zookeeper (stateful metadata store) and three Kafka brokers.

```
docker compose \
  -f ~/dataengineering/kafka/docker-compose.yaml \
  up -d --remove-orphans
```

The `docker-compose` file wires up a fully functioning Kafka cluster (using the `mde` Docker network from previous chapters), thus enabling access from your laptop and within your docker runtime. Now you can connect to the Kafka cluster from your Spark applications.

The containers running the Kafka cluster can be viewed using the following `docker ps` command. This will filter the running containers to the subset matching the `kafka_` prefix.

```
docker ps \
  --filter "name=kafka_*" \
  --filter "network=mde" \
  --format "table {{.Names}}¥t{{.Mounts}}¥t{{.Networks}}" \
  -s
```

The output will show the four containers running your shiny new Kafka cluster.

```
NAMES                   MOUNTS                  NETWORKS
kafka_kafka-2_1         kafka_kafka_2_...       mde
kafka_kafka-1_1         kafka_kafka_1_...       mde
kafka_kafka-0_1         kafka_kafka_0_...       mde
kafka_zookeeper_1       kafka_zookeepe...       mde
```

The three containers represent the cluster brokers, and the fourth container is Apache Zookeeper. Zookeeper is responsible for storing the Kafka cluster metadata (state) including the broker identifiers, the partition leaders, and replica locations across the cluster for each topic. You'll learn more about how topics work and create your first topic next.

Creating Your First Kafka Topic

Now that you have your Kafka cluster up and running, you can create your first *topic*, but before you do so, let's cover a few more traits, or behaviors, associated with Kafka topics. Then you can dive into creating your first one.

How Topics Behave

Topics at a high level behave like tables in a database, but act more like a distributed commit log of structured key/value pair records.

The general structure of the key/value-based Kafka record as viewed from Spark's perspective is shown in Listing 11-1. The format of the Kafka record (StructType) minimally contains a column named value that is encoded at rest as a byte array (binary), the topic where the data was read from, as well as the partition where the record resides at rest, and the offset (think of this as a pointer to the row location within the partition of a topic) of the individual record.

Listing 11-1. Kafka Record Structure Data Format

```
root
 |
 - key <binary>
 - value <binary>
 - topic <string>
 - partition <int>
 - offset <long>
 - timestamp <long>
 - timestampType <int>
 - headers <array<string, binary>>
```

Kafka records are sharded (or simply distributed) across the total number of partitions of an associated topic. Since the record key column is optional for each Kafka record, the default partitioning mechanism for a null key is to insert records in a round-robin fashion, which means records ends up with non-deterministic partition assignments for each record produced. This ultimately means that data won't be guaranteed to land on the same partitions upon repeated runs across the same dataset. However, by providing a non-null value for each record key, Kafka will use the key supplied as a partitioning mechanism (hash partitioning), enabling the deterministic routing of each record to a specific partition of a given topic.

Tip   For event streams, event data is either dependent on other events within a stream to tell the story of what happened (or is currently happening now), or each event can exist independently, telling a full story itself. Chapter 9 covered the concept of interdependent and independent event data, and you can use Kafka to optimize how events are emitted across partitions within each topic, allowing you to optimize for the interdependent or independent event use cases.

We've established that topics are a collection mechanism that act like a conduit, or channel, for a stream of events, but we have yet to cover what makes Kafka great. Each published record is stamped upon insertion with a broker side timestamp (in milliseconds) and acknowledged only after a successful write (commit). Along with the acknowledgement comes a guarantee that the topic will maintain a time-synchronous insert order on a partition-by-partition basis, essentially providing you with a sorted set of records by insert time.

Since the guarantee on the insert order is on the partition level only, this enables Kafka to manage the otherwise complicated distributed insert ordering task on a non-distributed, isolated subset of the topic. The partition-based topic architecture then in reality acts, at a lower level, more like a distributed file system, and like any file system, standalone Kafka doesn't enforce an explicit record schema to be applied for each value within a record.

By removing schema enforcement, Kafka topics can be incredibly flexible, but with flexibility comes the opportunity for corrupt or erroneous data to accidently flow downstream, and as a result it has become a common practice to associate a specific, binary encoded, event type (or types) to each topic.

Creating a Kafka Topic

Topic creation can be done using the administrative shell scripts that ship along with Kafka. Execute the command in Listing 11-2 to create your first topic.

Listing 11-2. Creating a Kafka Topic Using the Command-Line Utilities That Ship Out-of-the-Box with Kafka

```
docker exec \
  -it kafka_kafka-0_1 \
  /opt/bitnami/kafka/bin/kafka-topics.sh \
  --create \
  --if-not-exists \
  --topic com.coffeeco.coffee.v1.orders \
  --bootstrap-server kafka_0:9092,kafka_1:9092,kafka_2:9092 \
  --partitions 4 \
  --replication-factor 2
```

The result of running the command is a new Kafka topic named `com.coffeeco.coffee.v1.orders`. This topic will be conditionally created in the cluster (`if-not-exists`) and is configured with four partitions. For increased availability, each topic partition is replicated twice. You'll learn more about modifying topics in the next section.

Optional   Now that you've created your first topic, I bet you'd like to move ahead to doing something more with it. Feel free to skip ahead to Exercise 11-2. You can always come back here and learn more about how Kafka handles topic management and security.

Kafka Topic Management

Managing Kafka topics all boils down to understanding how the system behaves and learning how to use the admin tools that ship with Kafka. This section introduces you to some of the common commands and necessary skills required to operate Kafka.

Listing Kafka Topics

Listing the topics available in a cluster is essential to discovering topics. You can do this by using the `--list` operation on the `kafka-topics.sh` script, as shown in Listing 11-3.

Listing 11-3. Listing Kafka Topics Returns a Simple List of Topics

```
docker exec \
  -it kafka_kafka-0_1 \
  /opt/bitnami/kafka/bin/kafka-topics.sh \
  --list \
  --bootstrap-server kafka_0:9092,kafka_1:9092,kafka_2:9092
```

The `list` operation will print the names of all available topics in the cluster.

Describing a Topic

It is helpful from time to time to inspect the metadata associated with a given topic. This can assist in debugging issues you'll inevitably run into if you end up operating Kafka as part of your day-to-day responsibilities. Executing the command in Listing 11-4 will return the metadata associated with the partitions and the role each broker plays, thus maintaining topic consistency and availability.

Listing 11-4. Describing a Topic

```
docker exec \
  -it kafka_kafka-0_1 \
  /opt/bitnami/kafka/bin/kafka-topics.sh \
  --describe \
  --topic com.coffeeco.coffee.v1.orders \
  --bootstrap-server kafka_0:9092,kafka_1:9092,kafka_2:9092
```

The output of the `describe` command will be slightly different than the output shown next, given your cluster will have different assigned topic leaders and partition replica locations. Here is the modified output from my environment:

```
Topic: com.coffeeco.coffee.v1.orders
TopicId: eENgTyO1QUWOAe2gJzi_rw
PartitionCount: 4
ReplicationFactor: 2
Configs: segment.bytes=1073741824

Topic: com.coffeeco.coffee.v1.orders
Partition: 0    Leader: 2    Replicas: 2,1    Isr: 1,2
Partition: 1    Leader: 1    Replicas: 1,0    Isr: 1,0
Partition: 2    Leader: 0    Replicas: 0,2    Isr: 0,2
Partition: 3    Leader: 2    Replicas: 2,0    Isr: 0,2
```

The metadata from the `describe` command reveals how each partition is distributed across the three brokers. The broker IDs are 0, 1, and 2 and you can see that for each partition, there is a leader responsible for storing one copy of the partition, as well as a replica which is in sync with the leader. This capability allows for a partition leader to be lost, and for the in-sync replica to take over as the new leader, without missing beat.

Modifying a Topic

The configuration of a Kafka topic isn't set once and forgotten. In most cases, the initial configuration is only an estimation of the required total number of partitions and conditional configuration needs of a given topic. Luckily, changing (or altering) a topic can be done dynamically without interrupting the clients connected to a given topic.

You can modify the topic created earlier to increase its consistency and availability or general fault-tolerance by increasing the minimum number of topic replicas required to be considered healthy. The default value for `min.insync.replicas` is 1 and we will increase it to 2.

Altering Topic Configurations

To alter a topic after initial creation, you can use the `kafka-configs.sh` script to add or remove (delete) configuration properties for a topic. Execute the command in Listing 11-5 to change the minimum in-sync topic replicas.

Listing 11-5. Modifying the Configuration of an Existing Topic

```
docker exec \
  -it kafka_kafka-0_1 \
  /opt/bitnami/kafka/bin/kafka-configs.sh \
  --alter \
  --bootstrap-server kafka_0:9092,kafka_1:9092,kafka_2:9092 \
  --entity-type topics \
  --entity-name com.coffeeco.coffee.v1.orders \
  --add-config min.insync.replicas=2
```

With your topic reconfigured, each partition now requires two in-sync replicas. Can you see a potential problem here? Maybe having to do with the topic-replication factor? If a broker is lost (goes offline/unreachable/hosed), there will be additional strain put on the cluster, as the available brokers work to restore the lost partitions for any topic's partitions assigned to the lost broker. Secondary brokers will need to copy the lost topic partitions to get back to an in-sync (consistent) state with the leaders of those partitions. This can result in many file transfers, which can add network IO and CPU load across the cluster and reduce the availability of the affected topics. This can then result in client backoff for producers writing to the topic while regaining a consistent, available state.

To ensure your cluster can handle broker loss, you can change the replication factor of your topic from 2 to 3. This way, you can handle a single lost broker without having to scramble. This process is easier to do on topic creation, by adding –replication-factor 3, rather than using manual replica assignments like you will do next.

Increasing Topic Partition Replication

To ensure your cluster remains healthy with at least two in-sync replicas during the loss of a broker, you need to create a replication reassignment JSON file. This file can be produced using the information from the describe topic command. Just take the current assignments and add another broker ID to the replicas array. This will be used as a template for Kafka to increase the replication factor from 2 to 3 for the topic. The example in Listing 11-6 shows the replication reassignment's JSON format.

Listing 11-6. The Kafka Topic replication-reassignments.json File Format

```
{
  "version": 1,
  "partitions": [
    {
      "topic": "com.coffeeco.coffee.v1.orders",
      "partition": 0,
      "replicas": [2,1,0]
    },
    {
      "topic": "com.coffeeco.coffee.v1.orders",
      "partition": 1,
      "replicas": [1,0,2]
    },
    {
      "topic": "com.coffeeco.coffee.v1.orders",
      "partition": 2,
      "replicas": [0,2,1]
    },
```

```
    {
      "topic": "com.coffeeco.coffee.v1.orders",
      "partition": 3,
      "replicas": [2, 0, 1]
    }
  ]
}
```

With the replication file in hand, you'll need to copy it into one of your Kafka brokers so you can run the `kafka-reassign-partitions.sh` script from the broker. The copy command is shown in Listing 11-7.

Listing 11-7. Docker Copy Command to Upload the replication-reassignments. json to the Kafka Broker

```
docker cp \
  ./exercises/01_kafka_up_and_running/replication-reassignments.json \
  kafka_kafka-0_1:/opt/replication-reassignments.json
```

With the `replication-reassignments.json` file on the broker, you can now run the reassign partitions script. Use the same broker where you uploaded the `replication-reassignments.json` and execute the command in Listing 11-8.

Listing 11-8. Running the kafka-reassign-partitions.sh Script to Increase Topic Replication

```
docker exec \
  -it kafka_kafka-0_1 \
  /opt/bitnami/kafka/bin/kafka-reassign-partitions.sh \
  --bootstrap-server kafka_0:9092,kafka_1:9092,kafka_2:9092 \
  --reassignment-json-file /opt/replication-reassignments.json \
  --execute
```

As a result of running the replication reassignments, you can now handle the loss of one broker in your Kafka cluster without causing a major disruption in the behavior of your cluster. In a non-production scenario, it can sometimes be easier to just delete your topic and start from scratch, but understanding how to modify the behavior of a topic can teach you to solve problems pragmatically.

Tip   Another use case for reassigning partitions can come up in the case of Kafka broker skew. This is when one or more brokers are overburdened with too many partitions of a topic. This can slow down the producers and consumers connected to that broker (leader for their topic partitions) since more data and more connections are unevenly distributed across the broker.

Truncating Topics

Topics in Kafka will only retain records up to the configured retention milliseconds. Knowing this, you can truncate (clear all records) from a topic by altering the topic configuration. You'll need to first reduce the topic (record) retention period so Kafka begins to purge the topic records. After waiting for a minute or two, Kafka won't immediately delete the records, so simply add back a retention period that makes sense for your topic (one or seven days are common).

Reducing the Topic Retention Period

The first step uses the `--alter` command, adding the value of 1 second (1000ms) as the topic `retention.ms`. Execute the command in Listing 11-9 to change the topic retention.

Listing 11-9. Truncating a Kafka Topic to Purge All Records

```
docker exec -it kafka_kafka-0_1 \
  /opt/bitnami/kafka/bin/kafka-configs.sh \
  --bootstrap-server kafka_0:9092,kafka_1:9092,kafka_2:9092 \
  --entity-type topics \
  --alter \
  --entity-name com.coffeeco.coffee.v1.orders \
  --add-config retention.ms=1000
```

Now that the topic is being cleared, wait a minute or two and then increase the topic retention period.

Increasing the Topic Retention Period

Now that the topic is cleared (assuming you waited for a little while), you can now change the topic retention config back to a one-day retention period (see Listing 11-10).

Listing 11-10. Adding a One-Day Retention Period to a Kafka Topic

```
docker exec -it kafka_kafka-0_1 \
  /opt/bitnami/kafka/bin/kafka-configs.sh \
  --bootstrap-server kafka_0:9092,kafka_1:9092,kafka_2:9092 \
  --entity-type topics \
  --alter \
  --entity-name com.coffeeco.coffee.v1.orders \
  --add-config retention.ms=86400000
```

This technique can be used when you need to clear all records within a topic. When you create the CoffeeOrder generator in Exercise 11-2, you might want to clear the topic after testing things. This can also be used to test different failure modes with Kafka.

Deleting Kafka Topics

There are only two reasons that a Kafka topic should ever be deleted. The first reason is that the topic is end of life, and there are no producers and consumers of the topic. The second is because you are in a non-production environment, and there are no other teams (or services) relying on this topic. In fact, the Kafka brokers are configured by default to not allow topic deletion. This is because accidental topic deletion can be catastrophic to the network of data pipelines relying on access to the streaming source of data.

With the red tape and warnings out of the way, you can test deleting your topic by executing the delete command in Listing 11-11.

Listing 11-11. Kafka Topic Deletion Can Be Done Using kafka-topics.sh

```
docker exec \
  -it kafka_kafka-0_1 \
  /opt/bitnami/kafka/bin/kafka-topics.sh \
  --delete \
  --topic com.coffeeco.coffee.v1.orders \
  --bootstrap-server kafka_0:9092,kafka_1:9092,kafka_2:9092
```

The `delete` command will do a soft delete (marked in Zookeeper only) unless the Kafka brokers are configured with the `delete.topic.enable=true` on their brokers. You can modify your Kafka cluster to run with this dangerous configuration by updating the `docker-compose.yaml` file used to spin up the cluster. Locate the environment section for the Kafka brokers and add the environment variable in Listing 11-12.

Listing 11-12. Enabling Topic Deletion Using the delete.topic.enable=true Flag on Your Kafka Brokers

```
services:
  kafka-0:
    environment:
      ...
      - KAFKA_CFG_DELETE_TOPIC_ENABLE=true
```

This example shows how to modify the broker configuration in the `docker-compose.yaml` file for one of the three brokers. You can modify the other two broker configurations and then restart your Kafka cluster to have these changes take effect. Now you can do irreversible deletes, which is not recommended.

Securing Access to Topics

Securing access to Kafka and the associated cluster metadata, topics, and underlying physical data feels like a no brainer. However, like many things in life, security can sometimes fall through the cracks as a known issue for another day. In short, there are a handful of ways to secure access to your Kafka cluster, starting with basic firewall rules, moving to access control lists (ACLs), and finishing with fully encrypted end-to-end transport using SSL/TLS.

Firewall Rules

Firewall rules establish point-to-point connectivity from hosts within your colocation center or virtual private cloud (VPC) using Linux style iptables, or secure cloud-based access. This includes AWS security groups, which govern inbound and outbound connectivity and data transfer across IP addresses and between open ports. This is equivalent to locking your doors at night. It can keep bad actors out of your system unless they find a way to pick the locks.

Access Control Lists

Apache Kafka comes out-of-the-box with its own working ACL implementation called Authorizer. It can be used to secure Kafka beyond just firewall rules using access control list (ACLs). The ACL rules are stored in Zookeeper by default. ACLs are used to authorize and govern user (known as principals) access to operations across one or more topics or to the cluster itself. ACLs in Kafka are used like SQL Data Control Language (DCL) to grant specific sets of cluster-based operations restricting reads, writes, updates, and delete capabilities.

Enabling ACLs needs to be done across each broker, which can be done using `docker-compose.yaml`. Just like with enabling topic deletion, you can turn on Authorizer from the environment configs (shown in Listing 11-13).

Listing 11-13. Enabling Broker ACLs

```
services:
  kafka-0:
    environment:
      ...
      - KAFKA_CFG_AUTHORIZER_CLASS_NAME=kafka.security.authorizer.
        AclAuthorizer
```

Once the Authorizer has been turned on (which requires each broker to be restarted), you can add ACLs using the Kafka command line. Execute the command in Listing 11-14 to add user-based access control to your topic.

Listing 11-14. Adding Kafka ACLs for Topic-Level Authorization

```
docker exec -it kafka_kafka-0_1 \
  /opt/bitnami/kafka/bin/kafka-acls.sh \
  --bootstrap-server :9092 \
  --add \
  --allow-principal User:dataengineering \
  --operation read \
  --operation write \
  --topic com.coffeeco.coffee.v1.orders
```

Firewall rules, along with well-maintained ACLs, can go a long way to securing your Kafka cluster. If system-wide access is reviewed to keep things in check, and administrative super-user or "god-mode" access is not granted to insecure client-side software, you should be sailing smoothly.

End-to-End Encryption with Mutual TLS

Apache Kafka supports end-to-end SSL/TLS encryption to handle encrypted data in flight. This requires a shared certificate and private key pair that can be verified during the SSL/TLS handshake to ensure that data streaming from either end of the connection (client or server) is protected from bad actors.

Exercise 11-1: Summary

This first exercise introduced you to the ins and outs of running a local Kafka cluster. This exercise was intended to introduce the nomenclature and components of Kafka itself, and to get you oriented and familiar with operating a Kafka cluster. While we barely began to scratch the surface of all the various broker configurations, admin/ utility command-line operations, and general security and hardening, you now have a functioning Kafka cluster to use with Structured Streaming.

In the next exercise, you'll learn techniques for producing and consuming data from within Apache Spark using binary structured data (protobuf) as a follow-up to reading the OrderEvent data from Redis Streams in the last chapter.

Exercise 11-2: Generating Binary Serializable Event Data with Spark and Publishing to Kafka

In this exercise, you learn how to generate binary serialized structured data inside of Apache Spark. This is an essential skill for data engineers focused on interoperability with systems running outside of the data platform ecosystem. You discovered how data systems can be architected to be more resilient in Chapter 9 and learned that data is often encoded as binary avro or protobuf before being emitted through various API gateways. This technique helps to reduce bandwidth overhead (due to transport of binary data) and the gateways act as a safe and guarded means of routing event data (payloads) into various Kafka topics. This simple act enables downstream consumers to *reliably* fetch raw event data as it enters the data network with the knowledge that some checks and balances (aka initial authorization and data validation) have been completed.

Exercise 11-2: Materials

The exercise contains a Spark application as well as the Docker environment to run the application.

```
ch-11
├── applications
│   └── spark-kafka-coffee-orders
└── exercises
    └── 02_binary_serialization_in_spark
```

In this exercise, you learn to write an event generator that can produce an arbitrary number of CoffeeOrders that will be encoded using Google's Protocol Buffers (protobuf). We'll start with a look at the CoffeeOrder protobuf and move into the generator. This exercise concludes with you running the generator, while using Kafka's simple console consumer to watch the data as it arrives. This exercise shows the difference between event production and event consumption using Apache Spark.

CoffeeOrder Event Format

The CoffeeOrder protobuf definition probably looks familiar. It is essentially the StructType used in Chapter 10 to read data from com:coffeeco:coffee:v1:orders.
    The CoffeeOrder protobuf is shown in Listing 11-15.

Listing 11-15. The CoffeeOrder Protobuf Data Definition

```
message CoffeeOrder {
  uint64 timestamp  = 1;
  string order_id = 2;
  string store_id = 3;
  string customer_id = 4;
  uint32 num_items = 5;
  float  price = 6;
}
```

The CoffeeOrder message provides all the event information needed to track coffee orders for CoffeeCo. This sets the underlying binary data structure represented by the protobuf record. Interoperability between the Apache Spark DataSource APIs and Apache Kafka requires three steps:

1. You need to compile the protobuf data definitions into Scala case classes using the scalapbc compiler.

2. You need to include the scalapb-spark ScalaPB Spark library in the build.sbt application. This library provides the interoperability classes to handle the transformation between the protobuf native binary format and Spark's native binary format (Catalyst Rows).

3. You need to include the Apache Kafka Spark SQL library called spark-sql-kafka-0-10.

We'll go through these three steps while building the CoffeeOrder event generator.

Compiling Protobuf Messages

Moving ahead, let's compile the protobuf. This can be done with the standalone ScalaPB compiler. The compiler location can be added to your PATH, or exported as a simple environment variable, for example called SCALAPBC.

Tip   There is a longer set of notes in the protobuf directory README file including how to download and link the standalone ScalaPB compiler. Just look in the / ch-11/applications/spark-kafka-coffee-orders/data/protobuf/ README.md directory.

The compilation command is shown in Listing 11-16. The compiler uses the input path --proto_path, which is the directory where your protobuf definitions exist, and an output path --scala_out, which is the directory where you want the compiled Scala code to live. Execute the command in Listing 11-16 using /ch-11/applications/spark-kafka-coffee-orders as your working directory.

Listing 11-16. Compiling the CoffeeOrder Protobuf

```
$SCALAPBC/bin/scalapbc -v3.11.1 \
  --proto_path=./data/protobuf/ \
  --scala_out=./src/main/scala \
  coffee.common.proto
```

The compiler will parse the file named `coffee.common.proto` and generate Scala case classes, along with the deserializers and serializers needed to transform each class instance between their internal JVM object representation and the binary protobuf (wire format) representations. Because protobuf can be cross-compiled, the same `coffee.common.proto` could be compiled for interoperability among Node.js, Python, C++, Go, Scala, Java, and many more, using a language specific compiler.

Note   In Chapter 9 you learned about GRPC and how the framework is built to be interoperable between languages using a common binary data protocol, aka Google Protocol Buffers (protobuf). These binary messages interoperate natively with all the gRPC components. For companies already embracing the gRPC stack, it won't be a hard sell to use the same technology to support the data transport systems and ingestion services as well. If rich structured data makes it into the Kafka topics and other data streams, then everyone wins.

With our Scala classes generated, we can move on to the crux of the exercise, learning how to use Kafka as a transport device with Apache Spark. You'll use Spark as a conduit for event generation and you'll write these records using the Kafka DataFrameWriter. Given we need to produce data to be able to consume it, we will go through the process first before moving to consuming these records in Exercise 11-3.

Protobuf Message Interoperability with Spark and Kafka

You learned to use the Redis command line to generate new coffee orders in the last chapter. You'll be building a `CoffeeOrder` generator that will handle this process of generating random orders and encoding these records, and then you'll publish them into the new Kafka topic. It is unlikely you have much experience encoding binary data manually. It isn't something most people will need to do, so you can instead lean on some simple code to do just that.

Generators are a useful tool to have in your data toolbox and they can be used for several reasons, from generating stress tests and finding performance bottlenecks, to producing good and bad data to understand application behavior, and even helping to reproduce bugs exposed during incidents or investigations.

Adding the Protobuf and Kafka Spark Dependencies

Open the application named `spark-kafka-coffee-orders` in your favorite IDE. This application is set up using the same format as the applications in Chapters 7 and 10, so the layout should start to look familiar at this point in your journey.

Earlier in this section I mentioned there was a three-step process to using the protobuf. The first step was to compile the message definitions, the second includes the extension library for Spark called `sparksql-scalapb`, and the third includes the `spark-sql-kafka` library dependency. Open `build.sbt` in the root of the application directory, and you'll also see these Spark dependencies included under the `libraryDependencies`.

Listing 11-17. The Additional Application Library Dependencies Enabling Protobuf Interoperability with Apache Spark with Support for Reading and Writing to Apache Kafka

```
"org.apache.spark" %% "spark-sql-kafka-0-10" % sparkVersion
"com.thesamet.scalapb" %% "sparksql-scalapb" % scalaPbVersion
```

Toward the bottom of the `build.sbt`, you'll see two new blocks in the application sbt. These two blocks are `assemblyMergeStrategy` and `assemblyShadeRules`.

Escaping JAR Hell

When you are building your Spark applications, you will at times find yourself in what is known as JAR Hell. This is literally a cycle of trying to figure out why your application won't compile. To solve conflicts between the dependencies of various libraries, you can lean on the merge strategy (shown in Listing 11-18) to control how your JAR is built, what class files are added, and what files are deduplicated, removed, or renamed.

Listing 11-18. Assembly Rules Include the Ability to Select How to Solve
Conflicts when Constructing a Fat JAR

```
assembly / assemblyMergeStrategy := {
  ...
  case _ =>
    MergeStrategy.deduplicate
}
```

Merging gets you almost all the way out of JAR Hell; however, at times there are
issues outside of your control and that is where shading comes into play.

Take Apache Spark for instance. When running a Spark application there is already
a process running (literally Spark) that includes its own class dependencies. Given
these classes are loaded when Spark spins up, there is an opportunity to overrule your
application dependencies since your application is loaded into the Spark runtime.
You can use the process of shading (shown in Listing 11-19) to solve the problem with
colliding class versions by renaming the conflicting classes. This in turn will resolve the
runtime problems. `NoClassDefFoundError`'s issues are one of the many warning signs to
watch out for which stem from conflicts in your application classpaths.

Listing 11-19. Shading Can Be Used To Rename Classes That May Otherwise
Introduce a Conflict at Runtime

```
assembly / assemblyShadeRules := Seq(
  ShadeRule.rename("com.google.protobuf.**" -> "shadeproto.@1").inAll,
  ShadeRule.rename("scala.collection.compat.**" -> "scalacompat.@1").inAll
)
```

When combined, these two techniques come together to ensure your Spark
application can be deployed without the headaches caused by JAR Hell. Now you can
move on to the generator.

Writing the CoffeeOrderGenerator

The generator's package in the chapter materials has a working implementation of CoffeeOrderGenerator Essentially, the generator needs to do three things to be considered successful. The generator. needs to:

1. Generate a sequence of one or more CoffeeOrder events.

2. Encode the sequence events in a format that Spark understands..

3. Connect to Kafka and write the events to a topic.

Open the file named CoffeeOrderGenerator.scala. This singleton class. extends the SparkApplication trait from earlier in the book, enabling us to create an entry point for this new Spark application. In fact, we'll be reusing the other application helpers, thereby enabling the application to generate the SparkConf and SparkSession as well. This means the only thing the application needs to do is follow the three previous steps. Looking at the basic interface of the generator (shown in Listing 11-20), you can see that there isn't much going on.

Listing 11-20. The Core Interface of the CoffeeOrderGenerator

```
object CoffeeOrderGenerator extends SparkApplication {
  ...
  def generateCoffeeOrder(
    from: Instant,
    to: Instant,
    totalRecords: Int = TotalRecords,
    indexOffset: Int = IndexOffset): Seq[CoffeeOrder]

  override def run(): Unit
}
```

The generateCoffeeOrder method is the heart of the generator (see Listing 11-21). It is responsible for generating the actual CoffeeOrder. The method parameters provide a simple way to declare a window of time in which to generate your order events. Using from and to specifies the window boundaries in which to spread the generated CoffeeOrders across. Think of this like "I want to generate a specific total number (totalRecords) of order events spread across a time range identified by the bounds of from and to."

Listing 11-21. The Heart of the CoffeeOrderGenerator Creates an Arbitrary Number of Orders with a Minimal Amount of User-Provided Override Capabilities

```
def generateCoffeeOrder(...): Seq[CoffeeOrder] = {
  val random = new Random(totalRecords)
  val stepSize = (to.toEpochMilli - from.toEpochMilli)/1000
  (0 to totalRecords).map { index =>
    CoffeeOrder(
      from.plusMillis(index*stepSize).toEpochMilli,
      s"orderId${indexOffset+index+1}",
      s"store${random.nextInt(4)+1}",
      s"cust${random.nextInt(100)+1}",
      BoundRandom.nextInt(20)+1,
      BoundRandom.nextFloat()+1.0f
    )
  }
}
```

The method in Listing 11-21 uses the value of totalRecords as a seed for the random number generator. It enables the application to generate the same pseudo-random sequence of numbers for a unique seed. This means you can create the same sequence twice if the value of totalRecords is 15.

The method then uses the difference in seconds between the to and from values as the stepSize. The stepSize is used to spread the generated orders evenly across the time range provided to the method. You can see this at work in the lambda function within the Range iterator (0 to totalRecords). For each iteration, a new CoffeeOrder is generated using the loop index, the optional index offset, and the random number generators (random and BoundRandom).

Note   Notice that the random numbers are also scoped. For example, random. nextInt(4)+1 will select a random number between 0 (inclusive) and the specified value (4), exclusively.

This method returns a Seq[CoffeeOrder]. This checks off the first requirement of the generator (producing CoffeeOrder events). The generator itself is called in the run method. Listing 11-22 shows how the generator is incorporated into the run method.

Listing 11-22. The CoffeeOrderGenerator Calls generateCoffeeOrder to Produce an Arbitrary Number of CoffeeOrders That Can Be Emitted to Kafka

```
override def run(): Unit = {
  val until = LocalDateTime...
  val orders: Seq[CoffeeOrder] = CoffeeOrderGenerator
    .generateCoffeeOrder(
      from = until.minusHours(8).toInstant(ZoneOffset.UTC),
      to = until.toInstant(ZoneOffset.UTC))
  ...
}
```

The run method leans on the TotalRecords configuration to limit the number of orders produced.

```
lazy val TotalRecords = sparkConf.getInt("spark.data.generator.
totalRecords", 10)
```

This simple config lets you control how many coffee orders are generated when the application is run. By default, the generator is set to create records over an eight-hour window, ending at the current time (when you run the generator) in milliseconds. As a follow up, you could add another value to modify the window, or even to specify the from and to values for a specific moment in time.

After generating the sequence of orders, the run method uses the SparkSession (provided by the SparkApplication trait) to create a Dataset[CoffeeOrder] (see Listing 11-23). This is where the magic of the ScalaPB library comes into play. There is special import at the top of the generator, called import scalapb.spark.Implicits._ and these implicits provide the interoperability functions that convert between native protobuf and Spark's catalyst rows.

Listing 11-23. Generating and Transforming CoffeeOrders Between Native Protobuf to Spark's Catalyst Engine Can Be Handled Elegantly

```
val orderEvents = sparkSession.createDataset[CoffeeOrder](orders)
.map { order =>
  KafkaRecord(ByteString.copyFrom(order.orderId.getBytes),
  order.toByteString, KafkaTopic)
}
```

After generating the dataset, a map operation immediately occurs. The map function is used to iterate over each row in the dataset and transform each record into the binary format needed to write these events as Kafka records. KafkaRecord is another protobuf message type that was bundled in the coffee.common.proto along with the CoffeeOrder message.

```
message KafkaRecord {
  bytes  key   = 1;
  bytes  value = 2;
  string topic = 3;
}
```

The KafkaRecord message format provides the final piece of the Kafka producing puzzle, the essential API contract necessary to take the serialized CoffeeOrder into the Kafka ecosystem.

Note   Behind the scenes, Spark is doing some heavy lifting here. When map is called on the dataset, the value of each of the records is transferred from native memory (Tungsten) and each row is converted back into an JVM object. This sneaky line in the code generates a new protobuf message called KafkaRecord. Because we are using protobuf and Spark's Dataset API together, we are simply mapping (transforming) the CoffeeOrder protobuf into another protobuf named KafkaRecord. This functionality uses more memory because the Spark engine needs to allocate memory to serialization and deserialization the values. This is because Spark can't optimize the conversion from our CoffeeOrder to the KafkaRecord directly. This process can be optimized by creating a Spark SQL UDF. This is covered in Chapter 13.

Now that we have our orders generated and converted into the binary format required to publish records to Kafka from Spark, all that's left to do is write the records. Since the data encapsulated by the orderEvents is a typed dataset and not a DataFrame, we need to call select on the orderEvents to extract the columns required by the Kafka DataFrameWriter. This transformation process is shown in Listing 11-24.

Listing 11-24. Writing the Generated CoffeeOrders to Kafka Can Be Done Using the Same DataFrameReader Interface from the Spark's Sructured APIs

```
orderEvents
  .select($"key",$"value",$"topic")
  .write
  .format("kafka")
  .option("kafka.bootstrap.servers", KafkaBootstrapServers)
  .save()
```

Now that all the requirements of the generator are checked off, all that is left to do is compile the application and run the generator.

Running the Generator

You can run the generator by first compiling the Spark application, building the Docker container, and lastly running the generator.

Note   To run the end-to-end example, you need to have the Kafka cluster running. Follow the README under `ch-11/exercises/02_binary_serialization_in_spark` for extended directions.

Compile the application and create a new Docker container named `spark-kafka-coffee-orders` so you can run the generator.

```
sbt clean assembly &&
docker build . -t `whoami`/spark-kafka-coffee-orders:latest
```

To view the end-to-end example, you'll need to have two terminal windows open. The first will be running the Kafka console consumer. This is a command-line application that ships along with Kafka. It simply waits for new data to arrive on a Kafka topic and it prints the value of each record. You can start the Kafka console consumer application by executing the command in Listing 11-25.

Listing 11-25. Running the Kafka Console Consumer Application

```
docker exec -it kafka_kafka-0_1 \
  /opt/bitnami/kafka/bin/kafka-console-consumer.sh \
  --bootstrap-server kafka_0:9092,kafka_1:9092,kafka_2:9092 \
  --topic com.coffeeco.coffee.v1.orders
```

With the Kafka console consumer running in one terminal window, open another terminal window where you can run the Spark `CoffeeOrder` generator. Execute the `docker run` command in Listing 11-26 to run the generator.

Listing 11-26. Running the Generator to Produce Protobuf Based CoffeeOrders and Write Them Into Kafka

```
docker run \
  -p 4040:4040 \
  --hostname spark-kafka \
  --network mde \
  -v ~/dataengineering/spark/conf:/opt/spark/conf \
  -v ~/dataengineering/spark/jars:/opt/spark/user_jars \
  -it `whoami`/spark-kafka-coffee-orders:latest \
  /opt/spark/bin/spark-submit \
  --master "local[*]" \
  --class "com.coffeeco.data.generators.CoffeeOrderGenerator" \
  --deploy-mode "client" \
  --jars /opt/spark/user_jars/mariadb-java-client-2.7.2.jar \
  --driver-class-path /opt/spark/user_jars/mariadb-java-client-2.7.2.jar \
  --conf "spark.sql.warehouse.dir=s3a://com.coffeeco.data/warehouse" \
  --conf "spark.app.sink.option.kafka.bootstrap.servers=kafka_0:9092,kafka_
    1:9092,kafka_2:9092" \
```

```
--conf "spark.app.sink.kafka.topic=com.coffeeco.coffee.v1.orders" \
--conf "spark.data.generator.totalRecords=100" \
/opt/spark/app/spark-kafka-coffee-orders.jar
```

The generator will spin up and generate new coffee orders. If you are running the Kafka console consumer, you will see the binary representation of each order printed to the terminal window. You will see the binary representation of the `CoffeeOrder` rows being printed in the Kafka console consumer application (see Listing 11-27).

Listing 11-27. Using the kafka-console-consumer Will Enable You to View the Binary Serialized Records Being Published to Kafka in Real-Time

```
???orderId5store4"cust15(5,!??
?????orderId8store3"cust815?H??
?????orderId11store1"cust10(5????
```

Take a moment to pat yourself on the back. You now have a working solution for writing future event generators (and of course this one as well) using Apache Spark. All that is left is to connect the dots and learn how to consume the order events.

Exercise 11-2: Summary

You just learned how to assemble a fully functioning event generator in under 100 lines of code. While this is a cool exercise, it was intended to teach the important lesson of writing strictly typed (ridged) events in an interoperable way using Spark as the data producer. In the final exercise of the chapter, you learn how to reuse everything you learned during this second exercise to quickly wire up an event consumer that can be used to transform these CoffeeOrders into parquet data that can live in your data lake or data warehouse.

Exercise 11-3: Consuming Streams of Serializable Structured Data with Spark

The final missing piece in this 360-degree deep dive into Kafka and Spark is the process of reading (consuming) the serialized `CoffeeOrders` generated in Exercise 11-2. You'll be wading back into familiar territory with Spark's core APIs for Structured Streaming and learning how to use the `DataStreamReader`, with a Kafka twist provided by the `spark-sql-kafka-0-10` library.

Exercise 11-3: Materials

This exercise continues where Exercise 11-2 left off. To run the end-to-end application, you'll need to be running your local Kafka cluster as well as the MySQL (Hive Metastore) and MinIO (distributed data lake).

This exercise is a modified version of the application from Chapter 10. Remember the one that consumed order data from the Redis and wrote to the distributed table inside your MinIO data lake? The modified application flow for Exercise 11-3 is shown in Figure 11-1.

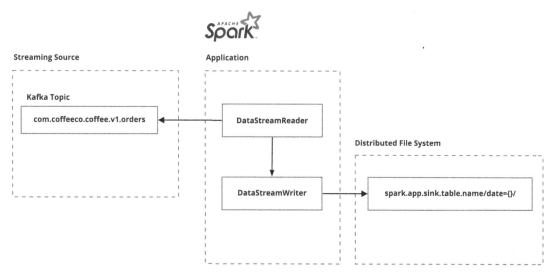

Figure 11-1. *The Spark application (Exercise 11-3) replaces Redis with Kafka; the rest of the application is just like the one from Chapter 10*

The application streams source data from Kafka by subscribing to a topic (in our example, it is `com.coffeeco.coffee.v1.orders`). The source topic is distributed across four physical partitions (which we configured in Listing 11-2). This means that the default number of partitions in the RDD source backing the Kafka DataFrame will also be four. Spark retains the partition information from the source to make the lineage of the data easier to follow back to a specific absolute partition. This is useful to recover all the data read between the starting and ending offsets for the data lost during a failure in your Spark application. This kind of recovery is necessary when an executor (node) in the Spark cluster goes offline. The application will have to recover all data encapsulating the state of your distributed computation. The way in which the stateful application operates using Kafka is shown in Figure 11-2.

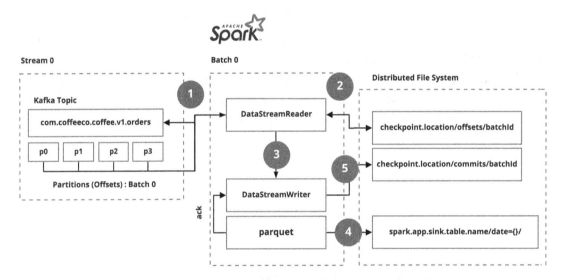

Figure 11-2. *The flow for each micro-batch in the Structured Streaming application*

For each micro-batch, the Spark application driver (1) queries Kafka to figure out if there is any new data to consume, and if there is any new data, the starting and ending offsets to be assigned for each topic partition are written ahead (2) to the application checkpoint directory. This process enables the application to pick back up from the exact location(s) of a Kafka topic if there is any processing failure. The write-ahead log (WAL) is important for ensuring that all data is processed, and that recovery is done in a deterministic fashion. The topic offset ranges are then passed to the application executors (3) as topic assignments. This enables the executors to fetch data in a distributed way through explicit assignments for a specific micro-batch in your Spark application. The application then transforms the Kafka binary data into datasets and processes the data. When the processing is complete, the results are then written (4) to a distributed table, and the results of successfully writing the output are committed back to the checkpoint directory (5). This end-to-end process using the WAL, processing, and committing back creates a completed micro-batch. Let's look at the process of building this now.

Consuming Binary Data from Apache Kafka

Open the SparkKafkaCoffeeOrdersApp, which is located in the same project as the CoffeeOrderGenerator. Let's start with the Kafka DataStreamReader and move backward from there.

Just like with the Redis Streams application, we are using the inputStream block (shown in Listing 11-28) to configure and bootstrap an instance of the DataStreamReader.

Listing 11-28. Creating a Generic Config Driven DataStreamReader

```
lazy val inputStream: DataStreamReader = {
  sparkSession
    .readStream
    .format(inputStreamFormat)
    .options(sparkConf
        .getAllWithPrefix(
          s"$StreamConfigSourcePrefix.option.").toMap)
}
```

The result of calling inputStream does in fact bootstrap an instance of the DataStreamReader for our Kafka connection. This is an example of using the Spark core APIs with a pure config-driven approach. For example, the settings in Listing 11-29 enable the application to connect from within the Docker network to the Kafka cluster (as long as the containers are running) and subscribe to the com.coffeeco.coffee. v1.orders topic, starting at the earliest available offsets per partition within the topic.

Listing 11-29. Minimal Configuration to Configure the Kafka DataStreamReader

```
--conf "spark.app.source.format=kafka" \
--conf "spark.app.source.option.subscribe=com.coffeeco.coffee.v1.orders" \
--conf "spark.app.source.option.kafka.bootstrap.servers=kafka_0:9092" \
--conf "spark.app.source.option.startingOffsets=earliest"
```

Once the configuration is applied to the `inputStream`, the resulting object (in Listing 11-30) is crafted to connect to Kafka, but it's flexible enough to handle any additional configuration needed to customize the Kafka connection or behavior.

Listing 11-30. The Resulting InputStream Object Created by Reading the External Application Configuration

```
sparkSession
    .readStream
    .format("kafka")
    .option("kafka.bootstrap.servers", "kafka_0:9092")
    .option("subscribe","com.coffeeco.coffee.v1.orders")
    .option("startingOffsets","earliest")
```

The way this works is all bound to a special method on the `SparkConf`. The beauty of using the `sparkConf.getAllWithPrefix` method is that you can modify the behavior of the `DataStreamReader` without needing to go back to the source code for each minor change. We revisit this pattern again in Chapter 13.

The common Kafka Topic properties are shown next.

Topic Subscription

For example, you can subscribe to multiple topics at the same time (enabling your application to read from many locations). This technique can be used to process multiple streams of events without having to run an application per stream. Just pass in one or more topic names in a comma-separated list.

```
.option("subscribe",
  "com.coffeeco.coffee.v1.orders,
com.coffeeco.customers.v1.events")
```

As an alternative, you can subscribe to a topic pattern. This technique enables your application to subscribe to all topics matching a prefix-based Regex. For example:

```
.option("subscribePattern","com.coffeeco.coffee.v1.*")
```

You can use `subscribe`, `subscribePattern`, or `assign`, which we will look at next.

Topic Assignment

Subscriptions do a lot of heavy lifting for your application. This includes figuring out what partitions are available and what offsets are required to assign to tasks during the operations within the lifecycle of the Spark application. At times you might find yourself wanting to assign a specific topic partition to an application as opposed to subscribing to all topic partitions. We won't cover this use case directly, but it is worth knowing it exists. For example, some topic partitions can become hot spots because upstream applications publishing to Kafka have accidently published records using empty strings as the record key. This will cause all records, keyed using an empty string, to be routed to the same partition, which can become a hot spot. You can preemptively mitigate this issue by enforcing non-empty strings, but you can also experience hot spots due to common behaviors in a system.

Throttling Kafka

We used the `stream.read.batch.size` configuration in Chapter 10 to control the rate in which the Spark application read (consumed) data from the Redis Stream. For the Kafka source, you can throttle the rate by providing a maximum number of records to read for each micro-batch. This is done using the `maxOffsetsPerTrigger` option.

```
"spark.app.source.options.maxOffsetsPerTrigger=1000"
```

This number is for the entire batch (across all partitions).

Handling Failures

Failure happens. What you do about it is up to the application. The Kafka source provides an option to handle this. When an application expects to pick up from the last successful batch, but the data (offsets of a partition of a topic) is simply missing (timed out of became corrupted), then the Spark application can simply fail rather than continue process data with a gap in it. Use the failOnDataLoss option to control this behavior.

```
--conf "spark.app.source.options.failOnDataLoss=false"
```

Splitting the Partitions

If you have four partitions, like with the com.coffeeco.coffee.v1.orders topic, but you want to further divide the dataset being processed by your Spark application, you can use the minPartitions setting. This means you can divide and conquer and distribute workloads across a variable number of executors (or processing threads) and achieve a higher throughput rate for your Spark application.

```
--conf "spark.app.source.options.minPartitions=8"
```

This would double the default number of partitions (for our four-partition topic).

You will get a chance to use these additional configurations after going through the rest of the exercise. Now back to the source code.

From Kafka Rows to Datasets

The KafkaOrderTransformer class is responsible for deserializing the byte array (value) from the Kafka DataFrame and transforming it into a CoffeeOrder. This is the reverse process you went through before, when building out the CoffeeOrderGenerator. The class is shown in Listing 11-31.

Listing 11-31. Converting the Kafka DataFrame into a Typed Dataset

```scala
class KafkaOrderTransformer(spark: SparkSession)
  extends DatasetTransformer[CoffeeOrder]
    with Serializable {
  import scalapb.spark.Implicits._

  override val encoder = typedEncoderToEncoder[CoffeeOrder]
  override def transform(df: DataFrame): Dataset[CoffeeOrder] =  df
    .map(_.getAs[Array[Byte]]("value"))
    .map(CoffeeOrder.parseFrom)
    .as[CoffeeOrder]
}
```

This transformer inherits from a typed version of the `DataFrameTransformer` from Chapters 7 and 10.

```
trait DatasetTransformer[T] {
  val encoder: Encoder[T]
  def transform(df: DataFrame): Dataset[T]
}
```

This then provides the glue to inform Spark how to operate with the data coming from the transformer, in this case the `CoffeeOrder` itself.

Moving to the `run` method in Listing 11-32 on the main application, you'll see how the `inputStream` and the `KafkaOrderTransformer` play together to produce the input to `streamingQuery`.

Listing 11-32. Consuming Serializable Structured Data from Kafka with Apache Spark

```
override def run(): Unit = {
  import scalapb.spark.Implicits._
  super.run()
  val inputSourceStream = KafkaOrderTransformer(sparkSession)
    .transform(inputStream.load())
  val writer = outputStream(inputSourceStream)
  startAndAwaitApp(writer.start())
}
```

The only thing left here is a look at the `outputStream` method. The `startAndAwaitApp` method is left over from Chapter 10. The `outputStream` method is using the configuration pattern shown in the `inputStream` method (Listing 11-23) from earlier, and is shown in Listing 11-33.

Listing 11-33. Transforming CoffeeOrders into Parquet Data Partitioned by Date

```
def outputStream(ds: Dataset[CoffeeOrder]): DataStreamWriter[Row] = {
  ...
  val streamOptions = sparkConf
    .getAllWithPrefixs("$StreamConfigSinkPrefix.option.")
  val dataStream = ds
```

```scala
    .withColumn("date", to_date(
      to_timestamp($"timestamp".divide(1000).cast(LongType)))
    )
    .writeStream
    .format(outputStreamFormat)
    .queryName(outputStreamQueryName)
    .outputMode(outputStreamMode)
    .partitionBy("date")
    .options(streamOptions)

  triggerType match {
    case "continuous" if triggerEnabled =>
      dataStream.trigger(
        Trigger.Continuous(processingInterval))
    case "once" if triggerEnabled =>
      dataStream.trigger(Trigger.Once())
    case "processing" if triggerEnabled =>
      dataStream.trigger(
          Trigger.ProcessingTime(processingInterval))
    case _ =>
      dataStream
    }
  }
```

The final piece of the pie here is the transformation from the `Dataset[CoffeeOrder]` into a `DataStreamWriter`, which will write all available rows of data into the data lake, achieving similar results as the application from Chapter 10. Now all that is left to do is run the application.

Running the Consumer Application

Now that you have had a chance to go through the application, you can compile the application if you made any changes. Otherwise the output of building the application and creating the Docker container will be the same as before, so if you have been following along with the exercises, you should already have the consumer application compiled and containerized.

Listing 11-34. Running the CoffeeOrder Consumer Application

```
docker run \
  -p 4040:4040 \
  --hostname spark-kafka-coffee-consumer \
  --network mde \
  -v ~/dataengineering/spark/conf/hive-site.xml:/opt/spark/conf/hive-
  site.xml \
  -v ~/dataengineering/spark/jars:/opt/spark/user_jars \
  -it `whoami`/spark-kafka-coffee-orders:latest \
  /opt/spark/bin/spark-submit \
  --master "local[*]" \
  --class "com.coffeeco.data.SparkKafkaCoffeeOrdersApp" \
  --deploy-mode "client" \
  --jars /opt/spark/user_jars/mariadb-java-client-2.7.2.jar \
  --driver-class-path /opt/spark/user_jars/mariadb-java-client-2.7.2.jar \
  --driver-java-options "-Dconfig.file=/opt/spark/app/conf/coffee_orders_
  consumer.conf" \
  --conf "spark.sql.streaming.kafka.useDeprecatedOffsetFetching=false" \
  --conf "spark.app.source.option.startingOffsets=earliest" \
  --conf "spark.app.stream.trigger.type=once" \
  /opt/spark/app/spark-kafka-coffee-orders.jar
```

The provided README.md in the spark-kafka-coffee-orders application root directory provides a more complete version of the run command with all config overrides provided. The command (11-34) also uses an alternative config alongside the application which is made possible using the driver-java-options on Spark Submit. While the Configuration helper was always capable of reading an alternative config (thanks to the TypeSafe config library), this is the first example of using what I like to call the bring-your-own-config pattern at work.

```
--driver-java-options "-Dconfig.file=/opt/spark/app/conf/coffee_orders_
consumer.conf"
```

The way this alternative configuration is made available inside the container is through a small change to the Dockerfile.

```
FROM newfrontdocker/apache-spark-base:spark-3.1.2-jre-11-scala-2.12

COPY target/scala-2.12/spark-kafka-coffee-orders-assembly-0.1-SNAPSHOT.jar
/opt/spark/app/spark-kafka-coffee-orders.jar
# copy the app configs
COPY conf /opt/spark/app/conf
EXPOSE 4040
```

The COPY directive simply moves the configuration from the application root directory into the container at build time. This way you can move any relevant config to a more final resting place rather than relying on the spark-submit runtime configs. It is less error prone to do things this way.

Exercise 11-3: Summary

This exercise was the final step in showing how to work with serializable structured data within the Apache Spark ecosystem. You have completed the full cycle required to work with serializable structured data from message (protobuf) definition, to compilation, to random record generation, to writing (producing) and then reading (consuming) the messages in an end-to-end modular way. I hope this sparks some new ideas for how you can work with data in an interoperable way, without falling back to JSON or CSV.

Summary

Apache Kafka is a Swiss Army knife for the data engineer. It provides a reliable platform for working with streams of rich data in near real-time. While there are many ways to work with data within the Kafka and Spark ecosystems, it is worth mentioning that working with explicit types provides a rich data contract between these separate systems and can provide a much more solid and stable foundation for building reliable, and more importantly, interoperable data systems.

In the next chapter, we will be continuing our work with Structured Streaming, with a focus on more advanced aggregations and analytical processing. If you're ready, let's go.

CHAPTER 12

Analytical Processing and Insights

Throughout the book you've focused on building your data engineering toolbox and learning powerful techniques for processing and transforming data, both at rest and in stream. Parallel to learning Spark, you have also been hands-on learning to use containers and operate additional technologies that interoperate well in the Apache Spark ecosystem.

Given that the modern data engineer wears many hats and has many responsibilities spanning the wide spectrum of the data ecosystem, from data production and ingestion, data validation and cleaning, data transfer and loading, batch and stream processing as well as more analytical responsibilities like simple and complex aggregation, and even streaming analytical processing and insights, this chapter focuses on filling the knowledge gap related to preprocessing data for analytics and learning to use powerful aggregation operators.

Exercises and Materials

The chapter exercises introduce you to Apache Spark's rich analytical processing capabilities. To get the most out of this chapter, all the exercises can be run interactively using Apache Zeppelin.

This chapter is broken into the following two hands-on sections:

- Using common Spark functions for analytical preprocessing

- Analytical processing and insights engineering

Learning to use Spark's higher order functions interactively will equip you to write and operate analytical stateful structured streaming applications in Chapter 13.

© Scott Haines 2022

S. Haines, *Modern Data Engineering with Apache Spark*, https://doi.org/10.1007/978-1-4842-7452-1_12

Exercise Materials   The exercise materials for Chapter 12 are located at `https://github.com/newfront/spark-moderndataengineering/tree/main/ch-12`. The README file located in the `ch-12/exercises` directory will provide you with a full set of setup instructions.

Setting Up the Environment

Once again, we will be leaning on our old friend Apache Zeppelin, but this time you'll be building your own Zeppelin container called `zeppelin-spark`. You might be asking yourself why you would want to create a different container. I'm glad you asked.

Throughout the book you've been required to use *volume mounts* and *environment variables* to plug your own local Apache Spark into the Zeppelin container at runtime. Back in Chapter 10, you'll remember that we migrated to using a base Spark container to build consistent containerized Spark applications. This was necessary given the growing complexity of the local data platform and to ensure a consistent environment was available at runtime for our applications. Until now, Zeppelin required you to bring your own Spark, but this new container will marry the official Zeppelin container with the same base Spark container you've been using in the last few chapters.

Zeppelin-Spark Directory Layout and Container Dependencies

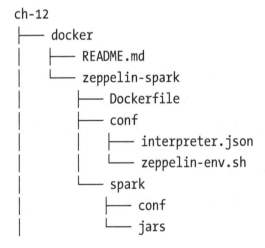

```
ch-12
├── docker
│   ├── README.md
│   └── zeppelin-spark
│       ├── Dockerfile
│       ├── conf
│       │   ├── interpreter.json
│       │   └── zeppelin-env.sh
│       └── spark
│           ├── conf
│           └── jars
```

From your terminal, change to the docker/zeppelin-spark directory. The *Dockerfile* located directly in the directory showcases the Docker multi-stage build pattern, essentially cherry-picking resources from the base spark container while also adding resources on top of the core apache/zeppelin container. The contents of the file are provided in Listing 12-1.

Listing 12-1. Adding a New File System Layer on Top of the Official Apache Zeppelin Docker Container That Brings Java 11 and Spark 3.1.2 as Well as Custom Configs and JARs to the Table

```
ARG spark_image=spark-3.1.2-jre-11-scala-2.12
FROM newfrontdocker/apache-spark-base:${spark_image} as spark
FROM apache/zeppelin:0.9.0

COPY --from=spark /opt/spark /opt/spark
COPY --from=spark /usr/local/openjdk-11 /opt/openjdk-11
COPY ./spark/conf/hive-site.xml /opt/spark/conf/
COPY ./spark/conf/spark-defaults.conf /opt/spark/conf/
COPY ./spark/jars /opt/spark/user_jars/
# add mariadb driver directly to the spark class path
COPY ./spark/jars/mariadb-java-client-2.7.2.jar /opt/spark/jars/

COPY ./conf/interpreter.json /zeppelin/conf/
COPY ./conf/zeppelin-env.sh /zeppelin/conf/
```

Multi-stage Docker builds allow you to create reliable instances of your favorite containers without the hassle of more complex builds. In the case of the zeppelin-spark container, we can extract Java 11 (JRE) as well as the Spark distribution directly from the apache-spark-base container (see Listing 12-2). In a similar spirit to using modules to build Spark applications, you can stack containers using FROM or use containers as a provider of resources.

Listing 12-2. Using the Multi-Stage Build Pattern Helps to Reduce the Final Weight Of Your Containers and Can Also Be Used to Reduce the Overhead of Managing Complex Containers

```
FROM newfrontdocker/apache-spark-base:${spark_image} as spark
FROM apache/zeppelin:0.9.0
COPY --from=spark /opt/spark /opt/spark
COPY --from=spark /usr/local/openjdk-11 /opt/openjdk-11
```

Building the container locally is an optional step. You can use the publicly available newfrontdocker/zeppelin-spark container, but building your own can be done simply by running the following build command (from the zeppelin-spark dir).

```
docker build . -t `whoami`/zeppelin-spark:latest
```

Next, you can go ahead and spin up the environment.

Spinning Up the Local Environment

Under the Docker directory, you'll find another directory named zeppelin. Copy the directory into ~/dataengineering/zeppelin-spark. As with prior chapters, the docker-compose.yaml file located in the directory provides you with all the wiring needed to spin up the environment.

```
docker compose \
  -f ~/dataengineering/zeppelin-spark/docker-compose.yaml up \
  -d
```

The compose process will bring up Apache Zeppelin (localhost:8080), MySQL (:3306), and the MinIO (localhost:9000) file systems.

Using Common Spark Functions for Analytical Preprocessing

This first exercise will be a hands-on tour using the more common higher-order functions (columnar expressions) available in the org.apache.spark.sql.functions package, focusing on the functions that will aid in preparing data for analysis.

Exercise 12-1: Preprocessing Datasets for Analytics

Open Zeppelin in your web browser (Localhost:8080), create a new note, and name it something memorable like 01_common_dataframe_functions. From this starting point, you'll learn to use functions that will help you modify and prepare data for analytical processing:

- Working with timestamps and dates
- Working with time zones

- Handling and replacing null column values

- Using case statements

- Using user-defined functions

Working with Timestamps and Dates

Working with time can be a difficult thing. You learned in Chapter 9 how clock drift and incorrect time zone information can cause problems and how issues can be generally resolved simply by standardizing on UTC and synchronizing clock drift on your servers using NTP. While standardizing can help future projects, you may find yourself needing to work with time in different ways or to correct and normalize timestamps.

Common Date and Timestamp Functions

Starting from your empty note (in Zeppelin), copy the code block in Listing 12-3 and run the paragraph. The SQL statement generates a row with columns representing the date and time information captured by Spark at runtime.

Listing 12-3. Using Spark SQL to Get a Sense of Time

```
%spark
spark.sql("""
SELECT current_timestamp() as ts,
current_timezone() as tz,
current_date() as date,
TIMESTAMP 'yesterday' as yesterday,
TIMESTAMP 'today' as today,
TIMESTAMP 'tomorrow' as tomorrow
""").show(6,0,true)
```

The output of the resulting row will be a snapshot of standard system date and time information collected when the code is evaluated.

```
-RECORD 0------------------------------
 ts        | 2021-09-12 00:54:03.691464
 tz        | Etc/UTC
 date      | 2021-09-12
```

```
yesterday  |  2021-09-11 00:00:00
today      |  2021-09-12 00:00:00
tomorrow   |  2021-09-13 00:00:00
```

The current_timestamp, current_timezone, current_date, and the TIMESTAMP constants yesterday, today, and tomorrow are all higher-order Spark SQL datetime functions. Next, we re-create the same output use the Spark DSL functions directly.

Applying Higher-Order Functions Using withColumn

Create a new paragraph. Inside the paragraph we will create a single 1x1 (row/column) DataFrame storing only a timestamp. Mimicking the current_timestamp expression, Listing 12-4 shows how you can wrap a simple Java Instant to replicate the current_ timestamp expression.

Listing 12-4. Create a DataFrame with a Single Row and a Single Column Storing a Timestamp

```
%spark
val tsDf = Seq(Instant.now).toDF("ts")
```

The simple technique in Listing 12-4 takes advantage of implicit conversions to encode a Scala Seq[Instant] as a Catalyst Row (DataFrame) with a TimestampType column. Using the tsDf DataFrame we can now add columns using the withColumn method on the DataFrame.

```
DataFrame.withColumn(colName, col)
```

The withColumn method adds a new column iteratively across all rows of a DataFrame. The col (Column) parameter is a powerful primitive that encapsulates a columnar SQL expression and can be used to enable user-defined functions to your applications. It is important to keep in mind that withColumn can only reference data from within adjacent columns of a row. There are other techniques that can be used to process all rows or a subset of rows to generate derived aggregates using window functions.

The code block in Listing 12-5 adds a column literal (which means "literally this column is exactly what you see") containing the time zone as well as the derived date using the column operation to_date.

Listing 12-5. Using withColumn to Add a Column Literal and a Derived Date Column Using to_date

```
import org.apache.spark.sql.functions._
import org.apache.spark.sql.types._

val dtInfoDf = tsDf
  .withColumn("tz",
    lit(spark.conf.get("spark.sql.session.timeZone"))
  )
  .withColumn("date", to_date($"ts"))
```

The result of the transformation is a new DataFrame representing the combination of two new columns added to the source DataFrame. Remember that because Spark operates lazily, no actual work will be executed until you arrive at a specific action. The operation from Listing 12-5 adds the promise to provide the new date column when the time comes to execute on the final series. The first column uses a special column named lit. The lit column wraps and encodes the underlying data type as a typed column using an implicit typed encoder (if available). The second column (date) is referred to as a derived column. When you use the to_date function and pass a source column ($"ts"), Spark will use the reference column (ts) to generate a DateType.

If you printed the schema of dtInfoDf at this point, you would see the following:

```
root
 |-- ts: timestamp (nullable = true)
 |-- tz: string (nullable = false)
 |-- date: date (nullable = true)
```

Tackling the rest of the missing columns needed to recreate the DataFrame from Listing 12-3 leaves the columns yesterday, today, and tomorrow.

Using Date Addition and Subtraction

The datetime functions date_sub and date_add, as well as a simple cast expression, can be combined to derive yesterday, today, and tomorrow, as shown in Listing 12-6.

Listing 12-6. Using Date Subtraction, Addition, and Data Type Casting to Mimic the Output from Listing 12-3

```
tsDf

  ...
  .withColumn("yesterday",
    date_sub($"date", 1).cast(TimestampType))
  .withColumn("today", $"date".cast(TimestampType))
  .withColumn("tomorrow",
    date_add($"date", 1).cast(TimestampType))
```

You can now create timestamps, derive dates from timestamps, and add or subtract dates. Rounding out the tour of datetime functions, we will be looking at the year, month, dayofmonth, dayofweek, and dayofyear calendar functions.

Calendar Functions

It is common to compare explicit periods (windows) of time using the calendar for analytics and insights. Trend analysis and timeseries forecasting are techniques that use statistics to measure the rate of change over time (deltas) between data points. Datasets can be partitioned (bucketed) and analyzed using Spark SQL to create aggregations broken down by seconds, minutes, hours, days, weeks, months, and even years using *fixed* or *relative* windows.

Fixed and Relative Windows   A *fixed window* is defined by an explicit start and end time. For example, yesterday is a window defined by the 24-hour period beginning at 00:00:00 and ending at 23:59:59. Fixed windows are typically used to compute changes across two or more datasets commonly computed hour over hour, day over day, week over week and month over month.

A *relative window* uses a non-fixed point in time to define one edge of a time-based boundary. This boundary can then be used to compute either the beginning or ending timestamp producing an arbitrary window to observe statistics. For instance, you can use relative time to isolate a dataset encapsulating the last 30 minutes rather than splitting an hour at a fixed 30-minute interval (06:00:00 – 06:29:59 | 06:30:00 – 06:59:59).

Using the code in Listing 12-7 as a reference, we will add five columns to the tsDf DataFrame.

Listing 12-7. Deriving the Year, Month, Day, the Day Within the Current Week, and the Day Within the Current Year Using a Single DateTime Column

```
tsDf
...
.withColumn("year", year($"date"))
.withColumn("month", month($"date"))
.withColumn("day", dayofmonth($"date"))
.withColumn("day_of_week", dayofweek($"date"))
.withColumn("day_of_year", dayofyear($"date"))
```

The final DataFrame represents the current date with columns expressing different observations generated from the initial timestamp column.

```
-RECORD 0-------------------------------
 ts          | 2021-09-12 01:04:39.04086
 tz          | Etc/UTC
 date        | 2021-09-12
 yesterday   | 2021-09-11 00:00:00
 today       | 2021-09-12 00:00:00
 tomorrow    | 2021-09-13 00:00:00
 year        | 2021
 month       | 9
 day         | 12
 day_of_week | 1
 day_of_year | 255
```

You now have a solid reference to go back to whenever you need a quick refresher about working with date and time. But what about the actual time zone?

Time Zones and the Spark SQL Session

Working with time-based data requires conversion between time zones for more reasons than just adherence to a common time zone like UTC. You can also use a time zone as a lens to view a dataset from the perspective of observers of specific events. This use case comes up when producing insights that are tied to specific geolocations and time zones.

Configuring the Time Zone

Spark defaults to using the local system time of its environment (your laptop or a remote server). Using the default system time can cause discrepancies when processing data. To ensure consistent behavior regardless of where the application is run, you can configure the default time zone using the config `spark.sql.session.timeZone`.

Modifying the Spark Time Zone at Runtime

The SparkSession handles timestamp conversions automatically for you globally. However, there may be times when you want to explicitly change the time zone used by a specific query. Create a new paragraph in Zeppelin and add the example code block in Listing 12-8, which shows how to dynamically set the time zone.

Listing 12-8. Observing Time Zone Changes in Spark's Output of Time Based on the SparkSession Time Zone Configuration

```
%spark
import java.time._
val ts = Seq(Instant.now).toDF("ts")
spark.conf.set("spark.sql.session.timeZone", "UTC")
ts.show(truncate=false) // utc
spark.conf.set("spark.sql.session.timeZone", "America/Los_Angeles")
ts.show(truncate=false) // pst
```

Running the paragraph, you will see first-hand how Spark's observation of time changes in step with the value of the time zone configuration. The output captures the reference to the single immutable timestamp (`ts`) as observed through the lens of the UTC and PST time zones, respectively.

```
UTC |2021-09-12 03:02:32.434387
PST |2021-09-11 20:02:32.434387
```

The ability to shift the observation of time using the Spark runtime config enables you to use a single source of immutable truth (UTC timestamps) for your backing data while simplifying how downstream applications compose queries for specific time zones. All this without the headache of creating multiple copies of a dataset to handle the conversion between time zones for timestamps.

Using Set Time Zone

You can also use SET TIME ZONE (shown in Listing 12-9) to directly set the time zone config using Spark SQL to switch between time zones dynamically.

Listing 12-9. Using SET TIME ZONE to Configure the Spark Session Time Zone

```
%sql
SET TIME ZONE 'America/Los_Angeles';
SELECT TIMESTAMP 'now' as now_pst;

%sql
SET TIME ZONE 'UTC';
SELECT TIMESTAMP 'now' as now_utc;
```

You can now declaratively change the time zone in Spark using the SparkSession. This simple configuration enables each application to select how time should be observed while querying and displaying time-based data. Just remember to use a default time zone for your Spark applications to ensure reliable, repeated results.

Seasonality, Time Zones, and Insights

Consider the following example. You are tasked with designing a system that automatically tracks and compares the relative changes in observed customer shopping behavior across goods being sold at CoffeeCo to produce insights on purchase trends. While testing that the system is working as expected, you stumble upon what at first appears to be an anomaly in sales of a particular item. Upon further investigation it turns out that the item is a *pumpkin spice latte*, and the sales numbers went from zero average sales per day, to customers buying these pumpkin lattes at records levels, as compared to all other items in category, in only a few days. Is this an anomaly? What other data might be necessary?

Some goods are only available for a limited amount of time, or at a specific time of year. A pumpkin spice latte is seasonal since it is considered a fall (season) beverage, but what is it about the seasons (holidays) that affect sales? What about emotions?

I grew up on the northeast coast of the United States, and for me, October always marked an observable change in the seasons. Leaves would fall from the trees, and the colder temperatures meant sweaters and jackets and the peace and quiet reflection of snow in New England. With this change came anticipation of the winter holidays, which meant cider (mulling spices) and pumpkin pie, and as an adult pumpkin spice lattes bring back fond memories. Is this unique only to me? Is there perhaps a correlation to higher sales on the east coast of the United States? Are sales driven by temperature or other weather patterns like snowstorms or rain?

While you might not have any emotional connection to fall in New England and you might hate pumpkin spice lattes, it is important to think about what kind of data can be useful to generate insights from analytical observations. Bottom line, think like a data detective or partner with a great data analyst to ensure that the work that goes into data collection, aggregation, and analysis can produce novel insights that can be used to drive experiences that make happy customers.

Timestamps and Dates Summary

Thinking about the different angles, lenses, or views you can derive from a single timestamp provides you a path toward event-based insights. You can calculate information about the time of year (seasonality), whether a date falls on a weekend or weekday (specific to the geolocation of an event, or even the local time zone), or you can check if the event occurred on a holiday, or even if the date is within proximity to any other meaningful historic date. Associating what happened and correlating data to real-world behaviors is a driving factor behind many successful analytics and insight initiatives. If you want any more proof, search online for "diapers beer correlations" to see how retailers derive insight from customer transactions to move commonly purchased items closer together to increase sales across seemingly irrelevant items. We will now look at some techniques for preparing data for analysis.

Preparing Data for Analysis

Data comes in all shapes and sizes. Unfortunately, that also means that data may be missing (null) for reasons outside of data corruption or user error. For example, null values in a dataset don't necessarily need to be filled in with default values at rest, since that may only prove to increase the overall footprint of a given dataset.

Additionally, depending on how long a dataset has existed and whether data is being actively collected and appended to that dataset, there is a high probability that the underlying schema of the data has changed slowly over time. Slowly changing schemas are one of the common factors behind null values as older records need to be coerced into a new common schema (StructType) to be processed side by side with newer data items from the dataset (table).

Outside of simple missing values, you are also aware that datasets are often transformed and joined with other datasets to materialize more specific views that cater to different data domains to assist in answering more specialized questions (queries).

We will look at the process of conditionally filling null values and using case statements (columnar expressions) to generate derived columns that can be used to conditionally label (tag), bucket (partition), and generally solve common data problems.

Replacing Null Values on a DataFrame

Filling in missing data conditionally can be done using na.fill on a DataFrame. Create a new paragraph (in Zeppelin) and copy the code block in Listing 12-10 into it. Then run it. The example creates a simple DataFrame representing four orders of various prices from registered and unregistered customers of CoffeeCo. Here I chose to represent unregistered customers with a null customer_id column.

Listing 12-10. Creating a DataFrame with Null customer_id to Use na.fill

```
%spark
val nullOrdersDF = Seq(
  (1, null, "decafe", 3.12),
  (2, "cust123", "pour_over", 5.15),
  (3, "cust234", "latte", 3.89),
  (4, "cust345", "special_pour_over", 6.99)
)
.toDF("order_id", "customer_id", "item_name", "price")
nullOrdersDF.show()
```

417

After running the paragraph, the resulting DataFrame will look like the following:

```
+--------+-----------+----------------+-----+
|order_id|customer_id|item_name       |price|
+--------+-----------+----------------+-----+
|1       |null       |decafe          |3.12 |
|2       |cust123    |pour_over       |5.15 |
|3       |cust234    |latte           |3.89 |
|4       |cust345    |special_pour_over|6.99 |
+--------+-----------+----------------+-----+
```

You can now use the na.fill operation to replace null values with a common value, across one or more columns, and return a new clean DataFrame. Using Listing 12-11 as a reference, test filling the column customer_id with the value "unknown".

Listing 12-11. Replacing Null Values from the Column customer_id with the Value Unknown

```spark
%spark
val nonNullOrdersDF = nullOrdersDF
  .na.fill("unknown", Seq("customer_id"))

nonNullOrdersDF.createOrReplaceTempView("order_nonnull")
nonNullOrdersDF.show(truncate=false)
```

The non-null (cleaned) DataFrame (after using na.fill) created as a result of calling na.fill Listing 12-11, which is saved into a new reference variable as well as a temporary view named order_nonnull. The result of the operation looks like the following:

```
+--------+-----------+----------------+-----+
|order_id|customer_id|item_name       |price|
+--------+-----------+----------------+-----+
|1       |unknown    |decafe          |3.12 |
|2       |cust123    |pour_over       |5.15 |
|3       |cust234    |latte           |3.89 |
|4       |cust345    |special_pour_over|6.99 |
+--------+-----------+----------------+-----+
```

Filling null values with a specific constant like unknown can help to declare the meaning of a column rather than requiring other engineers to guess the columns intentions.

As an alternative, if the table was being managed in the Spark SQL Catalog, then being the good data steward you are, you could go the extra mile and describe the intentions of the null column in the Hive Metastore. Then other data teams can consistently work with the null values in the same way.

When preparing data for analysis, you will more likely than not find yourself needing to add a derive columns as a function of one or more conditional expressions. Rather than creating a specialized user-defined function (UDF), you can simply use SQL case statements and Spark SQL when/then/otherwise expressions. Case statements work in a similar way to a pattern matching block or a traditional switch statement.

Labeling Data Using Case Statements

Case statements provide a mechanism for adding new columns based on the result of one or more chained Boolean expressions (predicates). You can use case statements to *tag*, *categorize*, or *label* data while it is being processed. The semantics of the case statement changes depending on if you are using the Spark SQL functions on a DataFrame or if you are using SQL expression on a table or view.

Case Statements on the Dataset

Create a new paragraph in the Zeppelin notebook. You will be transforming the target DataFrame nonNullOrdersDF (Listing 12-11) by adding two columns named is_registered and order_label in an effort to create a labeled order. The code block in Listing 12-12 shows how to use *when, otherwise expressions*. The first column is_registered tests if the order has a known or unknown customer_id, while the order_label expression uses boolean and numeric condition evaluation to conditionally flag an order as VIP, rush, or normal.

Listing 12-12. Using When, Then, Otherwise to Create Complex Derived Columns on Your DataFrame

```
%spark
nonNullOrdersDF
  .withColumn("is_registered",
    when(col("customer_id").notEqual("unknown"), true)
    .otherwise(false)
  )
  .withColumn("order_label",
    when(col("is_registered").equalTo(true).and(col("price") > 6.00), "vip")
    .when(col("is_registered").equalTo(true).and(col("price") > 4.00),
    "rush").otherwise("normal")
  )
  .show(truncate=false)
```

As discussed earlier, the first new column (`is_registered`) represents the registered status of a customer. The columnar expression is a predicate that checks that the evaluation conditions are met. *When* the value of the column is not equal to unknown *then* we will set the column value to *true, otherwise* it falls back to setting the column value as *false.*

The second derived column uses the value of `is_registered` to conditionally decorate a customer's order as a function of their registration status and the price of their order. The result is a column with three possible values: normal, rush, or vip. The columnar expression uses more complex conditional branching by introducing the concept of expression *and-ing* (which could also have been an or condition). The easiest way to think about when-then-otherwise expressions is to think about typical conditional branching.

```
if (row.is_registered) {
  if (row.price > 6.00) "vip" else "rush"
} else "normal"
```

Let's look at how to accomplish the same goal using Spark SQL.

Case Statements on a Spark SQL Table

The order_nonull temporary view can be used along with a nested inner select statement to accomplish the same results shown in Listing 12-12. Create a new Zeppelin paragraph, copy the SQL statement in Listing 12-13, and run it. You will see that the output is the same as Listing 12-12.

Listing 12-13. Using Interpreted SQL on a Spark SQL Table to Create Complex Derived Columns

```
%sql
SELECT *,
  CASE
    WHEN x.is_registered = TRUE and x.price > 6.00 THEN 'vip'
    WHEN x.is_registered = TRUE and x.price > 4.00 THEN 'rush'
    ELSE 'normal'
  END as order_label
  FROM (
    SELECT *,
    CASE
      WHEN customer_id != 'unknown' THEN TRUE
      ELSE FALSE
    END as is_registered
  FROM order_nonnull
) x
```

The Spark SQL case statement in Listing 12-13 provides you with the general framework for adding complex conditional expressions that can aid in tagging, bucketing, or in helping to find other patterns in a dataset.

Tip   The Spark SQL interpreter is a wonderful support tool when you simply want to let Spark do the work rather than providing the full set of functional transformations that are required to do the same thing on the DataFrame API. Ultimately, you can mix execution modes easily when you are using the DataFrame and Dataset APIs in batch mode (including via Zeppelin) while you are figuring things out.

The Case for Case Statements

Case statements provide a perfect mechanism for adding conditional expressions that can be used to declaratively tag, partition, or bucket, your datasets for analysis. For example, it may be interesting to know which customers spend more than $20 per order, maybe more than $100. Maybe you want to tag customers who purchase uncommon goods because they may tend to act like the other customers who also purchase uncommon goods? By adding tags, such as order_type, to your data, you can provide the input features for the data scientists and machine learning engineers in your organization to do more with the data. This technique is a lightweight way of testing data theories. They help explore a dataset by separating data to find patterns and insights.

User-Defined Functions in Spark

User-defined functions (UDFs) enable you to encode complex procedures using traditional Scala functions wrapped in a Spark UDF. UDFs can even be registered with the SparkSession to extend the available functions in the Spark SQL Catalog.

We will walk through the process of converting the order_label expression introduced in Listing 12-12 into a UDF called coffee_order_label. The base function will take a String and Double and return a String. The Spark SQL UDF function wraps a traditional Scala function, function literal, or inline literal function.

Using Scala Functions in UDFs

Listing 12-14 creates a traditional Scala function (labeler) and uses Spark's udf function to wrap an applied labeler function call.

Listing 12-14. Encoding a Traditional Scala Function for Use with the Structured Spark APIs

```
%spark
import org.apache.spark.sql.functions.udf

def labeler(customerId: String, price: Double): String = {
  if (customerId != "unknown") {
    if (price > 6.0) "vip" else "rush"
  } else "normal"
}
val coffeeOrderLabeler = udf(labeler(_,_))
```

Using a traditional function is probably the easiest way to get started with UDFs. Depending on how you like to work, function literals provide you with another means of expressing your UDF.

Using Function Literals in UDFs

Function literals in Scala provide a means of passing a function as an argument to another function. In the example in Listing 12-14, there was a need to apply the function using the udf(label(_,_)) expression. Listing 12-15 shows how you can use a function literal to simplify using the *udf* operation at a slight cost to the scanability of the code block (for non-functional programmers).

Listing 12-15. Passing a Function Literal to the UDF

```
%spark
import org.apache.spark.sql.functions.udf

// udfs can use function literals
val coffeeOrderLabelerFunc: ((String, Double) => String) =
(customer_id: String, price: Double) => {
  if (customer_id != "unknown") {
    if (price > 6.0) "vip" else "rush"
  } else "normal"
}
val coffeeOrderLabeler = udf(coffeeOrderLabelerFunc)
```

Since function literals can be passed as method parameters, the udf operation can infer the required function parameter types (String, Double) and generate a lambda expression automatically. Lastly, we will look at providing an inline function to the udf operation.

Using Inline Functions in UDFs

The udf operation will wrap whatever function you pass into the udf method, so you can simply provide an inline function (as shown in Listing 12-16) and Spark will encode the function without the need to do anything more outside of Spark's scope.

Listing 12-16. Using an Inline Function (Lambda Expression) with the Spark SQL UDF Method

```
val coffeeOrderLabeler = udf(
(customer_id: String, price: Double) => {
  if (customer_id != "unknown") {
    if (price > 6.0) "vip" else "rush"
  } else "normal"
})
```

Now that we've covered how to create user-defined functions, we should see how they work and then learn how to use them.

How User-Defined Functions Work

The type of object returned by the spark.sql.functions.udf method is an instance of the abstract class UserDefinedFunction. Specifically, an instance of the concrete class SparkUserDefinedFunction. This object is responsible for translating between the JVM data types and the Spark data types using implicit encoders. What is more interesting is that the return type of calling apply on your udf is a Column. This is the magic at work enabling the UDF to interoperate with the structured Spark APIs at the column level. The UDF just turns out to be another columnar expression.

Using UDFs with the Spark DSL

The newly defined UDF can now be invoked using withColumn. If you haven't already, create a paragraph in your notebook and run the code block in Listing 12-16 to create the coffeeOrderLabeler. Then create another new paragraph and copy the block in Listing 12-17, replacing the when/otherwise statement in Listing 12-12 with the new coffeeOrderLabeler UDF.

Listing 12-17. Using UDFs and the Structured Spark DSL

```
nonNullOrdersDF
  .withColumn("is_registered", ...)
  .withColumn("order_label", coffeeOrderLabeler(
    $"customer_id", $"price"))
```

If you want to use your new UDF with Spark SQL you will have to register it so that it becomes part of the Spark SQL Catalog. Then the function can be used like any other SQL expression.

Registering UDFs for Spark SQL

Once you have a UDF, you can register it using the SparkSession udf method. This will add the function into the Spark SQL catalog.

```
// registers the udf for use with Spark SQL
spark.udf.register("coffee_order_label", coffeeOrderLabeler)
```

You can check that your UDF is registered by using the Spark SQL Catalog methods from the SparkSession.

Introspecting UDF Functions

It is useful to search for and test for the existence for your UDFs. After registering a new function, you can list all the available functions of the SparkSession by calling the listFunctions method from the Spark SQL Catalog (catalog).

```
spark.catalog.listFunctions.show()
```

Because the `listFunctions` method returns a `Dataset[Function]`, you can also query the dataset for functions by pattern (in case you can't remember the full name).

```
spark.catalog.listFunctions
  .where($"name".like("%coffee%"))
```

And if you already know the name of the function, for example with the case of the `coffee_order_label` UDF, you can check to see if the function exists on the current SparkSession.

```
spark.catalog
  .functionExists("coffee_order_label")
```

UDF Visibility and Availability

User-defined functions are loaded into the SQL Catalog for each instance of the SparkSession. For example, if your Spark application forks the SparkSession then there is no guarantee that a UDF will automatically become available. This is because UDFs, like temporary views, are only available by default for the lifecycle of the SparkSession that created them. Checking if a UDF exists before using it can help resolve future problems today (see Listing 12-18).

Listing 12-18. Testing for the Presence of a User-Defined Function After Forking a SparkSession

```
val newSession = spark.newSession
if (!newSession.catalog.functionExists("coffee_order_label")) {
  ... register the function and use it
}
```

In the case where UDF functions must be available globally, there is also a pattern for creating and registering permanent UDFS in the default Spark SQL context.

Creating and Registering Permanent UDFs in the Spark SQL Catalog

It is common to share a common set of UDFs between Spark applications. To do so, you must take an additional step to create, package, and distribute your common UDFs. For instance, Listing 12-19 shows how to create a concreate class named CoffeeOrderLabeler. The class extends the Hive UDF class, thereby enabling the implementing class to be registered in the Hive Metastore. This capability also allows the UDF function to be loaded from a distributed file system like HDFS or S3 to provide the functionality on-demand.

Listing 12-19. Extending the Hive UDF to Create a Reusable UDF Class

```
import org.apache.hadoop.hive.ql.exec.UDF;
class CoffeeOrderLabeler extends UDF {
  def evaluate(customerId: String, price: Double): String = {
    if (customerId != null && customerId != "unknown") {
      if (price > 6.0) "vip" else "rush"
    } else "normal"
  }
}
```

With the class created, you can register the Hive UDF function by using the CREATE FUNCTION statement on the *Spark SQL DDL*. See Listing 12-20.

Listing 12-20. Registering a Permanent Global Function in the Spark SQL Catalog

```
%sql
CREATE FUNCTION native_coffee_order_label
AS 'CoffeeOrderLabeler'
USING JAR 's3a://.../udfs/orders/jars/udfs.jar'
```

Now the UDF is not only registered in the Hive Metastore but there is also a path to the JAR location within the shared file system. Now back to using UDFs in Spark SQL.

Using UDFs with Spark SQL

You know the ins and outs of creating, registering, and checking your *user-defined functions,* so all that is left is using the function in your Spark SQL statements. Listing 12-21 shows a modified version of the Spark SQL statement from Listing 12-13.

Listing 12-21. Using a Registered UDF with Spark SQL

```
%sql
SELECT *,
  coffee_order_label(customer_id, price) as order_label
  FROM (
    SELECT *,
    CASE
      WHEN customer_id != 'unknown' THEN TRUE
      ELSE FALSE
    END as is_registered
  FROM order_nonnull
 ) x
```

The SQL statement in Listing 12-21 shows the simple usage of the *coffee_order_label* UDF. UDFs enable you to extend the Spark SQL functionality and add new vocabulary (expressions and operators) that can be reused for a wide variety of purposes. The technique can be useful for reusing non-Spark native libraries, but there is also a catch.

Regarding User-Defined Functions

As you've seen, user-defined functions allow you to define new column-based operations expressed as native functions for use with Spark SQL and the Spark DSL. UDFs can fill in the blanks for any operations that is hard to express, or simply can't be expressed natively using default Spark SQL functions. There is a catch, and it relates to how Spark encodes UDFs. Simply put, Spark can't optimize the code block in your UDF since it treats the internal operation as a black box. Keep in mind the complexity of your UDF operations so you don't find yourself unwittingly slowing down your Spark applications. When performance really matters, it is always better to use native Spark SQL functions, since they have been optimized for the Catalyst Engine.

Exercise 12-1: Summary

This first exercise was a deep dive into the common SQL functions that are commonly used when preparing data for aggregation and analytical processing. From working with timestamps and date functions, to declaring which time zone to use when processing datasets, to filling null values and using conditional case statements, to finally defining your own functions (UDFs) to extend the common capabilities available within the Spark SQL ecosystem. We turn our attention now to using common aggregation and analytical functions in Spark.

Analytical Processing and Insights Engineering

Capturing and collecting data is generally only useful when the data can be used to answer questions, surface trends, analyze customer behavior, assist in decision automation, or provide an additional set of eyes and ears into complex processes and operations. Consider the operational data needed to run a coffee startup like the fictious CoffeeCo. To compete with more established companies in the busy bespoke coffee space, and to maintain a similar level of customer satisfaction, everything must be managed well. From the atmosphere of each individual store, to the coffee that is being served, to the logistics of contracts and vendor negotiations regarding fair trade and sourcing of coffee beans, essentially all processes that drive the business forward require multiple points of view across myriad data points that can be analyzed to ensure smooth operations. This is where analytics, insights, and machine learning get together to create intelligent systems. This story typically begins with aggregating data.

Data Aggregation

Data aggregation is a technique used to analyze large amounts of data to produce reports or statistics relating to observed patterns within a dataset. Data aggregation plays an important role in art of exploratory data analysis (EDA), data mining, insight generation, machine learning, and artificial intelligence. Like so many other things, data aggregation begins with a dataset and that dataset is the result of data being reliably captured and consistently processed by data engineers.

Exercise 12-2: Grouped Data, Aggregations, Analytics, and Insights

Start Zeppelin back up if it isn't running and create a new note. Give it a name such as 02_grouping_aggregations_analytics. For the rest of the exercise, you'll use this note as you learn to use grouping and aggregation functions to create reports and experiment with complex analytical functions.

Note   This exercise reuses data created using the random CoffeeOrder generator from Chapter 11.

Relational Grouped Datasets

Analytical processing is enabled using the groupBy, cube, rollup, and pivot DataFrame methods. These methods transform a DataFrame into a RelationalGroupedDataset and expose a wide variety of analytical processing functionality. Along with this new identity comes an important change in Spark's processes paradigm. Spark will shift the way it processes data from using row-based iterators, which are primarily concerned with performing row-wide transformations on individual columns, to instead processing the data points contained in specific columns across all rows in a subset of the dataset partitioned by a grouped expression or alternative partitioning logic.

Columnar Aggregations with Grouping

In this section, you create a simple report that aggregates the total number of items sold by store (store_id) as a function of the year and the month of each transaction. Begin by creating a temporary view named coffee_orders and referencing the date partitioned from the silver.coffee_orders table (from ch-11). Listing 12-22 can be dropped into Zeppelin and run to wire up this relationship.

Listing 12-22. Preparing to Read and Analyze the Partitioned coffee_orders
Table by Creating a View That Can Be Used in Your New Zeppelin Note

```
%spark
val dbName = "silver"
spark.catalog.setCurrentDatabase(dbName)
val dbUri = spark.catalog.getDatabase(dbName).locationUri
val df = spark.read.parquet(s"$dbUri/coffee_orders/")
df.createOrReplaceTempView("coffee_orders")
```

Using the coffee_orders view as a starting point, the Spark SQL query in Listing 12-23
will create a new temporary view named coffee_orders_datetime that adds the year,
month, and day derived columns.

Listing 12-23. Creating a New Temporary View (coffee_orders_datetime) that
Adds the Derived Columns year and month to the coffee_orders View

```
%sql
CREATE OR REPLACE TEMP VIEW `coffee_orders_datetime`
AS SELECT *, year(date) as year, month(date) as month, day(date) as day
FROM coffee_orders
```

Using the new year and the month columns, you can now compute statistics
pertaining to the monthly (or even yearly) sales totals. The query in Listing 12-24 uses
native Spark SQL and the SUM aggregation operator to compute the total number of items
sold. This query can be run without a grouping clause to compute the total number
of items sold globally. With the addition of a *grouping expression* (group by), you can
easily create multiple distinct subsets based on the grouping columns year, month, and
store_id.

Listing 12-24. Reporting the Total Items Sold by Store Partitioned by Year
and Month

```
%sql
SELECT store_id, year, month, sum(num_items) AS total_items_sold
FROM coffee_orders_datetime
GROUP BY year, month, store_id
ORDER BY store_id asc
```

The result of this aggregation will look like the following:

```
+--------+----+-----+----------------+
|store_id|year|month|total_items_sold|
+--------+----+-----+----------------+
|  store1|2021|   10|           61955|
|  store1|2021|    9|          197421|
|  store2|2021|    9|          199030|
|  store2|2021|   10|           63842|
|  store3|2021|    9|          202279|
|  store3|2021|   10|           62134|
|  store4|2021|    9|          200684|
|  store4|2021|   10|           62705|
+--------+----+-----+----------------+
```

Next, we look at using the Spark DSL to create the same aggregation.

Aggregating Using the Spark DSL

We just looked at using GROUP BY and the column aggregation SUM operator. In this next example, we use a DataFrame to create the grouped aggregation. Create a new paragraph in Zeppelin and copy the code from Listing 12-25. This example shows how a DataFrame transforms into a RelationalGroupedDataset and then back into a DataFrame again, post-aggregation.

Listing 12-25. Using the RelationalGroupedDataset to Compute a Columnar SUM Aggregation

```
%spark
import org.apache.spark.sql.RelationalGroupedDataset

val coffeeOrdersDf = spark.sql("""
SELECT store_id, year, month, num_items
FROM coffee_orders_datetime
""")

val groupedRelation: RelationalGroupedDataset = coffeeOrdersDf
  .groupBy($"year",$"month",$"store_id")
```

```
val aggDf = groupedRelation
  .agg(sum($"num_items") as "total_items_sold")
  .sort(asc("store_id"))
```

The code block in Listing 12-25 creates a SUM aggregation to calculate the total sales for each store by year, month exclusively. The interesting difference between Spark SQL and the Spark DSL is the way the latter exposes each operation to the engineer. Logically it makes sense that you would need to follow these steps:

1. Using the year, month, and store_id columns, separate all records into disjoint sets (group by/grouping set).

2. Within each distinct set (year-month-store_id) add all the values from the column num_items (aggregate).

3. Lastly, reorder the results by ascending store_id.

This simple report observes only the final totals within each distinct group. What about the worst or best performing days? What about percentile data? It is often necessary to provide additional information that can help summarize the shape of the data in an aggregation. We will look at computing summary statistics next and you will be introduced to the additional aggregation operator's min, avg, percentile_approx, and max fields to represent the statistics underlying the final numeric total.

Computing Summary Statistics for Insights

Analytics is the practice of capturing key performance metrics (measures) over time, such as with our *total sales* numbers from the prior section. *Insights* is the art of explaining why a metric changed over time. Comparing analytics to observe changes for better or worse requires additional data to be captured (or derived) along with key performance metrics over time.

Being able to observe the changes to the shape of a dataset over time requires more detailed statistical data to be computed per observation (window).

Using Describe to Compute Simple Summary Statistics

The set of aggregation functions comprised of count, mean (avg), stddev *(standard deviation)*, min, and max are commonly used to produce summary statistics across numeric (and even non-numeric) columns. Summary statistics help measure the shape of an underlying numeric distribution or can let you know the lexicographical min/max for a character-based column. Spark ships with a simple means of quickly computing columnar summary statistics using the describe function.

```
coffeeOrdersDf.describe("num_items").show
```

```
+-------+------------------+
|summary|         num_items|
+-------+------------------+
|  count|            100102|
|   mean|10.48980040358834|
| stddev|5.772833965567237|
|    min|                 1|
|    max|                20|
+-------+------------------+
```

Using Agg to Compute Complex Summary Statistics

When exploring a dataset, describe can be used to calculate some basic statistics quickly and inexpensively. However, if you want to capture more advanced statistics like the median value (50th percentile), the interquartile range (the distance between the 25th and 75th percentiles), or other counts of the distinct number of values in a column, then you can move these calculations back into the aggs block. Create a new paragraph and add the block of code shown in Listing 12-26.

Listing 12-26. Computing Summary Statistics Enables You to Understand the General Shape of a Dataset with Respect to a Given Measure (num_items)

```
%spark
val summaryDf = coffeeOrdersDf
  .groupBy($"year",$"month",$"store_id")
  .agg(
    count($"num_items") as "count",
    sum($"num_items") as "total_items_sold",
    stddev_pop($"num_items") as "sd", //sqrt(variance)
    min($"num_items") as "daily_min_items",
    ceil(avg($"num_items")) as "daily_avg_items",
    max($"num_items") as "max_daily_items",
    percentile_approx($"num_items",
      array(lit(0.25),lit(0.50),lit(0.75),lit(0.99)), lit(95)
    ) as "percentiles"
  )
  .withColumn("daily_p25_items", $"percentiles"(0))
  .withColumn("daily_median_items", $"percentiles"(1))
  .withColumn("daily_p75_items", $"percentiles"(2))
  .withColumn("daily_p99_items", $"percentiles"(3))
  .withColumn("iqr",
    floor($"daily_p75_items" - $"daily_p25_items")
  )
  .sort(asc("year"),asc("month"), asc("store_id"))
```

The statistical aggregation in the example in Listing 12-26 mixes in all the columns from the describe method along with the addition of the 25th, 50th, and 75th percentiles using the percentile_approx aggregation function.

PERCENTILES

Percentiles are used to observe a discrete number that represents a boundary where all values in the dataset (sorted from least to greatest) fall on or below. The 50th percentile, which is known as the median, represents the midpoint in a sorted set of numbers where 50% of all values fall on or below that number. Percentiles are used to capture the shape of the distribution of data and one of the common distributions is the normal distribution, shown in Figure 12-1.

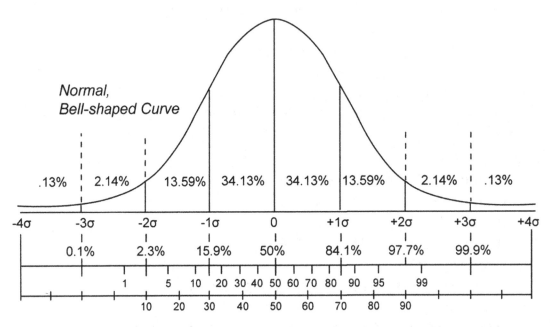

Figure 12-1. *The normal distribution is represented as what is called a "bell-shaped" curve*

Calculating a percentile requires you to sort all values from a given column, in ascending order, before you can use the size (total number of records) of the dataset to calculate the value for a given percentile. This can be costly in a very large dataset, and that is where the percentile_approx method comes into play.

```
percentile_approx(col, percentage, accuracy)
```

The `percentile_approx` function efficiently calculates the percentile(s) of a sorted sequence of continuous numeric data based on the provided percentage(s) parameter (which can be an array of doubles or an individual double). The accuracy parameter lets you define the acceptable error when generating the percentiles (the lower the number the lower the memory overhead but greater margin of error).

Percentiles can be used to create common bins of data to measure the frequency of a continuous numeric measure. Histograms, for example, are a graphical representation of the frequency (or tendency) for a data point to fall into a specific bin. Timeseries histogram analysis can be used to observe the change in the mass (percentage) of specific bins over time.

The aggregation from Listing 12-26 also introduces the numeric `ceil` (ceiling) function, which is used to round the average `num_items` up to the nearest whole number. If you wanted to round down to nearest whole number, you could use the `floor` expression.

Directly following the initial aggregation, we add columns using basic columnar expressions to extract the percentile data and to compute the interquartile range.

```
.withColumn("daily_p25_items", $"percentiles"(0))
.withColumn("daily_median_items", $"percentiles"(1))
.withColumn("daily_p75_items", $"percentiles"(2))
.withColumn("daily_p99_items", $"percentiles"(3))
.drop("percentiles")
.withColumn("iqr",
 floor($"daily_p75_items" - $"daily_p25_items")
)
```

The columns representing the 25th, 50th, 75th, and 95th percentiles use a technique that essentially extracts array values from the percentile's column in the DataFrame, and then drops the percentiles array from the final DataFrame since it is no longer of any use (columns are already extracted). Lastly, the interquartile range is calculated using the p75-p25, and this range helps to identify (statistically) the distance between the 25th and 75th percentiles, which is equal distance in a normal distribution. Being able to observe skews (to the left or right) in a dataset can be used to understand how measures change over time. The output of the aggregation in Listing 12-26 will look like the following:

```
-RECORD 0-------------------------------
 year                | 2021
 month               | 9
 store_id            | store1
 count               | 18951
 total_items_sold    | 197421
 sd                  | 5.777909019367008
 daily_min_items     | 1
 daily_avg_items     | 11
 max_daily_items     | 20
 daily_p25_items     | 5
 daily_median_items  | 10
 daily_p75_items     | 15
 daily_p99_items     | 20
 iqr                 | 10
```

Moving on. We will be looking at using rollup and pivot, which provide different ways of analyzing grouped data.

Using Rollups for Hierarchical Aggregations

Rollups offer another way of creating aggregations that are an extension of the groupBy known as grouping sets. The difference between a rollup and a groupBy is the way in which the data is analyzed. To generate a hierarchical aggregation, a rollup will be computed left to right across all columns represented within the rollup grouping set. Rollups enable you to create fine-grained reports that are represented as a tree-like structure. The topmost row represents the global aggregation of all data, and then for each column (left to right), an additional row is generated representing the subset of the data sliced by each distinct column value.

The code block in Listing 12-27 shows a concrete example of creating hierarchical aggregations.

Listing 12-27. Aggregating a Dataset Using Rollups Across the Grouping Set store_id, year, month

```
%spark
val rollupReport = ordersDf
  .rollup($"store_id", $"year", $"month")
  .agg(
    sum($"num_items") as "total",
    min($"num_items") as "min_daily",
    ceil(avg($"num_items")) as "avg_daily",
    max($"num_items") as "max_daily"
  )
  .sort(asc("year"),asc("month"),asc("store_id"))
  .limit(20)
```

The difference between the groupBy and the rollup can be seen when outputting the results of the rollupReport, as shown in Listing 12-28.

Listing 12-28. Rollups Compute a Hierarchial Aggregation Across a Grouping Set from Left to Right

```
+--------+----+-----+-------+---------+---------+---------+
|store_id|year|month|total  |min_daily|avg_daily|max_daily|
+--------+----+-----+-------+---------+---------+---------+
|null    |null|null |1050050|1        |11       |20       |
|store1  |null|null |259376 |1        |11       |20       |
|store2  |null|null |262872 |1        |11       |20       |
|store1  |2021|null |259376 |1        |11       |20       |
|store2  |2021|null |262872 |1        |11       |20       |
|store1  |2021|9    |197421 |1        |11       |20       |
|store2  |2021|9    |199030 |1        |11       |20       |
|store1  |2021|10   |61955  |1        |11       |20       |
|store2  |2021|10   |63842  |1        |11       |20       |
+--------+----+-----+-------+---------+---------+---------+
```

In the output in Listing 12-28, you will see that the global aggregation is followed by the first grouping column (store_id), then the next level is (store_id and year) followed by (store_id and year and month). This style of aggregation is good for exploratory reporting. Next, we look at pivots.

Using Pivots

Pivoting is a technique that enables complex aggregations across many rows, in a similar vein to the rollup and groupBy aggregations. The output of the pivot operation is a new dataset that has been *pivoted* around a specific column (such as an order_id or a month or year). This technique can be used for reporting, and for finding patterns across multiple rows of data to produce novel insights.

We'll explore creating a dataset that represents each individual items a customer ordered from the available menu items in a coffee shop. These menu items will then be associated to the column num_items on a CoffeeOrder. This new dataset will live in a table named coffee_order_items. This fine-grained order information would typically be available in a database or data lake for data mining and analysis purposes and could be used to figure out which category or typical items a customer commonly purchases.

In this case, we will be splitting orders items into two simple categories named beverage and food, and we will be using these identifiers to create categorical price-based rollup of the items purchased within an order, all using pivots.

Let's start by creating the menu_items table. Follow the code in Listing 12-29 to create the Item case class and generate a handful of menu items to complete the pivot tutorial.

Listing 12-29. Creating the menu_items view

```
%spark
import org.apache.spark.sql.types._
import org.apache.spark.sql._

case class Item(item_id: Int, name: String, category: String,
price: Double)

val menuItems = Seq(
    Item(0, "latte", "beverage", 3.46),
    Item(1, "americano", "beverage", 2.59),
    Item(2, "cappuccino", "beverage", 3.89),
    Item(3, "pour over", "beverage", 5.26),
```

```
    Item(4, "tea", "beverage", 2.99),
    Item(5, "water", "beverage", 2.99),
    Item(6, "cookie", "food", 2.99),
    Item(7, "salad", "food", 4.99),
    Item(8, "sandwich", "food", 6.99),
    Item(9, "popcorn", "food", 3.49)
)

val df = spark.createDataFrame(menuItems)
df.createOrReplaceTempView("menu_items")
```

The menu items can be saved for future reuse by using saveAsTable (see Listing 12-30).

Listing 12-30. Saving the Menu Items for Later

```
%spark
df
  .repartition(1)
  .write
  .mode("overwrite")
  .saveAsTable("silver.menu_items")
```

Now that you have some menu items to work with, let's use these items and associate them with a series of items (array of integers as menu_item_ids) for a specific CoffeeOrder. We will create the OrderItems structure to store this dataset.

Creating the Order Items

We need to create a new case class called OrderItems that stores an orderId, and a sequence of items purchased in an order. See Listing 12-31.

Listing 12-31. Creating Two Example OrderItem Rows and Saving Them Into a New View Named coffee_order_items

```
%spark
case class OrderItems(orderId: String, items: Seq[Int])
spark.createDataFrame(Seq(
  OrderItems("orderId1314", Seq(3, 7, 5, 0, 1, 1, 1)),
  OrderItems("orderId1696", Seq(3, 7, 7, 5, 0, 1, 1, 1, 2, 2, 8))
)).createOrReplaceTempView("coffee_order_items")
```

In Listing 12-26 from earlier, we looked at using array positions to extract an array field from our percentiles array (`$"percentiles"(0)`). This technique can be streamlined using the `explode` operation for variable length arrays. The reason I bring this up is because we will soon be using `explode` as an exercise to flatten the order items from a specific order and join each item with its menu item.

Using Array Explode to Create Many Rows from One

Working with array data isn't always cut and dried. It is easy to introduce bugs in the code when working directly with array index pointers, especially without guards in place. The `explode` functionality is a useful technique that extracts all data from a nested array, essentially flattening the nested structure by generating a new row per item. Let's see how this can be used to rotate an order across all items.

For reference, a single `OrderItem` row looks like this.

```
Row("orderId1314", array(3, 7, 5, 0, 1, 1, 1))
```

When we apply the `explode` function, as shown next.

```
%spark
spark
  .sql("select * from coffee_order_items")
  .selectExpr("orderId", "explode(items)")
  .show
```

We generate a new DataFrame that has a single menu item per row.

```
+-----------+---+
|    orderId|col|
+-----------+---+
|orderId1314|  3|
|orderId1314|  7|
|orderId1314|  5|
|orderId1314|  0|
|orderId1314|  1|
|orderId1314|  1|
|orderId1314|  1|
+-----------+---+
```

Using explode allows you to unpack data to create many rows from a single row. This technique is equivalent to using flatMap in Scala to generate a sequence of items from one. So how can we use explode and join to calculate the total cost of an order?

Using Array Explode and Join to Calculate the Order Total

Now that you have the pieces of the puzzle, you can mix the transformation (explode) with a join (order_items with menu_items) to create a new view that combines the order items with the price of each menu item in the order. Follow the code block in Listing 12-32 to generate the order_items_with_price view.

Listing 12-32. Using an Inner Join and the Explode Operator to Create a New View of Each Order That Combines the Price of Each Item in an Order

```
%spark
val coffeeOrderItemsDf = spark.sql("""
SELECT orderId, EXPLODE(items) as menu_item_id
FROM coffee_order_items
""")

val menuItemsDf = spark.table("silver.menu_items")
```

```
val joinedDf = coffeeOrderItemsDf
  .join(
    menuItemsDf,
    coffeeOrderItemsDf("menu_item_id") === menuItemsDf("item_id"),
    "inner"
  )
  .sort(asc("orderId"))
  .select($"orderId".as("order_id"), $"name", $"category", $"price")

joinedDf.createOrReplaceTempView("order_items_with_price")
```

Now that you have joined the order items along with the category and price of each item within an order, you will have a new view that looks like the following:

```
+-----------+----------+--------+-----+
|   order_id|      name|category|price|
+-----------+----------+--------+-----+
|orderId1314| pour over|beverage| 5.26|
|orderId1314|     water|beverage| 2.99|
|orderId1314| americano|beverage| 2.59|
|orderId1314|     salad|    food| 4.99|
+-----------+----------+--------+-----+
```

You can use a pivot operation to calculate the cost of items per sub-category of an order, which can be a useful tactic for observing trends in the purchasing behavior of customers of CoffeeCo.

Using Pivots to Calculate Price Aggregates Across Menu Item Categories on a Per Order Basis

Using the data from the order_items_with_price view (Listing 12-32), you can now run the SQL statement in Listing 12-33 to group by (order_id) and compute categorical sub-totals across the beverage and food menu categories.

Listing 12-33. Using Pivot to Compute Sub-Category Totals Across Orders

```sql
%sql
select * from order_items_with_price
  pivot (
    cast(sum(price) as decimal(4,1)) as category_total,
    count(name) as items
    for category in (
      'beverage', 'food'
    )
  )
```

The resulting aggregation pivots the data around the order_id, and then computes sub-aggregations on a named category. The results look like the following:

```
-RECORD 0------------------------------
 order_id                 | orderId1314
 beverage_category_total  | 19.5
 beverage_items           | 6
 food_category_total      | 5.0
 food_items               | 1
-RECORD 1------------------------------
 order_id                 | orderId1696
 beverage_category_total  | 27.3
 beverage_items           | 8
 food_category_total      | 17.0
 food_items               | 3
```

Pivoting is a powerful analytical operation that takes group by to the next level and can be used to create line-item reports across complex subsets of your data. The last and final operation we will look at is Sparks support of analytical window functions.

Analytical Window Functions

Spark offers a wide variety of analytical support and operations out of the box. This chapter has already covered the available grouped aggregations (groupBy, rollup) and explained how to take advantage of the powerful agg operator to support parallel aggregations. You learned to generate basic summations, to extract min and max

445

values, and compute percentiles. The last section taught you to use the powerful `pivot` operation to transpose (rotate) many rows of data to create complex sub-aggregations, but the one thing all these prior aggregation functions have in common is that they create an entirely new dataset. Windowing works a little different.

Windowing functions enable different analysis patterns than aggregation alone. The key difference is that grouped aggregations (`groupBy`, `rollup`, and so on) reduce many values down to a single value (per grouping set) and a window function adds an analytical value or aggregated result as a column for each row of the input dataset, preserving all rows.

For example, say we would like to create an ordered index (running count) of the orders by hour and across stores. To compose the query, we will be using `partitionBy` instead of `groupBy` since we are not computing a grouped aggregation. `PartitionBy` creates a window specification that can be used to optimally split and order a subset of a dataset for analytical processing. Let's look at using a Window function. The code block in Listing 12-34 generates a window that we can use to add new columns to our `coffeeOrdersDf`.

Listing 12-34. Creating a Window Specification to Compute the Hourly Transaction Number and the Cumulative Running Total Number of Items Sold Within the Window

```
%spark
import org.apache.spark.sql.expressions.Window

val hourlyOrders = coffeeOrdersDf
  .withColumn("timestamp",
      to_timestamp($"timestamp"/1000)
  )
  .withColumn("hour", hour($"timestamp"))

val window = Window
  .partitionBy("store_id","date","hour")
  .orderBy(asc("timestamp"))
```

```
hourlyOrders
  .withColumn("transaction_number", row_number.over(window))
  .withColumn("running_total",sum($"num_items").over(window))
  .limit(6)
  .show
```

The window represents a view of the dataset that is partitioned by store_id, date, and hour, and the resulting subset of the partitioned dataset is then sorted by timestamp in ascending order. This will let us calculate the total number of items sold, starting at the first transaction of the hour until the last. See Listing 12-35. For each transaction, we see the total number of items (num_item) increase for each row afterward by the value of the prior num_items.

Listing 12-35. The Output After Applying the row_number and sum window Aggregation Functions

```
+--------+----------+----+-----------+---------+------+
|store_id|      date|hour|transaction|num_items|totals|
+--------+----------+----+-----------+---------+------+
|  store1|2021-09-06|   3|          1|       11|    11|
|  store1|2021-09-06|   3|          2|       19|    30|
|  store1|2021-09-06|   3|          3|        3|    33|
|  store1|2021-09-06|   3|          4|       20|    53|
|  store1|2021-09-06|   3|          5|        5|    58|
|  store1|2021-09-06|   3|          6|       17|    75|
+--------+----------+----+-----------+---------+------+
```

Being able to add the transaction number (row_number) and a running total (totals) to the coffee orders is nice for tallying up sales. Other interesting things can also be accomplished, depending on what you want to analyze.

Calculating the Cumulative Average Items Purchased Difference Between Transactions

For example, say you wanted to compute the difference in the number of goods sold between transactions and simultaneously compute the cumulative rolling average difference in transactions per hour?

This can be done using the lag over window function. Lagging means to fall behind, and a lag operation uses the current_row and the provided column and the value of rows behind to look back to fetch a value from an arbitrary row in the dataset. In Listing 12-36, we use lag($"num_items",1) to look back one row so we can compute the absolute difference between items purchased between transactions. Lastly, we use the avg($"diff_last_sale") over the window in order to create our cumulative average.

Listing 12-36. Using lag(column,step) Over Window to Compute the Difference Between the Last Number of Items Sold and the Current Number of Items Sold Per Transaction

```
%spark
hourlyOrders
  .withColumn("transaction", row_number.over(window))
  .withColumn("totals", sum($"num_items").over(window))
  .withColumn("diff_last_sale",
    abs($"num_items" - lag($"num_items", 1).over(window))
  )
  .withColumn("avg_diff",
    bround(avg($"diff_last_sale").over(window),2)
  )
```

The result of computing the window functions (Listing 12-36) is the following output.

```
+-----------+------+--------------+--------+
|transaction|totals|diff_last_sale|avg_diff|
+-----------+------+--------------+--------+
|          1|    11|          null|    null|
|          2|    30|             8|     8.0|
|          3|    33|            16|    12.0|
|          4|    53|            17|   13.67|
|          5|    58|            15|    14.0|
|          6|    75|            12|    13.6|
|          7|    87|             5|   12.17|
|          8|    99|             0|   10.43|
+-----------+------+--------------+--------+
```

While the output has been truncated to the first eight transactions, you get the picture. You can achieve a lot by moving from `groupBy` (grouping sets) and into `partitionBy` (window) operations. By mixing and matching the analytical functionality of the window functions with the powerful aggregation capabilities, you can create sophisticated reports and generate truly novel insights.

If the last example didn't feel super relevant, consider that you are tasked with optimizing the ordering process for the CoffeeCo coffee shops and the data you are working with is related to finding (isolating) potentially problematic inefficiencies in the operations of each shop. Using the data captured and measuring the time an order is created to the time an order is completed, you can use the same lag over techniques, mixed with pivots, joins, and additional rollups to compute reports on the logistics and timings of menu items broken down by category and shop. It might be that the longest lines are due to items that are hard to make (fancy pour overs or food items) and that removing a single item from the menu could speed up all other transactions.

Exercise 12-2: Summary

This second chapter exercise exposed you to methods for using grouped aggregations and explained how to transform your data to create sophisticated reports that assist you in your data detective work.

Summary

Understanding how to capture, validate, clean, and store data reliably is one part of the data engineer's story. Being able to take advantage of the tools of the trade to efficiently perform your work is something that comes with time and practice. This chapter was a dive into using analytical data and learning to process that data to understand some of the work that goes into more advanced data transformations for analytics. Introducing you to how to clean data, switch time zones, replace nulls, create, and use user-defined functions led the charge of the first half of the chapter. We finished off exploring the aggregation and analytical processing methods and operations available within Apache Spark.

The next chapter is a little shorter and focuses on two simple structured streaming applications. The first is a coffee order metrics application that produces real-time reports that can be used for live dashboards and operations. The second application is more advanced, using the `flatMapGroupsWithState` operator. This chapter is the last chapter where we write code before looking at deployment patterns and monitoring.

PART III

Advanced Techniques

## CHAPTER 13

Advanced Analytics with Spark Stateful Structured Streaming

In the last chapter, you learned to use Apache Spark's powerful aggregation and analytics functions, from the `agg` operator that enabled powerful columnar aggregation capabilities directly off a grouped dataset, to the analytical window functions that allowed you to partition and analyze datasets using these unique windowing capabilities. This gave you the ability to look back (lag) or forward (lead) across many rows from your current position in an active iteration. You learned to use lag over to create row-by-row average deltas and similar techniques to create running cumulative totals.

This chapter focuses on creating stateful aggregations using Structured Streaming, continuing along the same grain as the last chapter. It teaches you how to move analytical processing and insight generation to the stream. It also expands upon the techniques and foundations established over the previous four chapters for creating reliable stateful applications, leveraging checkpoints, write-ahead logging, and process triggering. It also introduces the Spark `StateStore` for stateful streaming aggregations. You'll learn to build advanced streaming analytical applications.

Exercises and Materials

The chapter exercises and content take you through the following three sections:

- Stateful aggregations with Structured Streaming

- Arbitrary stateful operations using `FlatMapGroupsWithState`

- Testing stateful streaming applications using `MemoryStream`

© Scott Haines 2022
S. Haines, *Modern Data Engineering with Apache Spark*, https://doi.org/10.1007/978-1-4842-7452-1_13

Exercise Materials   The exercise materials for Chapter 13 are located at `https://github.com/newfront/spark-moderndataengineering/tree/main/ch-13`.

Let's begin with a continuation of the lessons learned in the last chapter, as you layer stateful aggregations on top of your Spark Structured Streaming knowledge.

Stateful Aggregations with Structured Streaming

Think for a moment about what you currently know about streaming data sources. You are likely thinking about the fact that streaming data sources are essentially an infinite source of data. Like rivers, streaming data sources continue to flow indefinitely into your application through the `DataStreamReader` until human or process-based intervention tamps down the flow or stops the source of data completely.

Now that you are thinking about a never-ending source of data, think for a moment about what you learned in the last chapter about how grouped aggregations (`groupBy`) work. You might be thinking that grouped operations act on a single DataFrame/dataset that defines the universal dataset (all known data at a particular moment in time) that is grouped (partitioned conditionally) and aggregated to produce analytical rows of columnar aggregations across subsets of a dataset. This is indeed correct. But what else is missing if, for example, you wanted to marry unbounded datasets across conditional groups over time? How would we group our data by time? What about windowing?

Creating Windowed Aggregations Over Time

Say you have an unbounded streaming data source like a Redis Stream or Kafka topic, and you would like to reliably ingest and analyze the data flowing through that data source continuously using Spark Structured Streaming. You know you can't store all data in memory forever, and you would also like to have some processing and accuracy guarantees. So what do you do?

For starters, you can use a technique called *windowing* alongside Spark Structured Streaming. Windowing allows you to automatically slice a static or streaming dataset into bins based on time-based intervals (such as every 15 minutes), or simply put, data is grouped into time-based bins automatically defined by a start and end time. Windowing, or time-based grouping, differs from the analytical window functions we discussed in the last chapter. The big difference is that analytical window functions preserve all rows in the dataset and add a new column based on the aggregation function, while windowed aggregations incrementally merge specific columns in aggregate based on the columns in the grouping set for the definition of the time-based window. Figure 13-1 shows the bounded data (windows 1-3) regarding the total number of coffee orders placed over time. Note that the windows are non-overlapping, meaning that the orders made within each 15-minute window are exclusive to each window. This way, each order is accounted for exactly once.

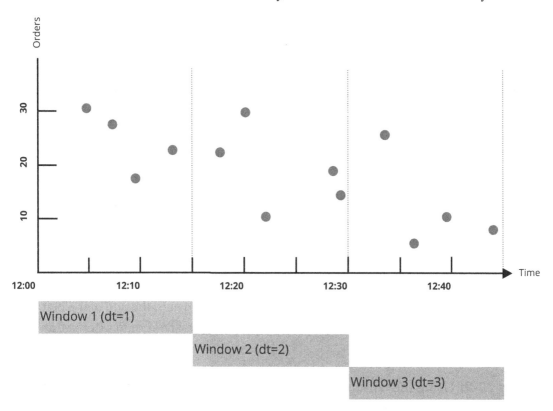

Figure 13-1. *Using windows to express dynamic groupBy functions for coffee orders over time*

Let's take this idea further with a concrete example.

Window Functions vs. Windowing

In the last chapter, you were introduced to using analytical window functions to process columnar data. You learned to use row numbers creatively, create cumulative running averages, and even compute aggregations based on the results of your analytical expressions. Listing 13-1 comes from Chapter 12 and is provided for reference. To recap, window functions transform input datasets by adding a new column, to all rows, representing the results of a specific analytical function. For example, you used these functions to compute cumulative running totals.

Listing 13-1. Columnar Window Functions as seen in Chapter 12, Listing 12-32

```
df.withColumn("running_total"),
  sum($"num_items").over(Window
    .partitionBy("store_id","date","hour")
    .orderBy(asc("timestamp"))))
```

Windowing is a columnar grouping technique that shares some commonality with analytical window functions, but under the hood they act differently. While both share the same name (window) and both work to partition datasets across conditional boundaries, there is a core differentiator. Mainly, a window expression is passed as an argument to the groupBy operator to enhance the capabilities of the grouping operator. Now data will be automatically partitioned into time-based buckets based on a common timestamp column (TimestampType) of the dataset.

Let's look at windowing in action. Listing 13-2 creates a *tumbling* window aggregation. Each window has an inclusive start time and an exclusive end time. This means that data will be split on specific time-based boundaries, kind of like boxes moving on a conveyer belt being filled as they move toward the end of the belt. The example in Listing 13-2 can be run directly in the spark-shell or in a Zeppelin notebook.

Listing 13-2. Using the window Function as a Grouping Column to Autmatically Create Windowed Aggregations

```
import spark.implicits._
import org.apache.spark.sql.functions.{col, window}

spark
  .sql("SELECT timestamp('2021-09-25 10:00:00' as ts, 200 as total,
'storeA' as storeId")
  .groupBy($"storeId", window($"ts","30 minutes"))
  .sum("total")
```

```
.select(
  $"storeId", $"window.start".as("start"),
  $"window.end".as("end"), $"sum(total)".as("total"))
.show(truncate=false)
```

The aggregation in Listing 13-2 is grouped by `storeId` and automatically aggregated into 30-minute windows curtesy of the timestamp column (`ts`) of the event data and the *window expression*.

The output of the windowed aggregation for reference is shown next. For each exclusive 30-minute window, order totals are accumulated across the total number of unique `storeId`s.

```
+-------+-------------------+-------------------+-----+
|storeId|start              |end                |total|
+-------+-------------------+-------------------+-----+
|storeA |2021-09-25 10:00:00|2021-09-25 10:30:00|200  |
|storeB |2021-09-25 10:00:00|2021-09-25 10:30:00|120  |
|storeA |2021-09-25 10:30:00|2021-09-25 11:00:00|350  |
|storeB |2021-09-25 10:30:00|2021-09-25 11:00:00|220  |
+-------+-------------------+-------------------+-----+
```

The window function used in Listing 13-2 is a columnar expression that transforms a timestamp column and converts it into a derived column for use in the `groupBy` operation. This allows us to automatically land data into specific time-based buckets for aggregation.

USING THE WINDOW FUNCTION

The `window` function enhances the way the `RelationalGroupedDataset` operator works to support automatic bucketing for time-based aggregations using *tumbling*, *sliding*, and *delayed* window capabilities.

The window Function from the org.apache.spark.sql.functions Package

```
def window(
  timeColumn: Column,
  windowDuration: String,
  slideDuration: String,
```

```
startTime: String): Column {
  withExpr {
    TimeWindow(timeColumn.expr, windowDuration, slideDuration, startTime)
  }
}.as("window")
```

The function takes four parameters (`timeColumn`, `windowDuration`, `slideDuration`, and `startTime`) and provides you with new ways to slide and dice your data across time. You looked at tumbling windows before, now let's look at sliding and delayed windows.

Sliding Windows

The `slideDuration` parameter of the `window` function enables the creation of sliding windows. Sliding windows are simply overlapping time-based windows evaluated at specific increments (such as a 30-minute window evaluated every 15 minutes). Unlike with tumbling windows, each row can land inclusively in the overlapping windows.

For example, by changing the `groupBy` window expression in Listing 13-2 to include a 15-minute slide duration, you would create two time-based buckets instead of the one.

```
groupBy($"storeId", window($"ts","30 minutes", "15 minutes"))
```

Now instead of a single row aggregate, you will have two row aggregates from a single input row.

```
+-------+-------------------+-------------------+-----+
|storeId|start              |end                |total|
+-------+-------------------+-------------------+-----+
|storeA |2021-09-25 09:45:00|2021-09-25 10:15:00|200  |
|storeA |2021-09-25 10:00:00|2021-09-25 10:30:00|200  |
+-------+-------------------+-------------------+-----+
```

Since the input row timestamp was 10:00:00, the data lands inside of two sliding 30 minutes. The first window starts at 09:45 and ends at 10:15. The second window starts at 10:00 and ends at 10:30. This is also known as a *staggered* window.

Delayed Windows

The window `startTime` parameter allows you to delay the computation for your derived windows. Change the `groupBy` from Listing 13-2 to use the following.

```
.groupBy($"storeId",
  window($"ts","30 minutes", "15 minutes", "5 minutes"))
```

This changes the starting offset for each aggregation by five minutes, and then creates a sliding window every 15 minutes.

```
+-------+-------------------+-------------------+-----+
|storeId|start              |end                |total|
+-------+-------------------+-------------------+-----+
|storeA |2021-09-25 09:35:00|2021-09-25 10:05:00|200  |
|storeA |2021-09-25 09:50:00|2021-09-25 10:20:00|200  |
+-------+-------------------+-------------------+-----+
```

Now that you understand the mechanics of *windowing*, let's look at a more complete example.

Windowing and Controlling Output

Given you just learned how the windowing process works, the next logical step is to look at applying this technique to streaming datasets. Given what we know about the unbounded nature of streaming data, it will also be wise to look at the controlling processes at work that inform Apache Spark when an aggregation operation is considered complete.

Say for example that you have to create a streaming application that ingests CoffeeOrder data to compute tumbling store performance metrics every 30 minutes forever. Listing 13-3 sets up the basic blocks of the StreamingQuery. You create a DataStreamReader (Redis) that aggregates (groupBy + agg) a stores (storeId) performance as a function of the count of total orders, the sum of the number of items sold, the total revenue as a function of the sum of the cost of an order, and the number of average items sold within the window.

Listing 13-3. Creating a Stateful Streaming Aggregation

```
spark
  .readStream
  .format("redis")
  ...
  .groupBy($"storeId", window($"timestamp", "30 minutes"))
  .agg(
    count($"orderId") as "totalOrders",
    sum($"numItems") as "totalItems",
```

```
  sum($"price") as "totalRevenue",
  avg($"numItems") as "averageNumItems"
)
.writeStream
...
.start()
```

Consider this example as a blueprint for a stateful aggregations app. Data flows in from Redis (Chapter 10) and you then transform the input stream data into a grouped aggregation (Chapter 12). The grouping set is now also enhanced with the windowing technique you learned in this chapter.

Essentially, almost everything you need to create this streaming application are in place, except for *two critically important missing pieces* that control how Spark will output these windowed aggregates. First, an output mode on the DataStreamWriter (writeStream) is missing, which defines how the aggregated window data will be emitted downstream, and secondly, there is also no way to inform Spark to stop processing data in aggregate (otherwise known as watermarking). Without an output mode and a watermark, Spark can't define the end of the conveyer belt, and your tumbling windows will keep riding in memory until you run out of memory.

Let's look at these two important processes now.

Streaming Output Modes

The output behavior of a Structured Streaming application is controlled by the output mode setting on the streaming sink (DataStreamWriter). There are three available values: complete, update, and append.

Complete

All rows of the streaming DataFrame/dataset will be written to the sink every time there is an update. This mode is commonly used when running tests locally and using the ConsoleSink (introduced in Chapter 10) or the MemorySink, which you'll learn more about at the end of this chapter.

Update

Only rows that were updated will be written to the sink. If the query does not contain any aggregations, the update mode acts like append. When writing a new streaming aggregation, it is helpful to be able to see the end results of each batch as it is merged (updated) with the prior state. Or for systems where you want to send each significant update (to a dataset) downstream, you can use this mode of operation to send multiple versions of each record on change.

Append

Only new rows will be written to the sink. This mode works in conjunction with the Structured Streaming state-based checkpoints, write-ahead logging and stage-level commits. Specific to stateful aggregations, append mode controls the exactly-once semantics regarding when streaming aggregation is considered complete and can be effectively sent downstream (when the data tumbles off the state-based conveyor belt and into the downstream sink).

Append Output Mode

Changing the example code from Listing 13-3, we can augment the writeStream to use append-only mode.

```
.writeStream
.format("parquet")
.options(...)
.outputMode(OutputMode.Append)
.start()
```

Now the streaming aggregation will only emit data when each aggregation is complete. But how does Spark understand when an aggregation is complete? If you were to run the code with output mode Append, you would see the following exception raised.

```
org.apache.spark.sql.AnalysisException: Append output mode not supported
when there are streaming aggregations on streaming DataFrames/DataSets
without watermark;
```

What exactly is a watermark?

CLOSE OF BOOKS AND WATERMARKING

There is a concept in the data industry called "close of books." Given streaming data has a limited window of time when it is still considered *fresh* and can be acted upon, there is a threshold (watermark) in which applications processing the event stream (Spark Structured Streaming or others) can ingest and process new data before calling it quits and dropping (or ignoring) any late arriving data. The watermarking process is also referred to as the "closing the books," since it is the point in time when the system emits its final processed state (for windowed aggregations) and sends the aggregations to be further refined downstream, written to reliable storage, distributed tables, or elsewhere.

The watermarking process is shown in Figure 13-2. Conceptually, if you consider that an unbounded stream of data is being ingested and processed, then for each micro-batch, the Spark engine will compute the min, max, average, and watermark values based on the timestamp column. The summary statistics of the entire state of the aggregation is considered when generating the stats, and each new micro-batch considers the watermark threshold before deciding to drop data that is considered latent.

Streaming Data (In-Order) ->

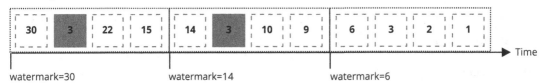

Figure 13-2. *The watermarking process can be thought of as a way to create a statistical time-based boundary. New event data being ingested that is smaller than the current watermark (latent) will be dropped based on the window interval plus the watermark interval*

For example, the processes in place used to monitor distributed systems in near real-time can't wait around for latent data and must act on the *available view* of the world as defined by the available data. Given streaming systems process data from many distributed upstream processes, data can get clogged just like people get held up in the queue to order or receive the coffee, or in their daily commutes to and from work (or wherever they are going) due to random traffic patterns.

Metric data is typically aggregated in the low seconds (on a source system/physical server / etc.) and emitted for processing in 5-10 second intervals. Each individual aggregation is then merged and joined with other data to compute a statistical view of a distributed system. Given that alerting systems use rules to establish when and how to alert a team or send events to another autonomous system, watermarking enables these systems to ignore data that no longer is of any interest and that could potentially trigger a false alarm (say if data shows up 30 minutes late due to bad network or some other problem).

Watermarks for Streaming Data

You just learned about the close of books and watermarking. We live in a world of rules and regulations, and the same can be said of Spark. To enable our streaming aggregations to properly emit new *tumbled window* aggregates, we need to mark our own close of books with a watermark (Listing 13-4) that is tied directly to a physical timestamp in our dataset.

Spark will ignore latent data (based on a pre-defined timestamp column) by automatically dropping it on the floor. In many use cases, dropping latent data is the best available option, rather than creating a new bucket to aggregate what most likely will only be a small number of latent records, but what is worse is that latent data can break any *exactly-once semantics* of Spark's append output mode.

Let's look at adding the watermark to the example from Listing 13-3. We add the watermark to the input stream, and Listing 13-4 shows how to add a 30-minute watermark to the Redis input stream.

Listing 13-4. Applying a Watermark to the Stateful Aggregation Example

```
spark
  .readStream
  .format("redis")
  ...
  .withWatermark("timestamp", "30 minutes")
  .groupBy($"storeId", window($"timestamp", "30 minutes"))
  ...
```

The relationship between the watermark (30 minutes) and the tumbling window (30 minutes) is used by Spark to control the append output mode process with exactly-once semantics. Essentially, close of books will occur one hour (30m window + 30m watermark) after the first event arrives, establishing the aggregate bucket.

Next, let's take the theoretical concepts we just covered and apply watermarking and windowing alongside a complete end-to-end stateful streaming aggregation application in Exercise 13-1.

Chapter Exercises Overview

This chapter's exercises reuse the CoffeeOrder event from earlier chapters. You will learn to read a stream of orders and generate store revenue aggregates, which are essentially store performance numbers across variable-length windows (five minutes in the examples).

Reading an unbounded stream of CoffeeOrders into your application, you will learn to transform this input dataset into a format that can be aggregated over time. This will result in a new dataset that represents tumbling window aggregates.

Input Data Format (CoffeeOrder)

```
+-------------+-------+-------+----------+--------+-----+
|    timestamp|orderId|storeId|customerId|numItems|price|
+-------------+-------+-------+----------+--------+-----+
|1632549600000| order1|  storeA|     custA|       2| 9.99|
|1632549660000| order2|  storeA|     custB|       1| 4.99|
|1632549720000| order3|  storeA|     custC|       3|14.99|
|1632549730000| order4|  storeB|     custD|       1| 3.99|
|1632549740000| order5|  storeA|     custE|       2| 6.99|
+-------------+-------+-------+----------+--------+-----+
```

Output Windowed Store Revenue Aggregates

```
+--------+--------------+------+---+-------+-------+-------+
|store_id|        window|orders|its|revenue|p95_its|avg_its|
+--------+--------------+------+---+-------+-------+-------+
|  storeA|{2021-... 06:00|    4|  8|  36.96|      3|    2.0|
|  storeB|{2021-... 06:00|    4|  7|  29.96|      3|   1.75|
|  storeA|{2021-... 06:05|    2|  4|   9.98|      2|    2.0|
|  storeB|{2021-... 06:05|    1|  1|   3.99|      1|    1.0|
+--------+--------------+------+---+-------+-------+-------+
```

Let's get started.

Exercise 13-1: Store Revenue Aggregations

Open the application located at /ch-13/applications/stateful-stream-aggs/ in your favorite IDE. The first application covered is an example of a stateful Structured Streaming application that automatically aggregates CoffeeOrder events across time-based windows to produce buckets of store performance metrics.

Start by opening SparkStatefulAggregationsApp and looking at the object signature.

object SparkStatefulAggregationsApp extends **SparkStructuredStreamingApplication**
[*DataFrame, Row*]

You'll notice a new application trait is being introduced in this chapter, called SparkStructuredStreamingApplication. The trait provides additional methods to use while building your Structured Streaming applications. Let's look at the new trait, and then we can return to the application.

Structured Streaming Application Trait

This new trait is shown next and extends the SparkApplication trait that was introduced in Chapter 7. There are a few new and noteworthy convenience methods added to speed up how you build your streaming applications, shown in Listing 13-5. This interface provides a reusable framework for simplifying how to write Structured Streaming applications using the core Spark classes DataStreamReader, DataStreamWriter, and StreamingQuery.

Listing 13-5. The SparkStructuredStreamingApplication Trait Interface Methods

```
trait SparkStructuredStreamingApplication[T, U]
extends SparkApplication {

  def streamReader: DataStreamReader
  lazy val inputStream: DataStreamReader = streamReader

  def outputStream(writer: DataStreamWriter[U])(implicit sparkSession:
  SparkSession): DataStreamWriter[U]

  def runApp(): StreamingQuery
  def awaitTermination(query: StreamingQuery): Unit
  def awaitAnyTermination(): Unit
}
```

As a refresher, Structured Streaming applications are composed by assembling a query plan that begins at a streaming data source and ends at a streaming data sink. With each stateful processing cycle, the application continues to read new batches of streaming data, transforming and processing each batch and sending results downstream or into a reliable state store (for aggregations).

The new trait wraps the stream reader and writer (DataStreamReader, DataStreamWriter) into reusable functions, thus providing additional mechanisms to use a config-driven approach that lends itself nicely to testing end-to-end applications (which you will see later in this chapter).

Let's look a little closer at the streamReader and outputStream functions, as they are evolutionary step-changes to the inputStream and ouputStream concepts used in earlier chapters.

Stream Reader

The streamReader method uses the prefix spark.app.source.* within the Spark config to be used when dynamically building a new DataStreamReader instance.

```
def streamReader: DataStreamReader = sparkSession
  .readStream
  .format(sparkConf.get(sourceFormat, sourceFormatDefault))
  .options(sparkConf.getAllWithPrefix(
    sparkConf.get(sourceStreamOptions)).toMap)
```

This technique enables your applications to easily switch an upstream data source used by the DataStreamReader. For example, recovery or replay jobs used to regenerate a final dataset may use a different upstream data source than what is used for the live application, such as selecting S3 instead of Kafka for the recovery. Additionally, this strategy can be used as a forcing function to prevent the hard-coding of values in the application source code. But be careful about reducing the flexibility of an application too much. This is why the trait provides methods that can be used and further decorated by the extension class.

Output Stream Decorator

The ouputStream method is shown in Listing 13-6 and is used to decorate a DataStreamWriter with common behaviors like triggering and partitioning. It streamlines the way Structured Streaming applications configure outputStreams with respect to the outputMode, columnar partitioning, and triggering.

Listing 13-6. The outputStream DataStreamWriter Decorator

```
def outputStream(writer: DataStreamWriter[U])(implicit sparkSession:
SparkSession): DataStreamWriter[U] = {
  import appConfig._
  val conf = sparkSession.conf
  ...
  Seq(OutputModeOption, PartitionByOption, TriggerOption)
  .foldLeft[DataStreamWriter[U]](writer)((w, config) => {
    ...
  })
  .format(SinkFormat)
  .queryName(QueryName)
}
```

In a nutshell, the outputStream method in Listing 13-6 uses the Spark runtime configuration to decorate an input DataStreamWriter. By opting to use the DataStreamWriter interface, the outputStream method doesn't need to concern itself with data lineage since that is a concern of the application. All the method is required to do is add a format, queryName, outputMode, and the optional triggers and output partitioning on top of the provided writer.

Conditional Object Decoration

Functional programming languages like Scala are immutable first, designed to reduce accidental side effects common with pass-by-reference languages. For this reason, we use the `foldLeft` operation to conditionally decorate the `DataStreamWriter` interface. The `foldLeft` operation traverses the list of config items to reliably generate the writer config.

Looking at the inline function from the `outputStream` method shown in Listing 13-7, you'll see that each config value from the initial sequence (`outputMode`, `partitionBy`, and trigger) is used to conditionally decorate the input writer object (`w`) before being passed along to the next value and the next checks.

Listing 13-7. The outputStream foldLeft Inline Function

```
(w, config) => { config match {
  case OutputModeOption => w.outputMode(sinkOutputMode)
  case PartitionByOption if sinkPartitionBy.nonEmpty =>
    val partitionColumns = SinkPartitionBy
      .split(SinkPartitionBySeparator).map(_.trim)
    if (partitionColumns.nonEmpty)  w.partitionBy(partitionColumns:_*) else w
  case TriggerOption if triggerEnabled => triggerType match {
    case TriggerProcessingTime =>           w.trigger(Trigger.ProcessingTime
    (ProcessTimeInterval))
    case TriggerOnce => w.trigger(Trigger.Once())
    case TriggerContinuous if sinkFormat.equals("kafka") =>
      w.trigger(
Trigger.Continuous(processTimeInterval))
    case _ => w
  }
}}
```

Together the code blocks in Listings 13-6 and 13-7 come together to create an extensible `DataStreamWriter` decorator.

Now let's move back to the application code and see how the trait is used in action.

Spark Stateful Aggregations App

Now that you've seen how the trait works, you can see how the application uses the new capabilities of that trait. Switch back to the SparkStatefulAggregationsApp source code so you can look at how the new underpinning help simplify the application architecture.

Let's begin our journey with the runApp method in Listing 13-8 and work our way backward through the app.

Listing 13-8. The runApp method Returns a StreamingQuery

```
override def runApp(): StreamingQuery = {
  val conf = sparkSession.conf

  val processor = StoreRevenueAggregates(sparkSession)
  val pipeline: Dataset[Row] = processor
    .transform(inputStream.load())
    .transform(processor.process)

  val w = outputStream(pipeline.writeStream)

  conf.get(sinkToTableName, "") match {
    case tableName if tableName.nonEmpty =>
      w.toTable(tableName)
    case _ => w.start()
  }
}
```

The runApp method in Listing 13-8 uses an instance of the StoreRevenueAggregates class as a proxy class to transform and process an input data stream.

```
val pipeline: Dataset[Row] = processor
    .transform(inputStream.load()) // DataFrame
    .transform(processor.process) // Dataset[Row]
```

Let's unpack the aggregation pipeline steps:

1. processor.transform(inputStream.load). The first transformation takes a DataFrame and returns a DataFrame. Behind the scenes, this transform is converting a timestamp column from UNIX epoch milliseconds into a TimestampType.

2. `DataFrame.transform(processor.process)`. The second `transform` method feeds the transformed DataFrame from Step 1 through the `process` method of the `StoreRevenueAggregates` class using a functional DataFrame transformation. This is where we add our `watermark`, `groupBy`, `window`, and `agg` operations to enable our aggregation stream.

The result of the pipeline is a DataFrame, aka `Dataset[Row]`), that can be used with the `outputStream` decorator function. The output stream is then conditionally started using `toTable` or the `DataStreamWriter` directly based on the application configuration.

Let's unpack the `StoreRevenueAggregates` class to see what this application is doing.

Streaming Aggregations

Open the `StoreRevenueAggregates` class located in the `processors` package. The class in Listing 13-9 is simply a `transform` and `process` method which is required by the `DataFrameTransformer` and `WindowedDataFrameProcessor` traits.

Listing 13-9. The StoreRevenueAggregates Class

```
class StoreRevenueAggregates(val spark: SparkSession) extends
DataFrameTransformer with WindowedDataFrameProcessor with Serializable {

  override def transform(df: DataFrame): DataFrame = df
    .withColumn(timestampColumn, to_timestamp(
      col(timestampColumn).divide(lit(1000))
      .cast(TimestampType)))

  override def process(df: DataFrame): DataFrame = {
    import spark.implicits._
    df
      .dropDuplicates("orderId")
      .withWatermark(timestampColumn, watermarkDuration)
      .groupBy($"storeId", groupingWindow())
```

```
  .agg(
    count($"orderId") as "totalOrders",
    sum($"numItems") as "totalItems",
    sum($"price") as "totalRevenue",
    percentile_approx($"numItems", lit(0.95), lit(95))
      as "numItemsP95",
    avg($"numItems") as "averageNumItems"
  )
}
}
```

The class implementation shown in Listing 13-9 is simple to follow. We'll begin with the `transform` method and move onto the `process` method next.

The Transform Method

The `transform` method takes an input DataFrame representing a `CoffeeOrder` and transforms the value of the input timestamp column from a `BigInteger` to a `TimestampType`. This action is necessary to remove the extra precision (`divide(1000)`) before casting the value to a timestamp that can be used by Spark with our window and watermark. If the division doesn't occur, the timestamp will flip around and become unusable.

The Process Method

The `process` method is where the lion's share of the application logic takes place.

We begin by dropping any duplicate rows using the `dropDuplicates` method on the `orderId` column. This process ensures that only the first row for each unique orderId is processed in the final aggregation. It is worth pointing out that the `dropDuplicates` method only works while an aggregation is in memory.

Next, we add a *watermark* to ensure the streaming query will ignore latent data as defined by our `timestampColumn`. *Remember that the watermark column must match the window timestamp column for watermarking to take effect.* With this in mind, we apply the familiar `groupBy` operator from the last chapter, along with the new window function introduced earlier in the chapter to produce a streaming grouping set. Afterward, we apply the aggregation functions introduced in Chapter 12 to the streaming dataset.

That's it. The application can now be configured to create arbitrary time-based aggregations using windows. We will break down what Spark is doing behind the scenes in Exercise 13-2, as we re-create the aggregation using the powerful typesafe FlatMapGroupsWithState functionality provided on the KeyValueGroupedDataset.

Exercise 13-1: Summary

This first exercise took you through the process of building an end-to-end application to reliably process streaming data and generate stateful windowed aggregations. Next, we'll be looking at writing an application using arbitrary stateful computations, which are a set of low-level operators used to extend stateful processing of grouped data to typed user-defined functions (UDFs).

Typed Arbitrary Stateful Computations

Apache Spark offers a set of rich low-level capabilities for streaming datasets. The capabilities are intended to be used to solve more complex problems than Spark SQL operations and DataFrame DSL alone can handle.

Think back to the *user-defined functions* (UDFs) you created in the last chapter. Those functions worked on a row-to-row basis, using one or more columns as input to a DataFrame UDF to create new columns. Arbitrary stateful computations take the basic UDF functionality to the next level with the additional capabilities for managed state and grouping set specific timeouts. All this behavior is hidden behind the KeyValueGroupedDataset.

KeyValueGroupedDataset

Recall how the agg operator of a RelationalGroupedDataset is only available after calling groupBy on a DataFrame. Datasets also have additional operators only available after calling groupByKey on a KeyValueGroupedDataset.

For example, Listing 13-10 transforms an inputStream into a KeyValueGroupedDataset grouped by the storeId of the order.

Listing 13-10. Converting a Dataset into a KeyValueGroupedDataset to Expose the Arbitrary Stateful Computation Operators

```
import spark.implicits._
val encoder = Encoders.product[CoffeeOrder]
val groupedDS: KeyValueGroupedDataset[
  String, CoffeeOrder] = inputStream.load().as[CoffeeOrder].groupByKey(_.
  storeId)
```

The prior example uses as[CoffeeOrder] to transform a DataFrame into a specific dataset that's ready for arbitrary stateful processing. The next step is to use the mapGroupsWithState or flatMapGroupsWithState operators. Let's look at these two operators now.

Iterative Computation with *mapGroupsWithState

The mapGroupsWithState and flatMapGroupsWithState methods takes a UDF function and define a key and an input value type. They also define the data types for the *stored state,* and the *final output data type* resulting from the arbitrary computation. They also take the parameters for the data type encoders for the managed state and the final output type. The final parameter timeoutConf describes how Spark should manage the internal state for these iterative computations.

```
def [flat]mapGroupsWithState[S, U](
    func: MapGroupsWithStateFunction[K, V, S, U],
    stateEncoder: Encoder[S],
    outputEncoder: Encoder[U],
    timeoutConf: GroupStateTimeout): [Dataset[U]|U]
```

The signatures of the functions you pass into the operators are shown next.

MapGroupsWithState Function

```
def func(key: K, values: Iterator[V], state: GroupState[S]): U
```

The udf function for mapGroupsWithState provides you with a key and an iterator of values (Iterator[V]) that can be used to create an initial state record or to update a previous version of a row stored within the state store. The output of the function is a reduction from one or more input rows (V) down to a single output row (U).

473

FlatMapGroupsWithState Function

The `flatMapGroupsWithState` function is almost identical to `mapGroupsWithState`. The only difference is the output type, which is an `Iterator[U]` rather than a single instance of `U`.

```
def func(key: K, values: Iterator[V], state: GroupState[S]): Iterator[U]
```

The UDF function provided to the *mapGroupsWithState operator by your state function is called for each micro-batch. You can guarantee that all available rows will be provided to your function split by your grouping key. This enables you to use the provided iterator of matching row values (V) to create a new state store record or to merge your new values into the prior state in an iterative fashion.

Exercise 13-2 looks at a practical example.

Exercise 13-2: Arbitrary Stateful Computations on Typed Datasets

Open the `SparkTypedStatefulAggregationsApp`. This application is implemented with the same trait as the application `SparkStatefulAggregationsApp` from Exercise 13-1.

SparkTypedStatefulAggregationsApp

The second application of the chapter yields a similar result to the first. You will produce store revenue aggregations over time. The difference this time around is that you will be using the `flatMapGroupsWithState` operator and defining your own grouping key that will replace the `groupBy(col, window(...))` functionality from the first exercise. The entire wrapper application is shown in Listing 13-11.

Listing 13-11. The SparkTypedStatefulAggregationsApp

```
object SparkTypedStatefulAggregationsApp
  extends SparkStructuredStreamingApplication[
    Dataset[CoffeeOrderForAnalysis], CoffeeOrderStats] {
  ...
  override def runApp(): StreamingQuery = {
    import sparkSession.implicits._
```

```
  val conf = sparkSession.conf

  val processor = TypedRevenueAggregates(sparkSession)
  val pipeline = processor
    .transform(inputStream.load().as[CoffeeOrder])
    .transform(processor.process)

  conf.get(sinkToTableName, "") match {
    case tableName if tableName.nonEmpty =>
      outputStream(pipeline.writeStream)
      .toTable(tableName)
    case _ => outputStream(pipeline.writeStream).start()
  }
}

run()
}
```

Beginning again inside the runApp method, you'll notice that aside from the proxy class TypedRevenueAggregates and the transform method taking a typed CoffeeOrder rather than a DataFrame, there isn't much difference between the application in Exercise 13-1. The main differentiator resides in the mechanics surrounding how the aggregations themselves behave, so let's open the TypedRevenueAggregates class from the processors package and start walking through the code.

TypedRevenueAggregates

There is a good amount of boilerplate code in the application, but let's begin by looking at the transform method, which is presented in Listing 13-12. The method takes a CoffeeOrder dataset and transforms it into a CoffeeOrderForAnalysis dataset.

Listing 13-12. Transforming One Dataset into Another in Preparation for Arbitrary Stateful Computation

```
override def transform(ds: Dataset[CoffeeOrder]): Dataset[CoffeeOrderFor
Analysis] = {
  ds
    .withColumn(timestampColumn,
```

```
      to_timestamp(
col(timestampColumn).divide(lit(1000)).cast(TimestampType)))
    .withColumn("window", groupingWindow())
    .withColumn("key", sha1(bin(hash(col("storeId"),col("window.start"),
    col("window.end")))))
    .as[CoffeeOrderForAnalysis]
}
```

There is a lot going on in that function. First things first, we reuse the timestamp conversion technique from Exercise 13-1 along with the groupingWindow method that was first seen in Exercise 13-1. Rather than adding the window function into our groupBy though, we add the window directly onto the dataset. Lastly, we compute a strong hashing function on the storeId, window.start, and window.end columns of what is now a DataFrame. This provides us with a strong composite key that we can use with groupByKey to process the streaming dataset.

The case class used to encode the transformed dataset is shown next.

CoffeeOrderForAnalysis

```
case class CoffeeOrderForAnalysis(
  timestamp: java.sql.Timestamp,
  window: Window,
  key: String,
  orderId: String,
  storeId: String,
  customerId: String,
  numItems: Int,
  price: Float
) extends Serializable
```

Given the transform method adds the timestamp, window, and key columns, we can then convert our updated dataset using .as[CoffeeOrderForAnalysis].

Listing 13-13 looks at the process method and unpacks what it is up to.

Listing 13-13. Transforming One Dataset Into Another in Preparation for Arbitrary Stateful Computation

```
override def process(ds: Dataset[CoffeeOrderForAnalysis]):
Dataset[CoffeeOrderStats] = {
  ds
    .dropDuplicates("orderId")
    .withWatermark(timestampColumn, watermarkDuration)
    .groupByKey(_.key)
    .flatMapGroupsWithState[CoffeeOrderStats, CoffeeOrderStats]
    (OutputMode.Append(), GroupStateTimeout.EventTimeTimeout)(stateFunc)
}
```

The process method is where we set up the logical plan for our arbitrary stateful computation. We begin by dropping duplicates based on the orderId of our coffee orders. This way we don't have to concern ourselves with bad aggregations based on an upstream emitting a second or third event for the same order (which does happen). We apply our watermark to the timestamp column we added in the transform method of our class, and unlike Exercise 13-1, we don't need to use the same timestamp column with our watermark as we use in our window.

This is because Spark enables our application to take over state management via the state timeout configuration GroupStateTimeout.EventTimeTimeout. Because of how the watermark is used with flatMapGroupsWithState, we only have the option of using Append mode (which doesn't necessarily stop our application from running in update mode; you will see why later).

CONFIGURING THE STATESTORE WITH GROUPSTATETIMEOUT

Spark provides three options for reliably managing your grouped state. The options are ProcessingTimeTimeout, EventTimeTimeout, and NoTimeout.

ProcessingTimeTimeout

ProcessingTimeTimeout sets the state timeout based on Spark's own internal clock. Your UDF function must set the timeout of the GroupedState to initially set, or push the timeout further, upon each update of the state.

```
state.update(storedValue)
state.setTimeoutDuration("5 minutes")
```

This example will add a timeout of five minutes to the updated state based on the internal Spark clock time.

EventTimeTimeout

EventTimeTimeout sets the state timeout based on the eventTime relative to the current global watermark of the StateStore. This option requires the withWatermark operator to function and requires append mode on your StreamingQuery.

```
state.update(storedValue)
state.setTimeoutTimestamp(state.getCurrentWatermarkMs(), "30 minutes")
```

This example would give any data stored in the state for this specific group an additional 30 minutes before being timed out.

NoTimeout

This configuration sets no timeout. This option can be used to continuously update state. As a word of caution, with each new unique key added to the state store, the managed state will continue to grow. At a certain point, your application may stop performing due to the size of the backing state.

Let's look at the state function (stateFunc) that has been passed to the flatMapGroupsWithState operator.

TypedStoreRevenueAggregates State Function

The stateFunc method is shown in Listing 13-14. This is the core routine enabling the arbitrary processing of data in your Spark applications. Let's walk through the main sections of the state function now, and then drill into things further.

Listing 13-14. The Function Responsible for Generating Store Revenue Statistics That Incrementally Update

```
def stateFunc(
  key: String,
  orders: Iterator[CoffeeOrderForAnalysis],
```

```scala
  state: GroupState[CoffeeOrderStats]): Iterator[CoffeeOrderStats] = {
  if (state.hasTimedOut) {
    val result = state.getOption
    state.remove()
    if (result.isDefined) Iterator(result.get)
    else Iterator.empty
  } else {

    val stats = ordersStats(orders)
    val stateHolder = state.getOption match {
      case Some(prior: CoffeeOrderStats) =>
        val avgNumItems = prior.averageNumItems/priorState.totalItems +
        stats.averageNumItems/stats.totalItems
        CoffeeOrderStats(stats.storeId, stats.window,
          stats.totalItems+prior.totalItems,
          stats.totalOrders+prior.totalOrders,
          stats.totalRevenue+prior.totalRevenue,
          avgNumItems)
      case None => stats }

    state.update(stateHolder)
    val timeout = stateHolder.window.end
        .toInstant.toEpochMilli
    state.setTimeoutTimestamp(timeout, "30 seconds")
    Iterator.empty
  }
}
```

Starting at the top of the function shown in Listing 13-14, we check the status of our state to see if it has timed out. As each micro-batch processing cycle is completed, the StateStore will mark the records to timeout efficiently on the next processing cycle. The next cycle occurs, and any state rows marked as timedOut will pass through the flatMapGroupsWithState function one last time. Let's look at the timeout operation in more detail in Listing 13-15.

Listing 13-15. StateStore records Will Time Out, Enabling Your Application to Remove the Record from the StateStore and Send it Downstream for Further Processing

```
if (state.hasTimedOut) {
  val result = state.getOption
  state.remove()
  Iterator(result.isDefined) Iterator(result.get)
  else Iterator.empty
}
```

It is important to note that if you skip removing the state from the `StateStore`, Spark won't remove it on your behalf. This would allow orphaned state rows to pile up, creating a healthy memory leak. Always remember to check your state if it is indeed timing out. Next, in the case where your state isn't timing out, you need to decide what to do with the current orders. This is where the `ordersStats` method comes into play.

```
val stats = ordersStats(orders)
```

The orderStats method reduces an `Iterator[CoffeeOrderForAnalysis]` down to a single `CoffeeOrderStats` object. Let's break down what is happening in the method shown in Listing 13-16.

Listing 13-16. The orderStats reduce Function

```
def orderStats(orders: Iterator[CoffeeOrderForAnalysis]):
CoffeeOrderStats = {
  val head = orders.next()
  val first = (head.storeId, head.window, 1, head.numItems, head.price,
  head.numItems.toFloat)
  val reduced =  orders.map(o => (o.numItems, o.price))
.foldLeft[(String,Window,Int,Int,Float,Float)](first)(
  (orderStats, order) => {
  val totalOrders = orderStats._3 + 1
  val totalItems = orderStats._4 + order._1
  (orderStats._1, // storeId
   orderStats._2, // window
   totalOrders, // totalOrders
```

```
  totalItems, // totalItems
  orderStats._5+order._2, //totalRevenue
  (orderStats._6+order._1.toFloat)/2) // averageNumItems
 })
 CoffeeOrderStats(reduced._1, reduced._2, reduced._3, reduced._4,
reduced._5, reduced._6)
}
```

The reduce operation uses a foldLeft process to reduce an arbitrary number of orders down into a single merged record using the first item of the orders iterator to act as the base object for iteration. For each incremental pass through the fold functionality, the prior stats are merged with each new set of statistics. Finally, when the fold operation returns, we are left with a tuple (String, Window, Int, Int, Float, Float) encapsulating the CoffeeOrderStats object.

Folds might not be your first choice for iterative processing, but I highly recommend them, given their flexibility to solve all sorts of problems.

Note   The iterative processing in the functional foldLeft operation in the orderStats method is a good mental model for thinking about how Spark handles aggregations across stateful batches. You essentially take multiple iterators (one for each stateful batch) and fold the values into each other to create a single record per batch. Depending on the aggregation heuristic at play, enough data is retained in order to merge the values of each reduce operation across the lifecycle of the stateful aggregation.

This leaves us with the final piece of the pie, the merge functionality across existing state, which is shown in Listing 13-17. The stateHolder variable is responsible for testing the GroupState object and looking for some prior state. If there is no prior state (None), then we simply return the stats object generated by the call to orderStats. However, if there is prior state kicking around, it is up to us to merge that into our more recent state.

Listing 13-17. Merging the Prior State of Our Aggregations with the Current Batch of Updates

```
val stats = orderStats(orders)
val stateHolder = state.getOption match {
  case Some(currentState: CoffeeOrderStats) =>
    CoffeeOrderStats(stats.storeId, stats.window,
      stats.totalItems + currentState.totalItems,
      stats.totalOrders + currentState.totalOrders,
      stats.totalRevenue + currentState.totalRevenue,
      currentState.averageNumItems/currentState.totalItems +
      stats.averageNumItems/stats.totalItems
    )
  case None => stats
}
```

In a similar way to how the `orderStats` method reduces an iterator of orders down into a single merged `CoffeeOrderStats` object, the merge process essentially creates totals through addition and creates an average of averages to ensure the average number of items sold per order continues to be valid.

Average of Averages

The one tricky thing here is keeping the average number of items accurate across merged batches. But using some basic math we can compute the average of averages and retain a fairly accurate average across many merges.

```
currentState.averageNumItems/currentState.totalItems +
    stats.averageNumItems/stats.totalItems
```

The trick is to add the average of both sides of averages.

Now that you've seen how the `TypedStoreRevenueAggregates` handles windowed aggregations across arbitrary stateful computations, let's look at some sample output for Exercise 13-2.

Final Aggregations: The Window Values Have Been Redacted to Fit the Screen

```
+-------+-----------+----------+----------+--------------+
|storeId|totalOrders|totalItems|totalRevenue|averageNumItems|
+-------+-----------+----------+----------+--------------+
|storeA |4          |8         |36.96      |2.25          |
|storeB |6          |3         |23.97      |1.5           |
|storeC |1          |1         |2.99       |1.0           |
|storeA |4          |2         |9.98       |2.0           |
|storeB |2          |2         |9.98       |2.0           |
|storeC |7          |2         |35.55      |2.0           |
|storeB |1          |2         |3.99       |2.0           |
|storeD |1          |9         |39.99      |9.0           |
|storeA |1          |1         |2.99       |1.0           |
|storeF |2          |12        |115.979996 |6.0           |
+-------+-----------+----------+----------+--------------+
```

Exercise 13-2: Summary

You were introduced to arbitrary stateful computations for Spark Structured Streaming. You learned how to use `flatMapGroupsWithState` to provide low-level incremental processing to the datasets stored in the state store for your computations. You learned how to use `GroupStateTimeout`.EventTimeout and learned about the other options, `GroupStateTimeout.ProcessingTimeTimeout` and `GroupStateTimeout.NoTimeout`, and you also learned how the watermark affects the way the state store records timeouts using event time and `OutputMode.Append`.

While you learned a lot about how to work with analytical data over the past two chapters, you have yet to learn how to unit test these complex operations. This last and brief introduction to the topic will leave you with a fully functional testing setup.

Exercise 13-3: Testing Structured Streaming Applications

Like any piece of software, your Structured Streaming applications aren't complete until you can prove that the logic works as intended. That can be a tricky situation, especially when it comes to ensuring code quality and performance while running localized unit tests. I am going to introduce you to the concept of the MemoryStream and MemorySink now, which will grant you a huge advantage when it comes to testing how your queries operate.

MemoryStream

MemoryStream comes from the org.apache.spark.sql.execution.streaming package and is intended for local testing.

```
case class MemoryStream[A : Encoder](
    id: Int,
    sqlContext: SQLContext,
    numPartitions: Option[Int] = None)
```

The added benefit of using memory streams rather than wiring up conventional test containers or local environments means your application code can be tested where your application logic begins. If you are using the Kafka DataSource, you don't often need to ensure that Kafka is working as expected, but you should be more concerned with the data that is being emitted from Kafka for processing in your Apache Spark applications. Listing 13-18 shows an end-to-end unit test for Exercise 13-1.

Listing 13-18. Testing End-to-End Structured Streaming Applications Using MemoryStream

```
class SparkStatefulAggregationsAppSpec extends StreamingAggregateTestBase {

  "StoreRevenueAggregates" should " produce windowed statistics" in {
    implicit val testSession: SparkSession = SparkStatefulAggregationsApp
  .sparkSession.newSession()
    import testSession.implicits._
    import org.apache.spark.sql.functions._
```

```scala
    implicit val sqlContext: SQLContext = testSession.sqlContext
    testSession.conf.set(AppConfig.sinkQueryName, outputQueryName)
    val outputQueryName = "order_aggs"

    // Group the iterator into groups every 6 events
    val coffeeOrders = TestHelper
      .coffeeOrderData()
      .grouped(6)

    val coffeeOrderStream = MemoryStream[CoffeeOrder]
    coffeeOrderStream.addData(coffeeOrders.next())

    val processor = StoreRevenueAggregates(testSession)
    val pipeline = processor
      .transform(coffeeOrderStream.toDF())
      .transform(processor.process)

    val streamingQuery = SparkStatefulAggregationsApp
      .outputStream(pipeline.writeStream)
      .start()

    // queue up all the data for processing
    coffeeOrders.foreach(orders =>
      coffeeOrderStream.addData(orders))

    // tell Spark to trigger everything available
    streamingQuery.processAllAvailable()

    val row = result
      .where($"store_id".equalTo("storeA"))
      .sort(desc("orders"))
      .collect().head

    row.getDouble(row.fieldIndex("avg_items")) shouldBe 2.0d
    row.getInt(row.fieldIndex("p95_items")) shouldBe 4d
    row.getDouble(row.fieldIndex("revenue")) shouldBe 36.96d
    streamingQuery.stop()
  }
}
```

Breaking down the unit test, we start off by reading in some CoffeeOrders that have been added to the TestHelper object in the test package. This is a sequence of orders that have been tuned to test running the same data through both the traditional groupBy/window-based aggregations of Exercise 13-1 as well as the flatMapGroupsWithState aggregations of Exercise 13-2. This way, we can test the behavior of each, side by side. So, looking at the memory stream:

```
val coffeeOrderStream = MemoryStream[CoffeeOrder]
coffeeOrderStream.addData(coffeeOrders.next())
val processor = StoreRevenueAggregates(testSession)
val pipeline = processor
  .transform(coffeeOrderStream.toDF())
  .transform(processor.process)
```

The neat thing you'll notice about the MemoryStream is that you can use the addData method to add another batch of data to the stream. This is the way you control how many records you process per artificial tick of the Structured Streaming test. You can also easily generate the correct "streaming" dataframe by calling toDF() on the coffeeOrderStream. This saves you the trouble of having to set up and integrate something like Redis or Kafka just for the sake of unit testing. Next, we add more batches of data and force Spark to process all available data.

```
coffeeOrders.foreach(orders =>
coffeeOrderStream.addData(orders))
streamingQuery.processAllAvailable()
```

This will force a full processing cycle, write-ahead logging, and committing of the final state afterward. Now for the testing part:

```
val row = result
  .where($"store_id".equalTo("storeA"))
  .sort(desc("orders"))
  .collect().head

  row.getDouble(row.fieldIndex("avg_items")) shouldBe 2.0d
  row.getInt(row.fieldIndex("p95_items")) shouldBe 4
  row.getDouble(row.fieldIndex("revenue")) shouldBe 36.96d
```

The test itself is simple. I am testing that the result of the aggregations for a specific record for `storeA` has the values `2.0d`, `4d`, and `36.96d` for the columns `avg_items`, `p95_items`, and `revenue`, respectively. You can do a lot more when you test your Spark applications, from testing the expectations for runtime configuration, to sophisticated event scenarios that ensure your application will last the test of time. And lastly, we stop the streaming query and let the test complete.

```
streamingQuery.stop()
```

Feel free to run the unit tests and watch how the application processes data and calculates the aggregations you've set up throughout the course of this chapter.

Exercise 13-3: Summary

There is an art to writing unit tests that catch bugs 9.99 times out of 10. It is always nice to feel confident in your applications capabilities before pressing the button and launching an application into production. One of the things I find to be more satisfying than writing good unit tests is the ability to debug and learn about the many classes at work that power the transformations and aggregations in the Spark engine itself. Lastly, unit tests are a great way to test an end-to-end idea using `MemoryStreams` and quickly put together a scrappy test application before going too deep into the weeds on a far-out idea. Remember that you are writing unit tests for yourself, for your application, and for your future self, or anyone who must maintain your code when you are gone. Remember to write tests that cover and wrap your application tightly enough that mistakes can be caught quickly, so that your application can have a life after you are on to your next set of projects.

Summary

This chapter covered a lot of ground. It covered many interesting areas with respect to analytics and streaming stateful aggregations. You learned about the magic window expression for `RelationalGroupedDatasets` that partners with the watermark to create exactly-once semantics for complex streaming aggregations. You also looked at using arbitrary stateful processing to create a system that worked like the grouping set with window expression capabilities in Exercise 13-1. The big difference was you got to work manually creating the equivalent of the `sum(col)` and `avg(col)` methods in your state function.

Although `mapGroupsWithState` and `flatMapGroupsWithState` can process any kind of data and be used to do things like track people entering or exiting a coffee shop or figuring out which shop is more congested (as a factor of the delta between orders being ordered and received, as well as the number of people who have entered but not exited the coffee shop), I thought it would be neat to showcase how to create windowed aggregations the hard way. Feel free to use your new knowledge and knowhow to create any of these two systems:

- Track people entering or exiting a coffee shop to figure out the total number of people in a shop at any given moment in time.

- Track the shop congestion as a factor of the people currently in the shop, along with the average delta between orders being ordered and received.

These projects can be done using the framework we've built this chapter. The next two chapters look at the process of deployments as well as monitoring and a little tuning. Let's get ready to release our applications into the wild.

CHAPTER 14

Deploying Mission-Critical Spark Applications on Spark Standalone

This is *day one*. You've probably heard people talk about *day one* with mixed emotions, ranging from very real excitement and joy to mild-frustration, and even anger stemming from high amounts of stress and anxiety. The initial live release of any software is bound to cause all sorts of feelings, since it's *the first* major release milestone. But with the high fives, and words of congratulations aside, the reason people call it day one is because it is the beginning, the first release of many, in a continuous development and support cycle. Day one is where the hard work really begins.

This chapter is dedicated to making sure your *day one* goes smoothly. You will learn about the various deployment strategies for your Spark applications and as well as some additional tips, tricks, and techniques to ensure you are prepared for wherever the tide will take you.

Deployment Patterns

Let's begin with the basics. You haven't technically *run* many Spark applications during this book. That doesn't mean the work you have done up until now is all for not, rather this yet another natural step in the process readying applications for deployment. To get to this point where you can deploy Spark applications, you had to first understand how to *build* a Spark application, which libraries to use, as well as what design patterns and best practices to weave together.

© Scott Haines 2022

S. Haines, *Modern Data Engineering with Apache Spark*, https://doi.org/10.1007/978-1-4842-7452-1_14

Ultimately, we had to come to a common understanding of what processes are at play (and at work), for instance, to control the runtime behavior of your applications through configuration, you needed to learn how Spark configuration can be assembled, merged, and overridden, to augment application behavior without needing to recompile. This technique enables you to control, for example, an application processing rate, trigger interval, checkpoint directory, and in a meta way (introduced last chapter), you can even plug different sources and sinks to leverage reading and writing to alternative locations.

Tip   It is common for debugging and replay (aka data recovery) purposes to run a second copy of a Structured Streaming application with overridden checkpoint locations and alternative sink configurations. This technique enables you to analyze and resolve problems without corrupting any stored state, or without writing duplicate records that could end up in downstream systems.

Being able to control the behavior of your application externally (config-driven) grants you more ways to operate mission-critical applications and can also reduce the mean-time-to-recovery (MTTR). In the case where problems can be solved through configuration and tuning versus fixing bugs in the source code.

Let's turn our attention back to the deploy process. To be able to successfully deploy applications, we needed to look at the core frameworks and build new layers on them. Essentially allowing us to crawl before we walk, walk before we run, and finally run before we *deploy*. So, what is the difference between *running* and *deploying applications*?

Running Spark Applications

Throughout the book you have built numerous nano- and micro-applications using the `spark-shell` and Apache Zeppelin. These execution modes encapsulate patterns for *running* Spark applications *dynamically*. By building up from a series of blocks (code) and orchestrating step-based routines (programs), you were able to create Spark applications on-the-fly through composition. The output (learnings) that can be converted into fully compiled applications (with tests!). By Chapter 7, we were writing full applications, and *running* them using a hybrid of containers and hosted local volume mounts (overlays) to run *Apache Spark on-a-box* (or clusterless) using `docker exec` and `spark-submit`.

The difference between *running* and *deploying* has to do with the presence of a *physical cluster*, or the lack of a single point of failure for the continuity of your mission-critical Spark applications. Let's look at the cluster mode options and learn to deploy Spark applications.

Deploying Spark Applications

Consistency. Reliability. Dependability. If you depend on something or someone, it can hurt to be let down. In a similar way, deploying your Spark applications, aka submitting your driver application, to run across an external compute cluster, provides your applications with the fault-tolerance and other essential capabilities to navigate the natural ebbs and flows of the torrents (streams or trickles) of data flowing through your apps. Need to scale up an application to handle increased traffic? Great, you can allocate more machine resources (either dynamically, or by bumping the number of cores in your static application config). The same goes with scaling down. You can deallocate or return machine resources back to the cluster when your application is sitting more idle.

There are many factors at play even before deciding where, how, and when to deploy your applications. Let's begin the process by looking at the Apache Spark *Resource Managers* and the part they play in the lifecycle of your structured streaming applications.

Spark Cluster Modes and Resource Managers

Apache Spark ships with many resource managers out of the box. The most popular (at the time of writing) being *Standalone, Kubernetes*, and *Yarn* cluster resource managers. So, what is a cluster resource manager?

Cluster resource managers are responsible for many things—the most important of which is the task of maintaining and managing the state of a cluster, as well as providing common capabilities (APIs) to communicate, schedule, track, assign, and distribute compute resources to be used by the Spark Driver applications running in the cluster.

At a high level, the resource managers monitor and observe the cluster (one or more network connected compute nodes), tracking the latest statistics from the physical compute nodes, checking the status and health of the workers (or executors), and essentially asking if the nodes (or processes) are still alive. As well as keeping a stateful ledger of the total schedulable real-estate in terms of physical nodes, CPU cores, and memory that is unscheduled enabling the cluster resource manager to control or delegate the approval or denial decisions when additional compute resources are requested by the driver applications. Because clusters are dynamic, new nodes (servers capable of running Spark processes) can be added to the cluster, or nodes can also be removed if they become unresponsive, fail health checks, or go offline for any other reason. By ensuring there is a common channel for Spark application(s) to communicate their needs, the resource manager is prepared to handle new resource allocation requests when existing resources are lost or the load on the application becomes more than the existing compute resources can tolerate without scaling outward and the application signals for help.

This all probably makes you think about CAP theorem and the issues described in Chapter 9 with the "thundering herd" problem and multi-tenancy. Distributed systems like Kubernetes, Yarn (Hadoop), and Spark Standalone provide mechanisms to delegate control of systems and services, and the Spark resource managers are all fault-tolerant, providing capabilities to maintain consistency and be reliable under various failure scenarios. Everything just boils down to the cluster resource managers being the foundation enabling your structured streaming applications to withstand a failure (like a server going offline) and being able to bounce back without the need for human interference (in most cases).

We'll look at Spark Standalone this chapter, and we will conclude with Spark on Kubernetes in the next chapter. If you are interested in the Yarn or Mesos resource managers, there is a lot of good information online that can help you get up in running in no time, but they are out of scope for the book.

Spark Standalone Mode

Running a Spark standalone cluster can be done with a minimum of three nodes for a simple test cluster. There is a cluster coordinator node (the master), and two worker nodes that can distribute the work of one or more Spark applications. Standalone mode enables you to submit your Spark Driver application (in client or cluster mode) to run distributed workloads across a subset of compute resources of the underlying (backing) Spark cluster.

While you can run with a minimum of three nodes, the cluster wouldn't be fault-tolerant since the coordinator (master) could go offline and the state of the cluster would disappear. It is essential for production use cases to run Spark Standalone in High Availability mode.

Spark Standalone: High Availability Mode

To be highly available, you must be fault-tolerant, which essentially means you need to be reliable and consistent under failure. Spark adopts a similar process as you saw with checkpointing for the state of the cluster (in Chapters 10, 11, and 13). The active master simply writes the state of the cluster to a reliable (durable) atomic file system (Apache Zookeeper), and the first Spark Master to come online and connect to Zookeeper wins the active state. Figure 14-1 shows the high-available (HA) setup for Spark Standalone, including the (1) Spark Masters, (2) the Spark Workers, and (3) Apache Zookeeper.

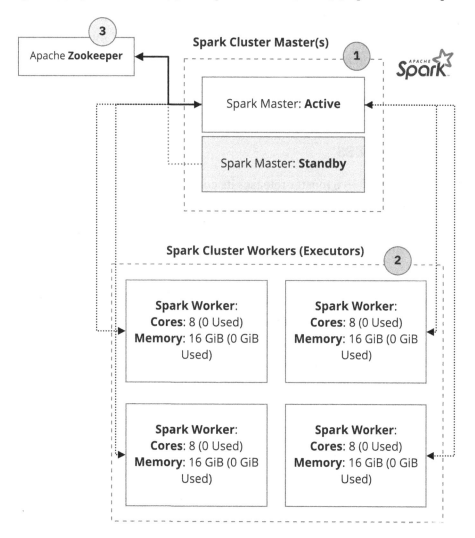

Figure 14-1. *Spark Standalone cluster mode*

The cluster diagram in Figure 14-1 shows a six-node cluster (not including the Zookeeper (zk) cluster, which would require at least three nodes for high availability due to the quorum-based architecture of Apache Zookeeper, so around nine servers are required for full high-availability). Starting at (1), the two Spark Masters are overseeing (watching) the cluster. One is in an Active state (consider this primary), while the other is waiting in *Standby* mode (secondary), on the sideline to be called into action when the current active (primary) master goes offline. The (2) workers (aka executors) are pictured at the bottom of Figure 14-1. When the worker services come online and register with the Spark Masters, the process effectively increases the total available compute resources of the cluster. The cluster can grow and shrink dynamically with workers coming and going, enabling the Spark cluster to handle network issues, be optimized for performance and cost, and for continuality of the cluster to survive other faults. Common behaviors like pushing updates and rolling out new services or replacement hosts (servers) in the cluster can be done with limited downtime to any running application.

The Failover Process

As soon as the current active master releases its connection to (3) the Zookeeper, the *standby* master will pick up the torch and switch from a *passive* to *active* state to keep the cluster alive (failover to the standby master usually takes around 1-5 minutes depending on your network and the size of the cluster).

Caution   It is worth pointing out that the cluster will only remain highly available (HA) after a failover event has occurred, for example the current active master goes offline, and Spark promotes the passive *standby* master to *active*. Monitoring for node failure and automating the process of deploying a replacement is important for the health of the clusters you operate.

If you do lose the cluster state, *don't panic*. Spark applications will continue to run without the need to communicate with the cluster coordinator (master) until the point where a given Spark application needs to acquire new executors (which run on the workers in Standalone mode). This gives you time to recover the cluster. So, if you are running your Spark (and not paying for a managed service), make sure you and your team experiments with loss of one or both masters and automates the process of keeping the cluster running under failure.

Deploying applications in Standalone cluster mode, by default, will allow the first deployed application to consume all resources in the cluster unless configured to take a limited number of CPU cores and memory. As you might imagine, individual applications hogging all the cluster resources can cause problems, but there are processes in place you can take advantage to ensure applications behave well together. All starting with the application deployments.

Application deployment begins with the driver application being submitted to a cluster, and before we go further, it is important to understand how a driver application runs in a distributed way. There are two main deploy styles available for the driver application controlled by the `spark.submit.deployMode` application setting.

Deploy Modes: Client vs. Cluster Mode

When you deploy a Spark Driver application, there are two options for the deploy style driven by the `spark-submit` configuration. We'll start with the *client* mode and move to the *cluster* mode.

Client Mode

Deploying a driver in client mode essentially means the Driver program will *run locally* on the machine, issuing the `spark-submit` command to the Spark Masters. For example, the configuration in Listing 14-1 comes from Chapter 11, and if we were to change the configuration to point to a live standalone cluster (`--master "spark://{master-hostname}:7077"`), we would in fact be declaring to the cluster resource manager that we only want to allocate executors to distribute the application workload, but that we've got the Driver program resource allocation already taken care of.

Listing 14-1. Deploying a Spark Application in Client Mode

```
docker run \
  ...
  -it `whoami`/spark-kafka-coffee-orders:latest \
  /opt/spark/bin/spark-submit \
  --master "..." \
  --class "com.coffeeco.data.SparkKafkaCoffeeOrdersApp" \
```

```
--deploy-mode "client" \
...
/opt/spark/app/spark-kafka-coffee-orders.jar
```

Figure 14-2 shows that the Driver program is running outside of the cluster in client mode, and that the executors are assigned to the remote application.

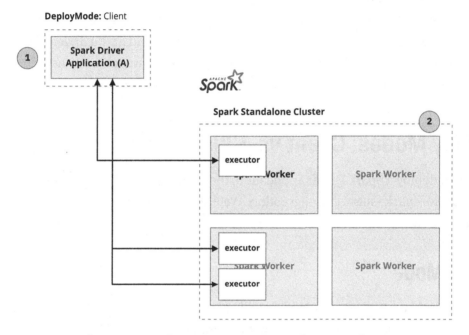

Figure 14-2. *Spark Driver application running in client mode*

There are many reasons for running the Driver outside of the Spark cluster. The top three reasons are to:

- Protect and isolate your Spark applications from disruptions caused by noisy neighbors or other bad behavior within the cluster.

- Safeguard secrets and credentials from cluster-wide access by restricting access to the drivers (outside) of the cluster, which is important for multi-tenant, shared clusters running workloads generated by many disparate teams with various pipeline needs.

- Provide a known host (endpoint) to connect to a driver application to view the Spark UI or to generate metrics for monitoring.

Cluster Mode

Deploying a Driver program in cluster mode essentially means that the application will run remotely inside of the Spark cluster being overseen by the Spark Master that received the inbound `spark-submit` command. For example, the configuration in Listing 14-2 is a revision of Listing 14-1. It replaces the deploy mode value, while also specifying the runtime resources (driver and executors), and the special `supervise` flag.

Listing 14-2. Deploying a Spark Driver Application to be Managed by the Cluster

```
docker run \
  ...
  -it `whoami`/spark-kafka-coffee-orders:latest \
  /opt/spark/bin/spark-submit \
  --master "..." \
  --class "com.coffeeco.data.SparkKafkaCoffeeOrdersApp" \
  --deploy-mode "cluster" \
  --driver-memory "2g" \
  --executor-memory "4g" \
  --total-executor-cores 4 \
  --supervise \
  ...
  /opt/spark/app/spark-kafka-coffee-orders.jar
```

As mentioned, when you deploy an application to the cluster, you run the Driver program *collocated* on a worker inside of the cluster. Figure 14-3 shows how the Driver application is collocated within the cluster, claiming a portion of the total available resources available on the assigned worker.

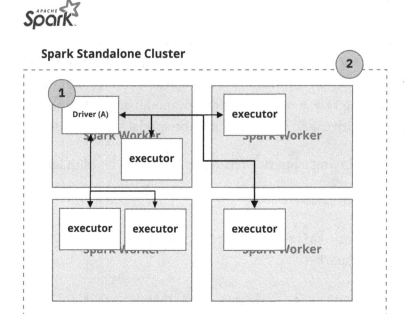

Figure 14-3. *Spark Driver application running in cluster mode*

Understanding the overhead of each collocated application running in a multi-tenant cluster is an important fact to remember given the SLAs for the Driver application running within the cluster. If there are no resources to run the Driver program (after a failure or because it is being deployed for the first time), then the Driver will wait in a Queued state.

Just like there were benefits to running the driver program outside of the cluster, there are also benefits to running drivers within the Spark cluster itself. Essentially, you reduce the surface area (in terms of resources) required to operate your streaming applications, or pipeline infrastructure, by removing the need to run external hosts or additional processes outside of the cluster. Additionally, for tested and mature Spark applications, cluster-based supervision, using the `--supervise` flag, enables applications to respawn in the case of death.

On the other side, if the Spark standalone cluster is dedicated to a common overarching mission, like processing CoffeeCo customer orders or merging store transactions and customer event data to create windowed analytic streams, then clusters can be spun up to solve similar problems, reusing approved access (abiding by data governance rules and regulations) authentication, authorization, and shared data access.

There are pros and cons of both deployment strategies (client mode and cluster mode). They allow you to use a strategy that fits the SLIs, SLOs, and SLAs for the respective application. Therefore, it is critical to understand how to share resources in a multi-tenant environment. We'll look at how Spark Driver applications can be deployed and distributed in a standalone cluster now.

Distributed Shared Resources and Resource Scheduling for Multi-Tenancy

Consider the Spark Standalone cluster in a similar fashion to a distributed operating system. You have a total number of *cores* (CPU), and a total amount of *memory* (RAM), as well as disk space on each *node*, worker (executor), and driver all being shared between the spark processes (*pids*), the operating system processes (also pids!), as well as any other services running on each server in the distributed cluster.

Tip   Using the Linux `top` command will allow you to see what processes are running and what resources they consume on each server. For the resource overhead on Docker, you can use the `docker stats` command to view the CPU, memory, and network consumption by running container.

Why does this matter? Glad you asked. Essentially you can't realistically expect to use 100% of the resources on any *node* within the cluster (*master, worker, driver*) without coming up empty handed, or resource constrained, either at the OS or at the application level. *A general rule of thumb is to preserve at least 1 CPU core and 1GB of memory for the operating system.* You must also weigh the cost of any other process being run outside of the scope of Apache Spark. For example, metric collectors, network profilers, and so on. Once you understand the overhead required for non-Spark processes, you can configure the worker behavior to play nicely and not be overtly greedy with the resources of a given server.

Controlling Resource Allocations, Application Behavior, and Scheduling

Controlling the number of resources allocated to each Spark application running in a multi-tenant (shared) Apache Spark cluster can be done using the Spark settings in Table 14-1 for your application at the time of deployment.

Table 14-1. *Controlling How to Allocate and Schedule Resources for Your Spark Application*

Property Name	Default	What This Does
spark.cores.max	(not set)	Sets a limit on the total number of cores that can be schedulable for your application at runtime. Otherwise, the application will take all available cores.
spark.task.cpus	1	Informs the scheduler of the true intentions of your Spark application. If you have additional CPU bound tasks that require thread pools or other IO, consider bumping this value to 2, or more depending on real CPU consumption per task.
spark.driver.cores	1	Defines how many cores (physical CPUs) are required to run your driver application. Set this number higher if your application is doing final processing on the Driver by way of calling collect() on your DataFrame/Dataset/RDDs.
spark.executor.memory	1g	Defines the soft limit for the memory allocated by the JVM process running the Spark Executor. This value is an allocation for the executor process, and in Standalone mode, this is not a hard limit: meaning the JVM can allocate more memory dynamically.
spark.executor.cores	ALL	This setting is important to note since your Spark application will take all available cores unless you specify otherwise. For multi-tenancy, you can share resources by setting this value to 1, 2, or N cores.

Table 14-1 contains a small subset of the more important values to pay careful attention to when deploying applications into a Standalone, shared (multi-tenant), Spark cluster.

Remember that Structured Streaming applications run forever (24/7/365), so setting resource limitations for each application becomes critically important because each application will allocate and hold the resources claimed when the application driver starts up. This means the total number of cores (physical CPUs) and memory (JVM allocated heap space) will be taken once when the application starts up (if running in static allocation mode).

For example, Figure 14-4 shows the flow of a new Spark Driver application requesting resources from the cluster. There are four nodes, each running as a Spark Worker process. Each of the workers has eight cores and 16 gigabytes (GB) of RAM, meaning that the entire cluster has a total of 32 cores and 64GB of memory to allocate across one or more Spark Driver applications. If the Spark application (A) had the following config:

```
spark.cores.max: 16
spark.executor.cores: 8
spark.executor.memory: 16gb
```

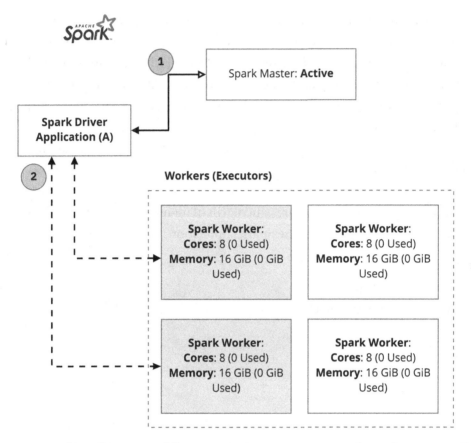

Figure 14-4. *Flow diagram of the negotiation processes at play when a new Spark application is deployed to a Spark cluster*

Then the driver would be assigned two executor instances (JVMs) that completely take over two of the four available worker nodes. Put another way, the configuration says, "I want to take two entire workers."

The sequence of events in Figure 14-4 show a Spark application being deployed into a cluster. It first negotiates with the (1) *active master* to receive *executor assignments* (2) based on the application settings (or cluster defaults) regarding total requested resources (cores/memory) to run the distributed workload. In the example, we are using the limit of 16 total cores, directing the cluster resource manager (active master in standalone mode) to assign the application two of the four total workers. So now 50% of the Spark cluster is allocated and 50% is free. Once the executors are assigned to a the Spark Driver application, managed via the SparkContext, the Driver application no longer needs to communicate with the master unless an executor is lost (server dies, or JVM runs out of memory and exits), or the spark. dynamic.allocation is enabled and the app requests or returns resources to the cluster.

This also means that a disruption to the assigned executors (say both workers go offline) is a failure scenario that the current cluster setup can easily handle. The driver application (A) needs only to ask the resource manager (spark active master) for new executors—a request that can be handled given there are two entirely free workers in the cluster.

If we look at the example in Figure 14-5, the diagram shows the process of a second Spark application joining a shared cluster. For the purpose of example, the second application will have the following application deployment settings.

```
spark.cores.max: 4
spark.executor.cores: 2
spark.executor.memory: 4gb
```

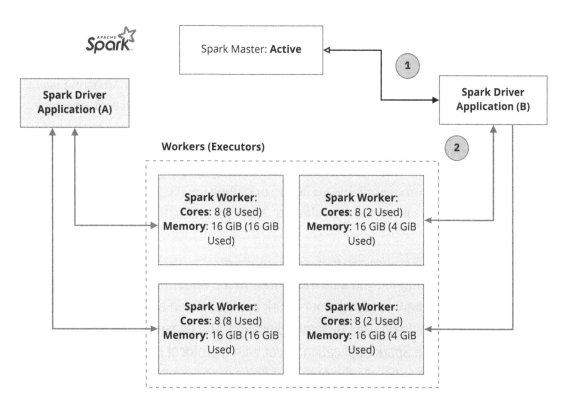

Figure 14-5. *Spark Driver application multi-tenancy on a shared cluster*

The second driver application will request resources from the two remaining workers in the cluster (that have schedulable resources to allocate). This will now leave us with executors scheduled on 100% of the workers. On the upside, there will be fewer wasted resources (since no core or memory goes wasted), and more than one application is running in the cluster, but on the flip side what would happen if a worker was lost now?

In the case where one worker is lost from the cluster in Figure 14-5, it depends on which worker was knocked out. There are two scenarios. In the first, the driver application (B) would be running solely on one worker because each executor is configured with the two cores and 4GB memory. To continue operating, new assignments would be made, and the workload moved to the only other worker with free resources available. That would be the worker where the application (B) already has executors running.

In the second scenario, if the driver application (A) loses a worker, then not all cores or memory required by the config executor (eight cores by 16GB memory) would fit in the cluster. This would leave application (A) running in a degraded state, but at least it would still be running.

When it comes to scheduling compute resources, it is also a good rule of thumb to have some head room to grow in the cluster. You can do this by adding a new node to the cluster when the cluster is more than 75% scheduled, by adopting auto-scale groups or cloud-based rules like you see with managed node groups on Amazon Elastic Kubernetes Service, or by changing the executor settings to be able to fit each executor in a different configuration. Having a 4-core x 8GB with the same 16 max cores would spread driver application (A) across more JVMs, but you could handle the loss of a worker without running in a degraded state.

With enough resources available for failure or scaling up, the process of dynamic allocation can enable a more elastic scalability. Next, you learn how to control this process.

Note   In the next chapter, you'll be introduced to the Kubernetes (k8s) resource manager and learn how to launch your Spark applications in an elastically scalable way, which again uses dynamic allocation. The dynamic allocation process provides the Spark application driver with a protocol to request and return additional resources to handle the ebbs and flows of the current of data being processed.

Elastically Scaling Spark Applications with Dynamic Resource Allocation

Spark applications have variable processing requirements composed of resources at an executor level (cores/memory/*GPUs) that change (think ebbs and flows) throughout their lifetime. To manage the variable needs of many applications running in parallel, dynamic allocation enables each Spark application to opt in and enable dynamic allocation, and to define appropriate settings to control how to release and reallocate additional resources. Table 14-2 lists the common spark configurations.

Table 14-2. *Enabling Elastic Scaling for Spark Applications Can Be Done Using Dynamic Allocation*

Property Name	Default	What This Does
spark. dynamicAllocation. enabled	false	Conditionally opts into dynamic resource allocation.
spark.shuffle.service. enabled	false	To release executors back to the Spark cluster, the Spark Workers must be running the *external shuffle service* as a prerequisite for dynamic allocation. This allows an executor to be released, even if it still holds some shuffle (data) required by the Spark application.
spark. dynamicAllocation. minExecutors	0	Set this value to 1 or higher in order to ensure your application isn't resource starved while waiting for resources to be allocated on available (open) workers.
spark. dynamicAllocation. maxExecutors	INF	Set this value to block an application from requesting all (open) executors in the cluster.
spark. dynamicAllocation. schedulerBacklogTimeout	1s	The driver application will wait up to this value before requesting new executors from the master when tasks are pending.

Dynamic allocation enables automatic management for elastically scaling workloads, which request and return resources from the cluster on-demand based on need. Depending on the priority or expectations of a job (or application), dynamic allocation can allow you to effectively manage the cost of running many co-located (multi-tenant) workloads across a shared compute cluster. However, depending on the various (dynamic) needs of each application, the priority of the workload needs to be taken into consideration when enabling dynamic allocation, since you can easily find yourself in a situation where many applications are waiting in an idle state for resources to become available in the cluster.

For Spark Structured Streaming applications, each micro-batch tends to process smaller amounts of data more frequently, which is different than the behavior of the typical large batch-based applications that read and process large swaths of data in one go. Yes, you can also run stateful batch jobs that use `Trigger.Once`, which will require thinking through the semantics of the application and the SLAs or priorities associated with the nature of the application.

Given the always-online nature of structured streaming applications, dynamic allocation and the external shuffle service can at times get in the way and hurt the performance of your mission-critical streaming applications. Sometimes the only thing getting in the way of the application performing at top speed is the internal application scheduler, which controls the parallelism and processing modes of internal (jobs) running within a Spark Driver application.

SCHEDULING MODES: FIFO VS. FAIR

By default, Spark applications run in *FIFO* (first-in, first-out) scheduling mode. Given that Spark Structured Streaming applications divide and conquer a continuous (unbounded) stream of one or more data sources, which are chunked and broken down through the processing paradigm of micro-batches (which yield deterministic, fault-tolerant, datasets), the Spark Driver application is hard at work ensuring that each stage of transformation across each micro-batch, which is defined by a `StreamingQuery`, aka `spark.readStream...` `writeStream.start()`, has priority based on the `QueryPlan` defining the mission of the streaming application.

Well, what if an application has more than one `StreamingQuery`? In the last chapter, you were introduced to the `SparkStructuedStreamingApplication` trait, which included the following two methods: `awaitTermination` and `awaitAnyTermination`.

```
def awaitTermination(query: StreamingQuery): Unit = {
  query.awaitTermination()
}

def awaitAnyTermination(): Unit = {
  sparkSession.streams.awaitAnyTermination()
}
```

In a basic Spark Structured Streaming application, there is commonly only one upstream source and one downstream sink. Therefore, calling awaitTermination on the main StreamingQuery will keep the Spark Driver application alive until the StreamingQuery terminates. This single application concern is perfectly fine for the FIFO mode scheduler, since there is only one primary concern.

However, the second you decide to introduce a second streaming query to the equation, you'll now have to wait in line for resources at each micro-batch, which doesn't seem fair now does it?

Luckily, Spark ships with a FAIR scheduler. This execution mode allows parallel sharing of resources. This allows for a more equal distribution of the total resources within an individual Spark application.

FAIR **Mode Scheduler**

You can opt into using the FAIR scheduler by setting the scheduler mode on the driver application configuration.

```
spark.scheduler.mode=FAIR
```

In addition to setting the scheduler mode, Spark requires your application to define one or more scheduler pools using an allocation file.

```
spark.scheduler.allocation.file="s3a:///path/to/fair-scheduler.xml"
```

The scheduler allocation file is a simple XML file defining one or more named pools with options to define schedulingMode, weight, and minShare for the respective pool.

- schedulingMode: This will default to FIFO, but can also be set to FAIR. FIFO will prioritize the first workload to be scheduled across the pool, letting additional work stack up in a queue. FAIR will distribute work evenly across the pool resources.

- `weights`: The pool weight declares the priority for scheduling across a given pool. This priority defaults to 1, and can be set higher to set a priority pool for scheduling (like a priority queue).

- `minShare`: Set the core allocations for a given pool. The scheduler will attempt to allocate a predefined number of cores to ensure each pool has the number of cores, before distributing any additional CPU core resources across the rest of the defined pools.

Listing 14-3. An Example fair-scheduler.xml Allocation File

```
<?xml version="1.0"?>
<allocations>
  <pool name="default">
    <schedulingMode>FAIR</schedulingMode>
    <weight>10</weight>
    <minShare>4</minShare>
  </pool>
  <pool name="queued">
    <schedulingMode>FIFO</schedulingMode>
    <weight>1</weight>
  </pool>
</allocations>
```

The example configuration in Listing 14-3 is intended to further define the way the scheduling modes work within the respective scheduler pools. Given the driver application utilizes the SparkContext for scheduling work, the FAIR scheduler can be used to modify the behavior of each SparkSession (which reuses the same Spark context) to govern and share the common pool of cluster resources allocated for a Spark application.

By default, Spark will assign all work to the *default* pool when `spark.scheduler.mode` is set to FAIR. By changing the default pool config, we can change the scheduling weights to prioritize the default (FAIR) pool, or we can carve out different pools for running lower priority workloads in FIFO or in FAIR mode. Using the `fair-scheduler.xml` file, you can guarantee that higher priority work will be scheduled in a FAIR way.

If there is lower priority work, or in the case of a shared environment (like Zeppelin), where you may have multiple concurrent sessions, you can assign one specific pool to be used for lower priority work. The example in Listing 14-4 shows how an ad hoc job can be spawned from an existing SparkSession.

Listing 14-4. Setting a Low-Priority Scheduler Pool for Non-Essential, Nice to Haves

```
// spark = SparkSession
new Thread {
  val other: SparkSession = spark.newSession()
  other.sparkContext.setLocalProperty(
    "spark.scheduler.pool", "queued")
  other.sparkContext.setJobGroup("batch.job", "runs in thread", true)
  other.range(100)
    .write
    .format("console")
    .save()
}
```

You might be wondering why you'd spawn a new session in a new thread. Well, there is a simple answer to that. The thread that sets the local property (setLocalProperty) on the SparkContext will use the scheduler pool assigned to it. If we are running a StreamingQuery on our main thread (current thread), then we'd want to carve out a second thread to run our non-essential work.

Consider revisiting this sidebar again after reading the "Spark Listeners" section. You can use the SparkSession provided in the SparkApplicationListener or QueryListener to create and schedule ad hoc queries, as shown in Listing 14-4. Given the rich capabilities granted to you by Apache Spark and the myriad connectors, you can leverage ad hoc processes to create reactive applications or control event-driven subroutines.

It is essential to have some sort of monitoring and observability for each Spark application that is running. Without understanding how things ran in the past, it is difficult to measure how applications change in the cycle of build and release. Hopefully, you begin capturing metrics for your application before you release them into the wilds of production for the first time (if they are mission-critical). On that note, there are myriad metrics available (out of the box) for measuring, recording, and monitoring the performance and behavior of your Spark applications. You'll learn about two of the more common Spark listener interfaces (SparkListener, QueryListener) and about the *Apache Spark REST endpoint for metrics*. Combined, these capabilities can be used to monitor the behavior of your Spark applications and enable reactive event-driven alerting.

Spark Listeners and Application Monitoring

There is a long-standing debate regarding what came first, the chicken or the egg. I can't answer that, but there is a subtle art when it comes to monitoring Spark applications and defining what metrics are essential for measuring the performance of your mission-critical Spark applications. Identifying the KPIs for each application changes depending on the *SLIs, SLOs, and SLAs* defined in your Spark application data contracts. The metrics help define and shape the performance characteristic of each application, and they let you understand what behaviors to expect (what is normal). Questions regarding metrics may be hard to initially answer given the potentially large application surface area, as well as the constituent parts, including the upstream (sources) and downstream (sinks), but it can help to break things down into groups, identify the sources of data necessary, and then go from there.

Broadly speaking, monitoring isn't just about understanding how your service operates but it is equally about being able to pin-point where in the system something goes wrong and to reduce MTTR (mean time to recovery) in the case of an outage or simply to better understand application to ensure optimal behavior.

Monitoring helps inform other decisions as well. If this is your first monitoring rodeo, it can also help to understand what we can monitor out of the box, especially when it comes to Spark. This begins with listening.

Spark Listener

The `SparkListener` class can be added to your driver application through the `SparkContext` object of the `SparkSession`, or via your application configuration using the `spark.extraListeners` option.

```
// spark = SparkSession
spark.sparkContext
  .addSparkListener(SparkApplicationListener())
```

The `SparkListener` is a prime candidate to use to tap into the runtime application metrics and general behavior of your Spark applications. Furthermore, if you need to take a more reactive approach to problems in-flight, the listeners can be a jumping off point to creating ad hoc event-driven controls.

Listing 14-5 extends the `SparkListener` class with a case class named `SparkApplicationListener`.

510

Listing 14-5. The SparkApplicationListener Case Class

```scala
case class SparkApplicationListener()
  extends SparkListener {
  val logger: Logger = Logger.getLogger(classOf[SparkApplicationListener])
  val session: SparkSession = SparkSession.getDefaultSession.getOrElse {
    throw new RuntimeException("There is no SparkSession")
  }

  override def onApplicationStart(applicationStart:
  SparkListenerApplicationStart): Unit = {
    super.onApplicationStart(applicationStart)
    logger.info(s"app.start app.name=${applicationStart.appName}")
  }

  override def onJobStart(jobStart: SparkListenerJobStart): Unit = {
    super.onJobStart(jobStart)
    logger.info(s"job.start jobId=${jobStart.jobId} jobStart.
    time=${jobStart.time}")
  }

  override def onStageSubmitted(stageSubmitted:
  SparkListenerStageSubmitted): Unit = {
    super.onStageSubmitted(stageSubmitted)
    val stageInfo: StageInfo = stageSubmitted.stageInfo
    logger.info(s"stage.submitted stage.id=${stageInfo.stageId}")
  }

  override def onStageExecutorMetrics(executorMetrics:
  SparkListenerStageExecutorMetrics): Unit = {
    super.onStageExecutorMetrics(executorMetrics)
    val em: SparkListenerStageExecutorMetrics = executorMetrics
    logger.info(s"stage.executor.metrics stageId=${em.stageId} stage.
    attempt.id=${em.stageAttemptId} exec.id=${em.execId}")
  }

  override def onTaskStart(taskStart: SparkListenerTaskStart): Unit = {
    super.onTaskStart(taskStart)
```

```scala
    val ts = taskStart
    val taskInfo: TaskInfo = ts.taskInfo
    // taskInfo has a wealth of status and timing details
    logger.info(s"task.start " +
      s"task.id=${taskInfo.taskId} task.status=${taskInfo.status} task.
      attempt.number=${taskInfo.attemptNumber} " +
      s"stage.id=${ts.stageId} stage.attempt.id=${ts.stageAttemptId}"
    )
  }

  override def onTaskGettingResult(taskGettingResult:
  SparkListenerTaskGettingResult): Unit = {
    super.onTaskGettingResult(taskGettingResult)
    logger.info(s"task.getting.result ${taskGettingResult.taskInfo.
    taskId}")
  }

  override def onTaskEnd(taskEnd: SparkListenerTaskEnd): Unit = {
    super.onTaskEnd(taskEnd)
    val te = taskEnd
    val taskInfo: TaskInfo = te.taskInfo
    val taskEndReason: TaskEndReason = te.reason
    val taskMetrics: TaskMetrics = te.taskMetrics

    logger.info(s"task.end task.type=${taskEnd.taskType} end.
    reason=${taskEndReason.toString} task.id=${taskInfo.taskId}" +
      s"executor.cpu.time=${taskMetrics.executorCpuTime}")
  }

  override def onStageCompleted(stageCompleted:
  SparkListenerStageCompleted): Unit = {
    super.onStageCompleted(stageCompleted)
    val stc = stageCompleted
    val stageInfo: StageInfo = stc.stageInfo
    //val stm: TaskMetrics = stageInfo.taskMetrics
    logger.info(s"stage.completed stage.id=${stageInfo.stageId} stage.
    completion.time=${stageInfo.completionTime.getOrElse(0)}")
```

```scala
  }

  override def onJobEnd(jobEnd: SparkListenerJobEnd): Unit = {
    super.onJobEnd(jobEnd)
    logger.info(s"job.end job.id=${jobEnd.jobId} end.time=${jobEnd.time}")
  }

  override def onExecutorAdded(executorAdded: SparkListenerExecutorAdded):
  Unit = {
    super.onExecutorAdded(executorAdded)
  }

  override def onExecutorMetricsUpdate(executorMetricsUpdate:
  SparkListenerExecutorMetricsUpdate): Unit = {
    super.onExecutorMetricsUpdate(executorMetricsUpdate)
  }

  override def onExecutorRemoved(executorRemoved:
  SparkListenerExecutorRemoved): Unit = {
    super.onExecutorRemoved(executorRemoved)
  }

  override def onEnvironmentUpdate(environmentUpdate:
  SparkListenerEnvironmentUpdate): Unit = {
    super.onEnvironmentUpdate(environmentUpdate)
  }

  override def onOtherEvent(event: SparkListenerEvent): Unit = {
    super.onOtherEvent(event)
  }

  override def onApplicationEnd(applicationEnd:
  SparkListenerApplicationEnd): Unit = {
    super.onApplicationEnd(applicationEnd)
    logger.info(s"application.ended end.time=${applicationEnd.time}")
  }
}
```

The case class in Listing 14-5 extends the `SparkListener` and provides a simple jumping off point for further metric exploration. The order of events matter. Each job starts and spawns one or more stages, which distribute tasks across the executors assigned to your application. Being able to track the rate of failed tasks or failed stages provides you with high-level eyes and ears into the runtime behavior of your application.

For a concrete example, you can observe the `SparkApplicationListener` directly from within your Spark unit tests to view the events generated as the application runs. Try turning on the listener, located in the `TypedRevenueAggregatesSpec` (from Chapter 13). Just uncomment the line and run the application unit tests to see things in action.

```
val testSession = SparkTypedStatefulAggregationsApp
  .sparkSession
  .newSession()
testSession.sparkContext
  .addSparkListener(SparkApplicationListener())
```

The `SparkListener` can be used to publish metrics using `stats.d`. Combined with your own in-house metrics systems or using paid services like DataDog, you can store important metrics pertaining to the runtime of your applications.

Tip   Application metrics are available from the Driver Application UI (port 4040 by default), using the metrics Rest API on the Spark application driver. Use the official "Spark Monitoring" documentation to see the various APIs available for consuming metrics through the driver and standalone master processes.

In addition to the main `SparkListener`, there is another useful class that can be used to add control flow into your applications, called the `StreamingQueryListener`.

Observing Structured Streaming Behavior with the StreamingQueryListener

The `StreamingQueryListener` observes the `SparkSession.streams` object and can be used to tap into all `StreamingQuery` instances managed by the driver application. The three observable states tracked by the `StreamingQueryListener` correspond to the `started`, `progress` (micro-batch completed), and `termination` of all streaming queries.

Listing 14-6 extends the StreamingQueryListener and provides example method implementations, listener interface method names, and event objects. To get you started, it shows how to tap into some of the event metadata to add enhanced logging.

Listing 14-6. The StreamingQueryListener Enables You to Observe the Runtime Behavior of Your Structured Streaming Applications

```scala
case class QueryListener() extends StreamingQueryListener {
  val logger: Logger = Logger.getLogger(classOf[QueryListener])
  val session: SparkSession = SparkSession.getDefaultSession.getOrElse {
    throw new RuntimeException("There is no SparkSession")
  }

  override def onQueryStarted(event: StreamingQueryListener.QueryStartedEvent):
  Unit = {
    logger.info(s"query.started stream.id=${event.id} stream.name=${event.
    name} " +
      s"stream.run.id=${event.runId} stream.start.time=${event.timestamp}")
  }

  override def onQueryProgress(event: StreamingQueryListener.QueryProgress
  Event): Unit = {
    val queryProgress: StreamingQueryProgress = event.progress
    val batchId = queryProgress.batchId
    val batchDuration = queryProgress.batchDuration
    val inputRowsRate = queryProgress.inputRowsPerSecond
    val processedRowsRate = queryProgress.processedRowsPerSecond
    val outputRowsPerBatch = queryProgress.sink.numOutputRows

    val progressSummary = queryProgress.prettyJson

    logger.info(s"query.progress stream.id=${queryProgress.id} stream.
    name=${queryProgress.name} " +
    s" batch.id=$batchId batch.duration=$batchDuration" +
    s" rate: input.rows.per.second=$inputRowsRate processed.rows.per.
    second=$processedRowsRate " +
    s" sink.output.rows.per.batch=$outputRowsPerBatch " +
    s"\n progressSummary=$progressSummary")
```

```scala
  }

  override def onQueryTerminated(event: StreamingQueryListener.QueryTerminated
  Event): Unit = {
    if (event.exception.nonEmpty) {
      logger.error(s"query.terminated.with.exception exception=${event.
      exception.get}")
      // make a decision on what to do
    } else {
      logger.info(s"query.terminated stream.id=${event.id} stream.run.
      id=${event.runId}")
    }
  }
}
```

The QueryListener class can be used to monitor any Structured Streaming application. Just like with the SparkApplicationListener, you can test using the QueryListener using the TypedRevenueAggregatesSpec from Chapter 13.

```scala
testSession.streams.addListener(QueryListener())
```

You now have a direct means to set breakpoints and insert creative control flow to react to the various progress or termination events that come your way. By using the test-driven approach to understanding the capabilities of the QueryListener, you can create better tests to drive unexpected behavior, all from the comfort of your IDE.

Ultimately, query observation is a capability that helps to ensure your application continues to operate as expected. By extending the listener you can collect important streaming metrics (input and processing rates) to help you understand how your application is running.

As a final note, SparkApplicationListener in Listing 14-5 and QueryListener in Listing 14-6 both run directly on the Spark Driver itself. Knowing this you can write hooks to Slack or PagerDuty in the case of bad behavior in your application, or by using the SparkSession handle (available in both example Listener classes), you can trigger event-based subroutines. For example, using the QueryListener you could actively halt all running queries and gracefully stop a bad Driver application, depending on the reason (exception) on the active query a process. The QueryListener hooks are shown in Listing 14-7 to provide you with an idea of how to tap into the events to control your applications. Exceptions that terminate the StreamingQuery may be recoverable. Otherwise, you may want to shut down the application.

Listing 14-7. Using the QueryListener to React to an Exception Received in a
StreamingQuery

```
// QueryListener

def resetTerminated(): Unit = {
  session.streams.resetTerminated()
}

def haltAllStreams(): Unit = {
  session.streams.active.foreach { query =>
    logger.info(s"stream.stop name=${query.name} stream.id=${query.id}")
    query.stop()
  }
}

def shutdown(): Unit = {
  haltAllStreams()
  session.sparkContext.cancelAllJobs()
  session.stop()
}

override def onQueryTerminated(event: StreamingQueryListener.
QueryTerminatedEvent): Unit = {
  if (event.exception.nonEmpty) {
    logger.error(s"query.terminated.with.exception exception=${event.
    exception.get}")
    if (exception.is.recoverable) {
      resetTerminated()
    } else {
      shutdown()
    }
  }
}
```

Listing 14-7 hooks into `StreamingQueryManager` and `SparkContext` to a reset any terminated streams or stop any additional streaming query. Then an explicit call cancels all active jobs (giving open and active jobs time to gracefully stop and clean up). The code finally asks the driver application to gracefully shutdown. Choosing when to halt an application can be instrumental in protecting downstream applications from breaking due to corrupt data or other unknown unknowns. It is often better to fail, and page the application owner(s), when an unrecoverable exception is encountered, knowing that it can (and probably will) cause a traffic jam in the short term, but that it can protect any downstream systems from data problems in flight.

There are many strategies to ensuring applications continue to perform no matter how choppy the waters become. Knowing the right patterns to keep the application from treading water takes patience and practice, but with the right tools at your disposal it can become easier to automate recovery and broadcast need-to-know events directly from applications (before they collapse).

Monitoring Patterns

Connectors to metric aggregators like Prometheus or DataDog can be woven into your Spark Listeners to create central points for observability, emitted by the driver application. This is better than monitoring and observing all executors, since exception information and executor metrics will ultimately be passed back to the driver, which can then be tapped into using the `SparkApplicationListener`.

The metrics emitted by your applications—using the Spark listener interfaces or the metrics collected internally and provided via JMX or REST interfaces by Spark—can be used to power dashboards and other alerting systems. Additionally, once you have operational metrics flowing, you can use the historic metrics of your applications to ensure they continue to perform as well, or better, with each new version released to production. These prior observations (historic metrics) help to ensure updates and upgrades to each application (such as bug fixes or simply results of any performance tuning) can be monitored and measured, keeping your SLAs, SLOs, and SLIs in check.

Spark Standalone Cluster and Application Migration Strategies

The Internet is full of best practices and suggestions. When it comes to running multi-tenant applications on Spark Standalone, it is best to have a strategy in place for migrating the applications as well as the cluster components to mitigate any downtime.

Anti-Pattern: Hope for the Best

Tooling for deploying and rolling (metered replacement) fleets of servers has been around for a while. One pattern that stops working quickly is the "roll and hope" approach for Spark Workers. If you are running Structured Streaming applications, you will continue to interrupt the running driver applications, and the applications can handle the intermittent pause, request new executors, assign, and rebuild process for any lost application state. The process of constant interruption, for example of a rolling frequency of 5% of workers every five minutes, can eat away at SLAs. Application disruptions can be handled naturally when machines come online or go offline, but it is often better to boot a replacement Spark Standalone cluster and migrate the applications (depending on latency and on your tooling).

Best Practices

1. Create a new cluster.

2. Migrate existing applications into the new cluster (start in reverse priority order for applications based on SLAs).

3. Shut down the prior cluster (if you no longer need the compute resource: go green).

This is a best practices list because Spark applications and the clusters they run on are updated to handle new versions of Apache Spark, Java, Scala, and other Python or system dependencies, as well as to patch application configurations or global settings on the Masters, Workers, and Drivers. It is best to understand how the changes can affect performance, so until you've become comfortable with the process, use the "canary in a coal mine" trick and migrate the lowest common denominator (lowest risk) applications first.

Next you are going to learn about the listeners available to you for monitoring the performance and behavior of your Spark applications.

Regarding Containers and Spark Standalone

Spark Standalone can run using containers, but it is beyond the scope of this book to cover. Essentially, you can use the Zookeeper container (which should be cached) from running the Kafka cluster in Chapter 11. Then you have to modify the environment variables (`spark-env.sh`) for the Spark Masters and Workers, assign each one a unique hostname (such as `spark-master1`, `spark-master2`, `spark-worker1`, `spark-worker2`, `spark-worker3`), or reuse the Docker network (`mde`) or create a new one called `spark`. This will enable cross-cluster communication between the components of the Spark cluster and your Spark Driver applications.

The official Apache Spark documentation covers how to "Configure Ports for Network Security," which can be used to view the port requirements to punch firewall rules. For example, a Spark Driver application requires inbound access to the Spark Masters on ports 7077 to submit (deploy) an application. The Spark Master also exposes an optional REST-based service that runs on port 6066, which can be used to accept remote jobs to run in the cluster. For the Spark Master, the UI runs by default on port 8080, the Spark Workers expose a UI on port 8081, and each Driver application runs a UI on port 4040. The Driver port can be turned off using `spark.ui.enabled: false` to save on processing overhead of the Driver, or to restrict how folks can access the logs of the application.

Managed Spark

This chapter discussed the architectural components and Spark classes used to construct a Spark cluster, and to tap into the operational metrics of your Apache Spark applications. You may be wondering why there was no discussion of using Apache Spark on Amazon EMR or Databricks to simplify the process of running Spark applications without all the headache. Using either platform can greatly reduce the level of effort required for getting up and running and should be considered in your short-term and long-term business planning. This book takes an open-source approach (in all ways) as a means of pulling back the curtain. The goal is to let anyone get to know Spark and learn about the pillars of the modern data platform for free (laptop and book aside).

Summary

It is more common for Spark Standalone clusters to operate using traditional Linux deploy and DevOps patterns, including bootstrap scripts for creating and laying out the file system, and Linux package managers like Yum and Apt-Get to scaffold the runtime environment. There are pros and cons to running multi-tenant server environments without clear boundaries between application processes, CPU allocations, and memory limitations (aka cgroups)—mainly that you can bypass containers and lean on traditional RPMs for managing what is installed at runtime.

Depending on the requirements of the applications your platform supports, there are often compounding challenges associated with running multiple versions of Java (supporting Java 8 and Java 11), multiple versions of Scala (2.12, 2.13), handling the native requirements for PySpark, or debugging which process is to blame for performance degradation when running in a world void of containers.

As a final note, although it is also possible to support multiple versions of Spark on the same box, it can be a management nightmare. It is easier to run a Standalone cluster with the common dependencies of the workloads that will run in the cluster and carve out isolated clusters based on the performance characteristics and scalability (fan-out, fan-in) requirements for the workloads that will run in the cluster. If this sounds like a lot of effort, it can be, but you can also automate a lot of the nuances using frameworks like Chef or Terraform.

Given this book has been written using a mostly container-driven approach and given that containers provide consistency and reliability from an operational perspective carrying over to the Spark runtime, it would make the most sense that the final point of our journey is to deploy Spark applications on Kubernetes. (*In fact, the applications you've been building since Chapter 7 can all run using Spark on Kubernetes.*) We'll be spending the rest of our time together learning to run mission-critical applications on Kubernetes.

CHAPTER 15

Deploying Mission-Critical Spark Applications on Kubernetes

Before diving headfirst into what *Kubernetes* (K8s) is, and how *Apache Spark* fits into the distributed K8s ecosystem, it is important to first begin by stating simply that *Kubernetes enables Apache Spark applications to run in isolation, pairing elastic scalability with the runtime consistency of containers, collocated in independent micro-environments called pods* (which you'll learn about soon). Ultimately, you can rely on consistent runtime environments, without having to deal with the pain and other hardships of multi-tenancy in a share-everything ecosystem. Rather, imagine each application running in its own isolated world, which at a high-level act similarly to the local environments you've run to power Spark applications using `docker-compose` on the local data platform we've been constructing throughout this book.

The devil is always in the details, but many of the common pain points that stem from running Apache Spark applications in production clusters (which we saw in the last chapter), such as issues with cluster-wide resource contention, the problems of scaling (up or down) large multi-tenant clusters, or at a high enough level even running one or more physical Spark clusters simply goes away. Instead, each Spark application is deployed and auto-wires, aka scaffolds (and operates) its very own isolated Spark cluster (on-demand), composed across pods that are orchestrated on top of the physical nodes underpinning your Kubernetes cluster. By removing the many moving parts required to run standalone clusters and manage the applications running therein, you can instead focus on the independent needs of each application and generalize the rest.

Kubernetes, with the support of the *Apache Spark Kubernetes Resource Manager,* powers your mission-critical Spark applications.

© Scott Haines 2022
S. Haines, *Modern Data Engineering with Apache Spark*, https://doi.org/10.1007/978-1-4842-7452-1_15

In this chapter you'll learn to deploy your Spark applications on Kubernetes. Furthermore, you'll see how to take advantage of the inherent idempotency provided by K8s deployments, coupled with resource requests and limitations (e.g., CPU and memory allocations). You'll also learn how Spark's dynamic allocation comes to the rescue to help add super-powers to your self-assembling Apache Spark applications. By the end of this chapter, you'll see first-hand how Apache Spark on Kubernetes comes together to power the myriad needs of any data platform and how Kubernetes fits into the modern data ecosystem. Let's get started.

Kubernetes 101

In their own words, "Kubernetes is an open-source system for automating deployment, scaling, and management of production-grade containerized applications." In a nutshell, Kubernetes is a reliable orchestration framework that provides distributed compute and resource scheduling as a service, built on the first principles of isolation, consistency, and reliability. It also happens to have a rich API that can be harnessed to enable smarter decision making and flexibility when it comes to operating applications and services at scale.

This book has gradually immersed you into the world of containers while teaching you to architect and build containerized applications with Apache Spark. You learned to run your applications using Docker, Docker Compose, and the Docker network to orchestrate local workflows. Each workflow required the setup and bootstrapping of curated, application specific environments, to support the myriad data engineering use cases tackled in the book. Use cases ranged from common data processing workflows like ETL/ELT, to job scheduling and orchestration, as well as architecting and building pluggable (composable), config-driven, Spark pipeline applications.

Each previous chapter acted as a guide on this journey, drilling into specific aspects of the data engineering ecosystem to lead you along a curated path through the most important (need to know) tools and technologies, mental models, and architectures to provide you with the skills necessary to tackle streaming problems with grace and finesse. All previous chapters have been leading here, as evolutionary steps along the journey to the K8s revolution, and our last chapter together.

This final chapter is broken into two main parts. In the first part, will get you started with Kubernetes locally. You'll learn to use the common components and resource APIs hands-on, while building your environment. Afterward, in part two, you'll learn to deploy your Apache Spark Structured Streaming applications on K8s. The chapter concludes with next steps and pointers so you can take everything you've learned locally up to the cloud and onward from there.

Exercise Materials   The exercise materials for Chapter 15 are located at `https://github.com/newfront/spark-moderndataengineering/tree/main/ch-15`.

Part 1: Getting Up and Running on Kubernetes

You can get started with Kubernetes locally in under an hour. Embracing the entire ecosystem and learning the myriad APIs and best practices can take a lot longer, but there are many books (and online tutorials and videos) available to walk you through the core APIs at a lower level. We will instead go over the essentials for getting started, touching upon the absolute need-to-know constructs, concepts, and paradigms. Let's get up and running.

Using Minikube to Power Local Kubernetes

Minikube is a simple-to-use environment (and shell) that runs a small, dedicated Kubernetes cluster locally in a virtual machine (Mac/Windows/Linux). Installation instructions are available online at `https://minikube.sigs.k8s.io/docs/start/`, or you can use `brew install minikube` if you are running on macOS. The following sidebar provides an additional overview of the basic minikube commands. You can come back to this section if you want to learn more. Otherwise, you can simply run `minikube start` to get up and running.

MINIKUBE BASICS

Once it's installed there are only a small handful of commands needed to operate minikube effectively. The following commands can be run in your terminal of choice.

minikube start

This command simply spins up a local single-node Kubernetes cluster and downloads and installs any additional dependencies on your behalf.

```
minikube start
```

By default, the minikube virtual machine starts with two (CPU) cores and 2GB of (RAM) memory. There are many options that allow you to customize your installation and specify how the K8s cluster will operate. For example, you can choose the version of K8s, the total memory to allocate to the virtual machine, and the number of CPU cores.

```
minikube start \
    --kubernetes-version v1.21.2 \
    --memory 16g \
    --cpus 4 \
    --disk-size 80g
```

When you are ready to stop working, you can choose to keep the virtual machine running and simply pause the processes in the k8s cluster, or you can choose to fully stop the cluster and reclaim your system resources.

minikube pause

Pause is a novel concept in minikube. Rather than fully shutting down your cluster, you can choose to pause (freeze) specific resources (governed by namespaces) or pause everything that is running. Pausing specific resources, such as local Kafka or Redis, can help when you need to see how your applications act when a service is lost.

```
minikube pause -A
minikube pause -n kafka,redis
```

minikube unpause

Any containers paused at a prior point in time can be unpaused.

```
minikube unpause -n kafka, redis
```

Using `pause` and `unpause` can be lifesavers if you have limited local system resources, and as an added benefit, they provide you with a simple mechanism to return to a specific environment setup. This is not unlike how `docker compose` was used to ensure system services were available for our Spark applications to run.

minikube stop

When you decide you want to simply stop everything and shut down the physical virtual machine, just use the `stop` command.

```
minikube stop
```

minikube status

From time to time, you will want to check the status of the minikube virtual machine. Just run the `status` command to check in on the environment.

```
minikube status
```

That concludes the quick tour of the minikube basics. Come back here if you need to revisit any of the basic commands.

Armed with your local (single node) Kubernetes cluster, you are now ready to begin the scaffolding process required to run your Spark applications.

A Hands-On Tour of the Kubernetes APIs

When working with any new tool, the primary objective is to understand, or simply become aware of, the essential moving parts. With Kubernetes, that means understanding the core services, concepts, and components. In this section, you will use the `kubectl` installed via minikube to interact directly with the Kuberenetes APIs.

Tip   If you want to support multiple versions of `kubectl` on your laptop (for testing purposes or otherwise), you can use `minikube kubectl --` to forward commands into the internal `kubectl` inside the minikube virtual machine.

Let's begin at the core, with the concept of a *node*.

Nodes

Nodes are central units of scalability in a Kubernetes cluster. Nodes independently register and join the cluster through the cluster manager (K8s master). Newly joined nodes increase the horizontal scale of the cluster by expanding the total schedulable resources (CPU cores and memory). Similarly, nodes that exit the cluster (shutdown or failed) take away from the total cluster resources, thus enabling loosely coupled elastically scalability.

Common Node Services and Responsibilities

Each node in the cluster is comprised of a minimum of at least three components: the *kubelet*, the *kube-proxy*, and the *container runtime*.

kubelet

The *kubelet* is an agent (running locally on every node) responsible for ensuring that the node can carry out its primary objective, which is to ensure all containers continue to run within their respective pods.

kube-proxy

The *kube-proxy* is an (agent) process that runs on every node in the cluster. It is responsible for maintaining the network rules for forwarding packets to and from the pods managed by the kubelet. This essential service enables cluster-wide communication between applications and services at the node level and participates in the service layer that enables external requests to be forwarded to a container running on a pod.

Container Runtime

The *container runtime* is a pluggable interface extending the *Kubernetes Container Runtime Interface* (CRI). It supports Docker and other popular container runtimes. Together the three components—the *kubelet, kube-proxy*, and *container runtime*—work to support the workloads running in a cluster. They do this by ensuring that each node remains loosely coupled to the cluster and is primarily focused on isolated local concerns, rather than cluster-wide concerns.

Viewing All Cluster Nodes

To list all available nodes in your cluster (there will be one by default with minikube), you can use the get *verb* on the *node resource* via the kubectl.

```
% kubectl get nodes

NAME        STATUS    ROLES                      AGE    VERSION
minikube    Ready     control-plane,master       64m    v1.21.2
```

Once you have the unique node *name* of a given node, use the describe *verb* to view the node's metadata.

```
% kubectl describe node minikube
```

The result of the describe node command provides you with all the available information regarding the node, including the node IP address (hostname), labels (think tags), any annotations, as well as the available and used node resources (core/memory/ etc.). I recommend revisiting this command again when more services are running in the cluster. After all, with only the one local node in the cluster, all pods and processes will be scheduled there. Next up, *namespaces*.

Namespaces

Namespaces provide the critically important role of protecting and isolating groups of services from user error, to separate concerns between many people (teams) running many workloads in a single federated Kubernetes cluster, and to control access to protected resources like namespace-scoped secrets and compute and storage allocations and prevent resource starvation.

Depending on the needs of your team or company, you can decide how you want to manage your production clusters. However, when running your local development cluster, I find it useful to separate services based on context and ownership. For example, databases and other data services live in the data-services namespace, while Spark applications live in another. By doing this, you effectively cordon off the dependent services needed to run your Spark applications from the namespace where the Spark applications run.

Tip   In minikube the namespace can also be used to pause or unpause all pods in the namespace.

```
minikube pause -n data-services.
```

Use this technique to observe how fault-tolerant your Spark applications are.

Next, you'll be creating the `data-services` namespace.

Creating a Namespace

You have two options when creating a namespace (and most resources and services in Kubernetes). You can use the `kubectl` directly or you can provide an external resource config file that can be applied.

Using the command line, execute the following `create` command.

```
% kubectl create namespace data-services
```

Success. You should see that the `data-services` namespace has been created. You can confirm this assumption by checking the available namespaces using the get *verb* on the `namespaces (ns)` resource.

```
% kubectl get namespaces
```

As a result, an ordered list of all active namespaces will be returned. Now, let's remove the namespace we just created, so we can re-create it again with a resource file.

Deleting a Namespace

Resources created with the Kubernetes API can be destroyed using the `delete` *verb* on the respective resource.

```
% kubectl delete namespace data-services
```

You'll be recreating the `data-services` namespace next, but it is an important step to learn to manage resources in a stress-free way since things can always be rebuilt.

Creating Kubernetes Resources with Apply

Open your favorite text editor (or IDE), and create a file named `namespace.yaml`, as shown in Listing 15-1.

Listing 15-1. namespace.yaml

```
apiVersion: v1
kind: Namespace
metadata:
```

```
labels:
  kubernetes.io/metadata.name: data-services
name: data-services
```

The namespace can now be created consistently using the apply *verb* on the kubectl.

```
% kubectl apply -f namespace.yaml
```

Deleting File-Based Resources   Any resource created with apply -f {file} can also be deleted using the same file.

```
kubectl delete -f namespace.yaml
```

Now you've seen two different ways to create and delete resources using the kubectl.

Note   You might be wondering why anyone would choose to use a file when the command line is faster (and easier). This decision can be filed under consistency. If you want to achieve good portability with your systems and services while also maintaining an ecosystem that runs just as *easily* locally as it will in the cloud, you need a *simple* way to rebuild (scaffold) an entire working environment, and that process starts with a durable blueprint.

Next, you'll learn the steps to deploy Redis into your minikube cluster using the newly minted data-services namespace.

Pods, Containers, and Deployments

We'll begin with a quick introduction to *pods* and follow up with an exercise. You'll see first-hand how a pod deployment works, and as a by-product, you'll learn what a pod is and how containers fit into the problem. Along the way, you'll learn to manage cluster resources at the container level. For the icing on the cake, you'll be introduced to stateful (durable) storage using persistent *volumes*.

These steps are all necessary as you work our way toward deploying your first Spark Structured Streaming application on Kubernetes. As you'll find out in part 2 of this chapter, Spark does a lot of magic on your behalf, but without understanding the constituent parts, the simplicity can come back to haunt you.

What Is a Pod?

Pods get their name from nature. A pea in a pod or a pod of whales. In Kubernetes, a *pod* is the smallest deployable unit of compute. It is important to think of a pod as a construct encapsulating an isolated environment for a group of one or more containers that provide runtime protection (think of the protective shell wrapping English peas) to the shared resources of the underlying group. Resources that include the sharing of a local (pod-centric) network (localhost/127.0.0.1) and local access to shared storage (volumes/ disk space), all with the runtime guarantee that a pod will only be successfully scheduled if all resource requests (CPU/memory/storage) and other specialized requirements are met completely (across all containers). This is required for a pod to be considered in a *healthy* state. The relationship of the pod to a node is shown in Figure 15-1. The common components (covered earlier in the chapter) are nodes, the kublet, the container runtime, and the kube-proxy, and they work together to run pods and route traffic to and from the containers of the pod.

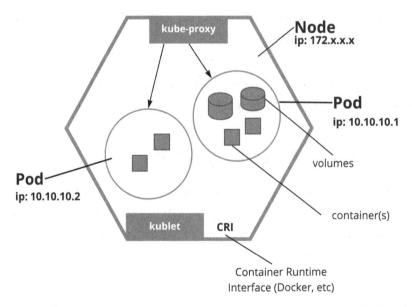

Figure 15-1. *Pods are scheduled to run on one or more nodes and allocate resources on the physical node, controlled by the kublet process*

Because each pod has its own localized scope, many pods can be scheduled to run on top of the same underlying node without the worry of collisions within the addressable network space. Furthermore, concerns regarding node-wide access to shared secrets or restricted files in the global user space can instead be managed at the pod level, making security and separation of concerns easier. Pods add a nice paradigm by separating workloads running across the shared cluster, as well as on the same physical node.

By design, containers running in a pod cannot break isolation and reach into other pods running on the same node. This would raise concerns from a security point of view and would also break the concept of loose coupling, since pods need to be aware of the other pods in the first place. Okay, so that is great, but to really appreciate pods, we'll create a pod spec to run Redis.

Creating a Redis Pod Spec

In Chapter 10, you were introduced to running Redis via docker compose. Let's revisit the Redis service from the chapter exercise's docker-compose.yaml and then convert the configuration into a Kubernetes pod spec. Listing 15-2 provides the Redis service from Chapter 10.

Listing 15-2. Docker Compose-Based Redis Service

```
services:
  redis:
    image: redis:6.2
    container_name: redis
    hostname: redis
    ports:
      - 6379:6379
    healthcheck:
      test: ["CMD", "redis-cli", "ping"]
      interval: 5s
      timeout: 30s
      retries: 50
    restart: always
```

Using the docker compose service configuration as our mental model, let's create a simple Redis PodSpec, and then launch the Redis pod. Listing 15-3 provides the configuration needed. Open your favorite editor, create a new file called redis-pod-basic.yaml, and add the contents from Listing 15-3.

Listing 15-3. Basic PodSpec to Configure and Launch Redis (redis-pod-basic.yaml)

```
apiVersion: v1
kind: Pod
metadata:
  name: redis-basic
  namespace: data-services
  labels:
    app: redis
    type: basic
spec:
  restartPolicy: OnFailure
  containers:
  - name: redis
    image: redis:6.2
    imagePullPolicy: Always
    ports:
    - containerPort: 6379
      protocol: TCP
    resources:
      limits:
        memory: "1G"
        cpu: "2000m"
      requests:
        memory: "500Mi"
        cpu: "1000m"
```

The Redis PodSpec in Listing 15-3 creates a uniquely named pod template. The reason that we call this a PodTemplate is because it can be used to launch one or more pods. Starting at the top, we begin by specifying the K8s apiVersion for our *pod resource* (kind: Pod). By declaring the API, Kubernetes is free to support and deprecate older APIs without forcing convergence.

metadata

Next, the metadata properties are used to uniquely assign a name to our pod (redis-basic) and to add configuration that can be used to schedule and locate our pod. In this case, we assign the data-services namespace and add some labels, which we can use later as a means of locating a specific subset of pods, or to isolate an individual pod within our namespace.

spec

Moving on to the most important config: the pod spec (spec) block itself. As mentioned, a pod encapsulates an isolated environment for one or more containers, and so the PodSpec is broken into many parts to allow for common configuration of shared resources, as well as container specifics configs. You'll see some overlap in the container specification between the docker-compose.yaml in Listing 15-2 and the Redis pod spec from Listing 15-3. The biggest differences being the granular control of the resource requests and limits, and the port configs.

Let's release this pod into the wild.

Scheduling the Redis Pod

Using kubectl we can apply the redis-pod-basic.yaml and let Kubernetes schedule our pod. Then we can ultimately reach convergence with our desired state of the world.

```
% kubectl apply -f redis-pod-basic.yaml
```

If all went as expected, you'll see the pod/redis-basic created message from the kubectl. Let's go find our pod.

Listing, Describing, and Locating Pods

From the command line, call get on the pods resource.

```
% kubectl get pods
```

You might be scratching your head at this point, wondering why no results are showing up (no resources were found in the default namespace). This is because Kubernetes (by default) will only list resources from a single namespace, which happens to be the default namespace. By adding the explicit namespace or by including the flag --all-namespaces, you can achieve the desired result.

```
% kubectl get pods -n data-services
NAME            READY    STATUS     RESTARTS    AGE
redis-basic     1/1      Running    0           44m
```

Now that you can find your pod, you can also gather more information regarding the pod itself by using the describe *verb* on the *pod resource* using the pod name.

```
% kubectl describe pod/redis-basic
```

The results of calling describe can be used to gather extended details about the pod itself, including the launch date, container, node, IP addresses, current observed pod conditions, and more.

Lastly, in the future case where you have many pods running, and you want to find a specific pod (or group of pods) you can lean on the metadata labels of the pod resource to find what you are looking for (across one or many namespaces). Listing 15-4 executes a universal search across all namespaces in order to find any pod with a matching label (app=redis) that is not the type pv (type!=pv).

Listing 15-4. Using Label Selectors to Locate Pods Across All Namespaces

```
% kubectl get pods \
  --all-namespaces \
  -l app=redis,type!=pv
```

The query in Listing 15-4 will return all matching pods. In our case there is currently only one.

```
NAMESPACE       NAME           READY    STATUS     RESTARTS    AGE
data-services   redis-basic    1/1      Running    0           78s
```

Our next order of business is to simply pop onto the new Redis pod and execute a few commands using the redis-cli.

Executing Commands on the Redis Pod

In the same way that you have grown comfortable using the docker exec command, you can literally transfer your skills directly to Kubernetes and the kubectl. The command in Listing 15-5 will connect you to the redis-basic pod and start the redis-cli.

Listing 15-5. Connecting to a Running Pod Using kubectl Exec

```
% kubectl exec \
  -n data-services \
  -it redis-basic \
  -- redis-cli
```

Get and set a key to make sure Redis is working as expected.

```
127.0.0.1:6379> set app redis-basic
OK
127.0.0.1:6379> get app
"redis-basic"
```

Exit the process (Command+C) so we can delete the pod.

Deleting the Redis Pod

You probably already know the command that is needed. Just like with the namespaces, you can simply use the delete *verb* on the pod resource to delete a pod by name.

```
% kubectl delete -n data-services pod/redis-basic
```

So why go through the effort to do all that work just to delete things? The answer resides in our expectations for a given service. In the real world, given that pods can be killed almost at random (due to the normal lifecycle of any cloud-native infrastructure), you'll ultimately need a reliable mechanism to ensure that all your services (including Spark applications) can be deployed, monitored, and restored in a consistent fashion with limited or zero human intervention.

We'll look at the Kubernetes deployment resource next, as we learn to create consistent and reliable deployments.

Deployments and Handling Automatic Recovery on Failure

You learned to configure a pod specification in the last section. Using a simple pod template, you can schedule and run a pod in the `data-services` namespace. Through this process you learned how a pod's metadata can be used to decorate a pod (using labels for label selection), and how to configure a simple single container spec. Using the skills you acquired for running the `redis-basic` pod, you'll learn next to generate managed deployments using the Kubernetes Deployment resource.

Creating a Kubernetes Deployment

A *deployment* is a special kind of K8s resource that introduces extended capabilities for managing and operating reliable deployments for a given pod specification. Managing reliability as a service, the Kubernetes cluster will supervise the resources configured and included within a deployment, ensuring common problems like node failures don't cause your services to come to a grinding halt.

Create a new file called `deploy-redis-memory.yaml` and add the contents of the deployment configuration in Listing 15-6 to the file. Then save it.

Listing 15-6. Deployments Provide Your Pod-Based Services with Runtime-Consistent Behavior Including Automatic Recovery from Failures, And Out-Of-The-Box Capabilities for Handling Rolling Updates for Replicated Services

```
apiVersion: apps/v1
kind: Deployment
metadata:
  name: redis-memory
  namespace: data-services
  labels:
    app: redis-memory
    type: canary
spec:
  replicas: 1
```

```
strategy:
  rollingUpdate:
    maxUnavailable: 100%
  type: RollingUpdate
selector:
  matchLabels:
    app: redis
    type: memory
template:
  metadata:
    labels:
      app: redis
      type: memory
  spec:
    containers:
    - name: redis
      image: redis:6.2
      volumeMounts:
      - name: redis-storage
        mountPath: /data
      imagePullPolicy: Always
      ports:
      - containerPort: 6379
        protocol: TCP
      resources:
        limits:
          memory: "1G"
          cpu: "2000m"
        requests:
          memory: "500Mi"
          cpu: "1000m"
    volumes:
    - name: redis-storage
      emptyDir:
        medium: Memory
```

You'll notice that the initial deployment config in Listing 15-6 wraps the pod spec that was introduced in Listing 15-3, with two small changes to introduce pod-based local *volume* storage. *Volumes* are shared resources in the pod, but rather than being immediately available to all containers, each container must declaratively add the volume using `volumeMounts`. Mounting (or attaching) volumes to a container also comes paired with the added benefit of enabling each container to create a custom mount path.

```
volumeMounts:
- name: redis-storage
  mountPath: /data
```

We'll cover `volumes` and `volumeMounts` later, but for now, take notice that the `volumes` spec is configured at the same level as the `containers` spec, while the `volumeMounts` are configured on a container-to-container basis. Now let's create the deployment.

Deployments and Resource Management

Using the deployment configuration from Listing 15-6, apply the config to create and kick off your deployment.

```
% kubectl apply -f deploy-redis-memory.yaml
```

Now that your deployment has been created, let's have a little fun. Rather than unpacking all the deployment details, let's instead investigate the deployment using the following four steps: locate, connect, observe, and delete.

Locate the Newly Deployed Pod

Find your newly deployed `redis-memory` pod using any of the techniques you've picked up in the chapter. When you locate it, you'll notice that the pod has a longer name than before. For example, mine was created as `redis-memory-78d89d7c87-zchc2`. If you want to understand why, check out the deployment metadata.

```
% kubectl get deployment/redis-memory -n data-services
```

Now, connect to the managed pod.

Connect to the Managed Pod

Connect to the managed Redis pod and repeat the steps from Listing 15-5. Remember to check that the keys were saved before you disconnect from the pod.

Observe Pod States and Behavior

In a second terminal window (preferably side-by-side to your existing terminal session), run the following command.

```
% kubectl get pods -n data-services --watch.
```

Watching the pods will enable you to observe real-time state changes for all the pods in the target namespace.

Delete the Managed Pod

Returning to your first terminal window, delete the Redis pod.

```
% kubectl delete -n data-services pod/redis-memory-78d89d7c87-zchc2
```

What happens?

You'll observe that the second you delete the managed pod, a new pod is immediately, in near real-time, scheduled to replace the deleted pod.

The output in Listing 15-7 shows the state change messages as observed on my cluster. The output shows my initial pod (`redis-memory-78d89d7c87-zchc2`) as it changes state from `Running` to `Terminating`. At the same time, a new Pod is created to ensure we always have at least one replica, given our deployment states that we desire replicas: 1. The replacement pod comes up initially in the `Pending` state, then `ContainerCreating`, and lastly `Running`. All of this happens in under five seconds locally. Unless the deployment is deleted, Kubernetes will keep your pod running forever.

Listing 15-7. The Observed State Changes When a Managed Pod (Behind a Stateful replicaset) Is Terminated

NAME	READY	STATUS	RESTARTS	AGE
redis-memory-78d89d7c87-zchc2	1/1	Running	0	35m
redis-memory-78d89d7c87-zchc2	1/1	Terminating	0	36m
redis-memory-78d89d7c87-p4hzl	0/1	Pending	0	0s

redis-memory-78d89d7c87-p4hzl	0/1	Pending	0	0s
redis-memory-78d89d7c87-p4hzl	0/1	ContainerCreating	0	0s
redis-memory-78d89d7c87-zchc2	0/1	Terminating	0	36m
redis-memory-78d89d7c87-p4hzl	1/1	Running	0	4s
redis-memory-78d89d7c87-zchc2	0/1	Terminating	0	36m
redis-memory-78d89d7c87-zchc2	0/1	Terminating	0	36m

Now for the bad news. While the deployment controller was able to restore the deleted pod, you'll find that there are zero keys in the Redis database. You can confirm this for yourself using the `redis-cli`. Well, why is that? Wasn't the pod supposed to be consistent and reliable?

This is in fact all by design. To declare each deployment idempotent and to support runtime fault-tolerance in a consistent way, pods and their underlying containers must use *immutable conventions*. Otherwise, runtime alterations (mutations) to any pod-local resource (such as adding new keys to the Redis database) will be lost when a pod is terminated or experiences irrecoverable failure. Given that pods, by convention, need to be capable of quickly moving around the cluster, it is a best practice for external configuration, secrets, as well as storage to be managed by Kubernetes to ensure consistency.

Let's complete our hands-on tour of Kubernetes by learning to persist data between deploys. Then we'll get our hands dirty with the Kubernetes Resource Manager for Apache Spark.

Persisting Data Between Deploys

Kubernetes provides a rich abstraction for file system (and memory) storage based on the concept of `volumes` and `volumeMounts`. You were introduced to these two concepts in the last section (in Listing 15-6), and now you'll learn to enable persistent storage that can survive pod failures.

To get started, we are going to use node-level storage, which persists on a Kubernetes node and prevents data loss when a pod is terminated. It is worth noting that if you delete your minikube virtual machine, you will lose your node-level data as well. From the command line, `ssh` to minikube and then create an empty directory located at `/mnt/data`.

```
% minikube ssh
$ sudo mkdir -p /mnt/data
```

With this step completed, we can move on to creating a persistent volume resource.

Warning   ssh access in managed production Kubernetes clusters (such as EKS/ AKS/GKE) is most likely not necessary. Consider restricting ssh access or turning it off completely to protect malicious actors from messing with your Kubernetes cluster.

Persistent Volumes (PV)

Local changes made in a container at runtime will disappear when the pod terminates, fails, or is otherwise replaced (rescheduled) due to normal lifecycle events (like a fresh deployment). As you've seen, this also includes any database records added to the local pod environment.

Fortunately, reliable storage can be provisioned for use in the Kubernetes cluster, using the local file system storage of the underlying Kubernetes' nodes, or the more loosely coupled practice of attaching managed cloud network file system storage. Consider for example the offerings of the major cloud vendors: Amazon's Elastic Block Storage, Microsoft's AzureFile/AzureDisk, and Google's Persistent Disk. Persistent volumes abstract away the storage providers from the actual consumers, enabling applications to request storage, and for Kubernetes to do the work of finding and claiming resources matching the storage requests.

Configuring a Persistent Volume (PV)

Create a file named persistent-volume.yaml and add the PersistentVolume resource configuration shown in Listing 15-8. This template can then be applied to the Kubernetes cluster to create a persistent volume located physically on the single node in your minikube cluster.

Listing 15-8. Creating a PersistentVolume for Redis (persistent-volume.yaml)

```
apiVersion: v1
kind: PersistentVolume
metadata:
  name: general-pv-volume
```

```
  namespace: data-services
  labels:
    type: local
spec:
  storageClassName: manual
  capacity:
    storage: 2Gi
  accessModes:
    - ReadWriteOnce
  hostPath:
    path: "/mnt/data"
```

The resource configuration in Listing 15-8 creates (cordons off) a 2GB volume
located in the /mnt/data directory of your minikube node. Apply the configuration to
create the new resource, then look at what you've created using the get and describe
verbs on the persistentvolume (pv) resource type to dive deeper into this new
resource type.

Create the Persistent Volume

```
% kubectl apply -f persistent-volume.yaml
```

View the Persistent Volume

```
% kubectl get pv general-pv-volume
NAME               CAPACITY   ACCESS MODES   RECLAIM POLICY
general-pv-volume  2Gi          RWO             Retain
```

Describe the Details of the Persistent Volume

```
% kubectl describe pv general-pv-volume
```

Armed with the general-pv-volume, our next step is to create a claim policy to
govern read/write access to the volume using the PersistentVolumeClaims resource.

Persistent Volume Claims (PVC)

PVs are storage-specific resources in your cluster. The PVC acts as a resource that claims a portion (or all) of a PV. Just like a node can be sliced up to run workloads of various sizes (controlled by pod specs) with respect to CPU cores and memory, PVCs represent a storage request of a specific amount of storage space, along with specific access modes to govern inclusive or exclusive read and write access to the underlying storage resource.

Configuring a Persistent Volume Claim (PVC)

Using the configuration in Listing 15-9, create a new file named `persistent-volume-claim.yaml`.

Listing 15-9. Creating a PersistentVolumeClaim (persistent-volume-claim.yaml)

```
apiVersion: v1
kind: PersistentVolumeClaim
metadata:
  name: redis-pv-claim
  namespace: data-services
spec:
  storageClassName: manual
  accessModes:
    - ReadWriteOnce
  resources:
    requests:
      storage: 500Mi
```

The PVC in Listing 15-9 configures a named claim (`redis-pv-claim`) requesting at least 500 Mi (MB) and the access mode `ReadWriteOnce`. This access mode allows a pod to claim exclusive access to the bound PV, which is the correct access mode for our Redis service. You'll probably also have noticed that the PVC doesn't mention anything about the actual PV (this is due to the loose coupling of storage provisioning and storage consumption; the details are simply abstracted away). Next, create the PVC resource, and then learn more about what you've created using the `get` and `describe` verbs on the `persistentvolumeclaim` (`pvc`) resource type.

Create the Persistent Volume Claim (PVC)

```
% kubectl apply -f persistent-volume-claim.yaml
```

Use Get to View the PVC

Checking on the status of the PVC shows if the claim has been bound to a PV, including the name of the bound volume.

```
% kubectl get pvc -n data-services
NAME             STATUS   VOLUME              CAPACITY
redis-pv-claim   Bound    general-pv-volume   2Gi
```

View the PV

Given that the PV to PVC relationship is bi-directional, you can also discover the PVC claim and the status of the claim (reference) through the PV.

```
% kubectl get pv redis-pv-volume
NAME               RECLAIM STATUS CLAIM
general-pv-volume Retain  Bound  data-services/redis-pv-claim
```

Tip   Kubernetes will use the `resources:requests:storage` value from the PVC to attempt to find a matching PV. In our case, we claim 2Gi when we only requested 500Mi. I recommend looking at dynamic provisioning for PVs as a follow-up if you are interested.

Enhancing the Redis Deployment with Data Persistence

Let's put the PV and PVC into action and modify our earlier Redis deployment from Listing 15-6. It turns out all we need to do at this point is modify the volumes configuration of the pod spec. Open your deployment config file from earlier and modify the volume type from `emptyDir` to `persistentVolumeClaim` (as shown in Listing 15-10).

Listing 15-10. Using a Persistent Volume Claim to Persist Data

```
apiVersion: apps/v1
kind: Deployment
...
spec:
  template:
    spec:
      volumes:
      - name: redis-storage
        persistentVolumeClaim:
          claimName: redis-pv-claim
```

Save the changes for the deployment and apply the updated deployment. Any data saved in your Redis database (stream objects, keys, etc.) will persist between deployments in a stateful and consistent way. Run through the process from Listing 15-6 and then delete your pod. Watch as your Redis service comes back up in the same state you left it.

Kubernetes Services and DNS

As a final step, let's add a Service definition for our Redis pod deployment. Services in Kubernetes enable service discovery within the scope of a single namespace (by default) allowing you, for example, to use Redis from within the `data-services` namespace, which could be used in powering Airflow or to act as a cache layer for MySQL.

Creating a Service

Create a file named `redis-service.yaml` and add the following Service configuration in Listing 15-11. Then apply the config using `kubectl apply -f`.

Listing 15-11. Defining a Kubernetes Service Resource for Redis

```
apiVersion: v1
kind: Service
metadata:
  name: redis-service
  namespace: data-services
```

```
spec:
  selector:
    app: redis
    type: memory
  type: ClusterIP
  ports:
    - name: tcp-port
      port: 6379
      targetPort: 6379
      protocol: TCP
```

The Redis service definition in Listing 15-11 creates a dynamic service pointing to port 6379 on the group of zero or more pods matching the `Service` selector (`app=redis,type=memory`). This abstraction enables you to define a loosely coupled service that leans on Kubernetes to fill in the endpoint(s) dynamically, using selectors. The bound endpoints are the pod's IP addresses for the subset of pods matching the selection criteria

Viewing Services (SVC)

```
kubectl get services -n data-services
NAME            TYPE        CLUSTER-IP      PORT(S)    AGE
redis-service   ClusterIP   10.106.24.62    6379/TCP   8d
```

We'll revisit services again in Part 2. For now let's recap what we've learned.

Part 1: Summary

The first half of this chapter introduced you to many of the core components, concepts, and APIs available for running cloud-native applications on Kubernetes. As mentioned, running Apache Spark on Kubernetes (using the official Spark Kubernetes Resource Manager) abstracts away (and protects you from) much of the runtime details and complexity required to get started with K8s. However, when moving from POC to production, it is always necessary to establish a solid foundation of not only the tools but of the environment where they run. That's why we walked through the steps to set up and configure the persistent Redis deployment.

As a follow-up exercise, test what you've learned in the first half of the chapter by bringing both MySQL and MinIO into your local Kubernetes cluster. You can use the earlier lessons in the book to reference everything you'll need, including using the `docker-compose` files and initialization steps. Get creative with how you solve these problems and use the process to take a closer look at what works and doesn't work as you build up your mental models.

Now the part we've all been waiting for—it is time to learn how to deploy your Apache Spark applications on Kubernetes.

Part 2: Deploying Spark Structured Streaming Applications on Kubernetes

With your Kubernetes cluster up and running and your `data-services` namespace carved out, we'll dive into the additional Kubernetes resources needed to power our Apache Spark environment.

Hands-On Content   The chapter material for this second part of the chapter is located in `ch-15/spark-on-kubernetes`. The complete walkthrough is available in the READMEs. Full solutions for running both MinIO and MySQL in the `data-services` namespace can prepare you for this next adventure.

Spark Namespace, Application Environment, and Role-Based Access Controls

There are a few essential resources required to enable your Apache Spark applications to run. In this section, we define the `spark-apps` NAMESPACE, and then you'll be introduced to roles, role-bindings, and service accounts.

Define the Spark Apps Namespace

The command in Listing 15-12 creates a new namespace for our Spark applications called `spark-apps`. Create this namespace.

Listing 15-12. *Defining the Spark Apps Namespace*

```
cat <<EOF | kubectl apply -f -
apiVersion: v1
kind: Namespace
metadata:
  labels:
    kubernetes.io/metadata.name: spark-apps
  name: spark-apps
EOF
```

With the `spark-apps` namespace defined, the next step is to establish some rules to govern the runtime capabilities of our Spark applications. Just like in the real world, you wouldn't give your house keys to any random stranger who asked, and the same goes in the Kubernetes world. Pods have known identities controlled by nonfungible mutual-TLS controlling authentication and authorization based on token/certification exchange between the pods and the Kubernetes API. A pod can assume a specific identity and receive special resource and access permissions governed by the service account, role, and role-binding resources.

Role-Based Access Control (RBAC)

Like establishing permissions in SQL systems using GRANT and REVOKE, role-based access control (RBAC) is used to define general capabilities and permissions for your pods as well as the real-world users accessing your K8s clusters.

Roles

Roles within Kubernetes are used to define common permissions at the API level (such as creating, modifying, or destroying resources).

```
% kubectl get roles -n {namespace}
```

Roles define a list of rules that are governed on the resource and are verb level within a specific namespace.

> **Note**   You can also define rules that govern access at the cluster-level, rather than just a specific namespace. These are called `clusterroles`: `% kubectl get clusterroles`

Service Accounts

Service accounts provide an identity mechanism in the Kubernetes API. This is equivalent to using IAM permissions on Amazon Web Services. Service accounts provide token-authorized access to specific resources in the Kubernetes cluster. By default, each namespace generates a common service account named `default`. This default identity is then automatically associated with any pod created in each namespace unless a different service account is specified.

```
% kubectl get sa -n {namespace}
```

Role Bindings

Role bindings connect a user, group, or service account to a specific role.

```
% kubectl get rolebindings -n {namespace}
```

Let's set up the `spark-controller` service account now, along with the associated role and role-binding.

Creating the Spark Controller Service Account, Role, and Role-Binding

Given that pods run within namespaces and given the fact that we want our Spark Structured Streaming applications to be able to supervise themselves—adding and removing executor pods when necessary—we need to create a special service account (`spark-controller`) that can be bound to our Spark application's pods. This will grant the pods capabilities to create and destroy themselves and manage additional resources within the namespace (so that we don't have to).

The multi-part resource definition in Listing 15-13 generates three new resources inside the `spark-apps` namespace. The Role (`spark-app-controller`), ServiceAccount (`spark-controller`), and RoleBinding (`spark-app-controller-binding`).

Listing 15-13. Creating Multiple K8s Resources Can Be Achieved Using the multipart --- Separator

```
cat <<EOF | kubectl apply -f -
apiVersion: rbac.authorization.k8s.io/v1
kind: Role
metadata:
  namespace: spark-apps
  name: spark-app-controller
rules:
  - apiGroups:
      - ""
    resources:
      - pods
      - services
      - configmaps
    verbs:
      - list
      - create
      - delete
      - get
      - list
      - patch
      - update
      - watch
---
apiVersion: v1
kind: ServiceAccount
metadata:
  name: spark-controller
  namespace: spark-apps
---
apiVersion: rbac.authorization.k8s.io/v1
kind: RoleBinding

metadata:
```

```
  name: spark-app-controller-binding
  namespace: spark-apps
subjects:
- apiGroup: ""
  kind: ServiceAccount
  name: spark-controller
roleRef:
  apiGroup: rbac.authorization.k8s.io
  kind: Role
  name: spark-app-controller
EOF
```

The technique in Listing 15-13 enables you to provide a single file declaring a set of resources. Leaning on the multipart resource definition pattern ensures that the various needs of each isolated deployment are met in a consistent way.

Tip   For complex deployments, or for bootstrapping entire environments, you can lean on multi-resource configurations like the Spark bootstrap config (Listing 15-13). In doing so, you can ensure that all your deployment dependencies (resources) are available and applied within the K8s cluster in the correct order before deployment. The name of the game is consistency. Two popular solutions have been adopted by the K8s community—helm and kustomize—to make it easier to manage deployment complexity. Helm is like a K8s package manager, enabling complex deployments to be installed and updated with ease using the command line. Kustomize, on the other hand, is a template-free configuration manager that makes it easy to merge (overlay) environment-specific configs on top of common deployment resources. This solution enables you to configure deployments for development, stage, and production environments while reusing common configuration to prevent the problems of configuration drift.

Now that you have applied the environment scaffolding, it is time to revisit the generic Structured Streaming application from Chapter 13 and prepare it for deployment in the K8s cluster.

Redis Streams K8s Application

You were introduced to Structured Streaming using Redis Streams in Chapter 10, as a lightweight (low overhead) streaming source. The initial application was built to show you how to reliably read unbounded streams of `CoffeeOrder` events, control the application triggering process (as a stateful batch or streaming micro-batch), and maintain the state of the application using checkpoints. The result was a simple, but consistent, streaming application that read and wrote `CoffeeOrders` into distributed tables in our data warehouse (MinIO).

In Chapter 13, we leveraged Spark Structured Streaming to generate time-based (windowed) aggregations, thereby simplifying the otherwise complex process of creating reliable streaming analytics data. We'll build on the generic aggregations application from Chapter 13 and create a new container that can be deployed and managed inside the minikube cluster.

To get started, open the `ch-15/applications/redis-streams-k8s` application in your favorite IDE or simply open the source code to follow along.

Building the Redis Streams Application

We won't be touching the source code of the application; it is ready to be built. There are just a few small adjustments that are worth pointing out from the container perspective.

First, the Dockerfile has been adjusted to copy the contents of the `user_jars` directory into our final container, which can be seen in Listing 15-14. This process is done for consistency, ensuring there are no unexpected missing dependencies or runtime gotchas.

Listing 15-14. Modifying the Dockerfile to Package Any Additional Runtime JARs and Base App Configurations

```
FROM newfrontdocker/${spark_base_version}
ARG app_jar

USER 1000


# copy and rename
COPY --chown=1000 ${app_jar} /opt/spark/app/jars/redis-streams-k8s.jar
```

```
# copy the config for the app
COPY --chown=1000 conf /opt/spark/app/conf
COPY --chown=1000 user_jars /opt/spark/app/user_jars
```

As mentioned, the changes to the Dockerfile were made to generate a consistent container, one that includes all required runtime JARs and common baseline configurations. The baseline configurations that ship in the container are the TypeSafe configurations from Chapter 13, the `fairscheduler.xml` file (introduced in the last chapter), and two executables (`spark-submit.sh` and `spark-submit-lite.sh`), which can be used to submit your application from within the Kubernetes cluster.

Tip   If you need similar common dependencies, those can be bundled into your base Spark container images. That way, each application doesn't have to repeat the process of BYO-JARs. Just make sure to add what is necessary for common use cases so the size of the image remains as small as possible.

Building the Application Container

The second change is to the container build process. You'll need to build your local application image using the minikube Docker environment. There is one small additional step, as shown in Listing 15-15, which boils down to a one-line `eval` statement.

Listing 15-15. Building Your Container for Use in the minikube Environment

1.   `sbt clean assembly`
2.   **`eval $(minikube docker-env)`**
3.   `docker build . -t mde/redis-streams-k8s:1.0.0`

Now that you have built the container, the next step is to deploy it.

Tip   It is faster while you are testing things locally to use the trick in Listing 15-15 to iterate quickly on your application. The alternative path is to build and push your image to DockerHub or an alternative container registry service like Amazon ECR.

Deploying the Redis Streams Application on Kubernetes

Thinking back to the journey of running the various Spark applications throughout the book, we often augmented the runtime environment using mounted configuration and runtime resources (JARs, secrets in `hive-site.xml`, and config) via `docker-compose`. To enable communication between specific containers, we also used the internal Docker network (`mde`). It makes total sense to apply similar strategies to our Spark applications on Kubernetes.

Earlier in the chapter, you learned about *services* and *volumes* (persistent and ephemeral). In preparation for deployment, you'll add a new service type to simplify cross-namespace DNS resolution with the services running in the `data-services` namespace. You'll also learn how to externalize environment variables and deployment-specific configuration, as well as secrets, like the `hive-site.xml` and the MinIO access/secret keys. We'll turn our attention to using volumes that extend beyond traditional file-based storage (directories and individual files).

Running the Code   From your terminal, switch your working directory to `ch-15/spark-on-kubernetes`.

Externalized Dependency Management

There are two important resource types that act like volumes (and can be mounted to a pod), but provide additional specialized capabilities. They are named `ConfigMaps` and `Secrets` and you will lean on them in your deployment.

External Configuration with ConfigMaps

The `ConfigMap` resource provides a solution for storing uniquely identified collections of data (files, binary data, etc.) scoped within a namespace. `ConfigMaps` are commonly used to provide shared environment variables (for `dev`, `stage`, and `production` environments, for example) and to provide isolated (deployment-specific) application resources.

Let's create a `ConfigMap` that can be used as our default Spark application configuration (`spark-defaults.conf`). From your terminal (within the `spark-on-kubernetes` directory), run the following command:

```
kubectl create configmap spark-redis-streams-conf \
  --from-file=spark-redis-streams/config \
  -n spark-apps
```

Magically, the contents of the config/spark-defaults.conf file are encoded into your spark-redis-streams-conf resource. To inspect the contents, you can run the following command:

```
kubectl get configmap spark-redis-streams-conf \
  -n spark-apps \
  -o yaml
```

The output essentially just shows you the contents of your spark-defaults.conf.

Secrets

Everyone knows you shouldn't commit usernames, passwords, or sensitive data into Git (or whatever version control system you are using. Otherwise, you may inadvertently leak sensitive data. The Secret resource type allows you to store sensitive data (within your namespace) that is guaranteed to travel over an encrypted (HTTPS) channel and is stored internally as a base-64 encoded string.

Securely Storing the Hive Metastore Configuration

The command in Listing 15-16 encodes the hive-site.xml that we've been using to access the MySQL backed Hive Metastore and wraps it into a secret resource called hivesite-admin that can be mounted safely onto our Spark application pod at runtime.

Listing 15-16. The hive-site.xml Wrapped as a Secret for Use Within Our Application Deployment

```
cat <<EOF | kubectl apply -f -
apiVersion: v1
kind: Secret
metadata:
  name: hivesite-admin
  namespace: spark-apps
type: Opaque
stringData:
```

```
hive-site.xml: |
  <configuration>
    <property>
      <name>javax.jdo.option.ConnectionURL</name>
      <value>jdbc:mysql://mysql-service.spark-apps.svc.cluster.
      local:3306/metastore</value>
    </property>
    <property>
      <name>javax.jdo.option.ConnectionDriverName</name>
      <value>org.mariadb.jdbc.Driver</value>
    </property>
    <property>
      <name>javax.jdo.option.ConnectionUserName</name>
      <value>dataeng</value>
    </property>
    <property>
      <name>javax.jdo.option.ConnectionPassword</name>
      <value>dataengineering_user</value>
    </property>
  </configuration>
EOF
```

You'll notice that `hive-site.xml` is stored in the `stringData` block. For binary files like `hive-site.xml` and `spark-defaults.conf`, you can't use simple key/value pairs, which makes the resource definition longer in size.

While we are here, it is worth also pointing out that `ConnectionURL` in `hive-site.xml` is pointing to a service domain name that doesn't currently exist in the `spark-apps` namespace.

```
jdbc:mysql://mysql-service.spark-apps.svc.cluster.local:3306/metastore
```

We'll be adding an `ExternalService` resource later to provide our alias to the `data-services` MySQL service.

Tip   Use the `ch-15/spark-on-kubernetes/data-services/README.md` to set up MySQL on K8s if you haven't tried your hand at it yet and want the solution.

Now that we have `hive-site.xml` encoded as a secret, the other sensitive information we've been working with is the MinIO access and secret keys. This is thankfully a simpler definition, using key/value pairs.

Securely Storing the MinIO Access Parameters

The resource definition in Listing 15-17 encodes the administration username and password information. Create the `minio-access` resource. Remember that secrets are controlled outside of your application source code, so you don't have to worry about accidently committing secrets to GitHub. Just reference known secrets and mount them to your pod for use in your container.

Listing 15-17. Creating the minio-access key/value Pair Secret for Use with the Spark Application Deployment

```
cat <<EOF | kubectl apply -f -
apiVersion: v1
kind: Secret
metadata:
  name: minio-access
  namespace: spark-apps
type: Opaque
stringData:
  access: minio
  secret: minio_admin
EOF
```

With the two secrets created, you can now use them in your deployment specifications using `volumes` and `volumeMounts`.

Viewing Secrets in the Namespace

If you want to view the encoded secrets, you can use the following commands:

```
% kubectl get secrets -n {namespace}
% kubectl get secret/{name} -n {namespace} -o yaml
```

The next order of business is to add the `ExternalService` configurations required to run our Spark Structured Streaming application.

Cross Namespace DNS and Network Policies

Fun fact: namespaces provide a natural protective barrier for cross-namespace communication. To enable crosstalk between namespaces, for example in an actual production Kubernetes cluster, you need to create a policy using an special service configuration called ExternalName, along with a NetworkPolicy. Combined, this enables applications running in an isolated namespace to make requests to services running in another namespace. For example, to enable your Spark applications to communicate with the data-services namespace.

When running on minikube, we are going through the motions here since the network is shared on our single node. Let's define the service pointer to the data-services MySQL service (shown in Listing 15-18).

Listing 15-18. The ExternalName Service Provides a Local Fully Qualified Domain Name (FQDN) Internal to Your Namespace that Resolves to an External Service in Another Namespace

```
cat <<EOF | kubectl apply -f -
kind: Service
apiVersion: v1
metadata:
  name: mysql-service
  namespace: spark-apps
spec:
  type: ExternalName
  externalName: mysql-service.data-services.svc.cluster.local
  ports:
  - port: 3306
    name: tcp-port
EOF
```

Why would you need to use this? As things go, internally managed services routinely switch ownership between teams over time. If a new team is picking up ownership of some services and switching namespaces, you can update a single resource and migrate your applications or chain ExternalName resources between namespaces.

Ultimately, as things grow in scale, you'll find yourself eventually migrating to a service mesh such as Istio, but since you are starting out, remember to keep things simple (there are already many new components and services to keep straight).

As a follow-up exercise, create additional policies for Redis and MinIO, following the pattern from Listing 15-18. When you are ready, it is time to deploy.

Controlling Deployments with Jump Pods

To begin, you'll deploy the Spark application via the `mde/redis-streams-k8s:1.0.0` container wrapped in a simple "jump" pod. If you come right out of the gate at full speed, it is easy to forget something and fall into what is known as a "crash loop." You might find it necessary to repeat this slow, controlled deployment flow for the first few applications that follow any new deployment patterns. This process is also helpful when debugging applications, since you can check common troublemakers like missing configurations prior to automating everything an official stateful deployment.

From the `ch-15/spark-on-kubernetes/spark-redis-streams` directory, open `deployment-manual.yaml` (shown in Listing 15-19) to view the pod specification. Pay special attention to the `volumes` and `volumeMounts` environment variables to see how the `Secrets` and `ConfigMaps` are mixed into the pod, and how the external DNS values are mixed into the specification.

Listing 15-19. The Spark Jump Pod (spark-redis-streams-app) Spec

```
apiVersion: v1
kind: Pod
metadata:
  name: spark-redis-streams-app
  namespace: spark-apps
  labels:
    app: spark-redis-streams
    type: canary
    version: "1.0.0"
spec:
  restartPolicy: OnFailure
  # The Jump Pod runs as root
```

```yaml
  securityContext:
    runAsUser: 0
    runAsGroup: 0
  serviceAccountName: spark-controller
  volumes:
    - name: spark-submit-conf
      configMap:
        name: spark-redis-streams-conf
    - name: hive
      secret:
        secretName: hivesite-admin
  containers:
    - name: redis-streams-app
      env:
        - name: SPARK_CONTAINER_IMAGE
          value: mde/redis-streams-k8s:1.0.0
        - name: REDIS_SERVICE_HOST
          value: redis-service.spark-apps.svc.cluster.local
        - name: REDIS_SERVICE_PORT
          value: "6379"
        - name: MINIO_SERVICE_HOST
          value: minio-service.spark-apps.svc.cluster.local
        - name: MINIO_SERVICE_PORT
          value: "9000"
        - name: MINIO_ACCESS_KEY
          valueFrom:
            secretKeyRef:
              name: minio-access
              key: access
        - name: MINIO_SECRET_KEY
          valueFrom:
            secretKeyRef:
              name: minio-access
              key: secret
      volumeMounts:
```

```
    - mountPath: /opt/spark/conf/spark-defaults.conf
      name: spark-submit-conf
      subPath: spark-defaults.conf
      readOnly: false
    - mountPath: /opt/spark/conf/hive-site.xml
      name: hive
      subPath: hive-site.xml
      readOnly: true
  image: mde/redis-streams-k8s:1.0.0
  imagePullPolicy: Never
  command: ["tail", "-f", "/dev/null"]
  resources:
    limits:
      memory: "2000Mi"
      cpu: "500m"
    requests:
      memory: "1000Mi"
      cpu: "250m"
```

The pod spec in Listing 15-20 securely mounts the secrets and configurations that you applied to the spark-apps namespace. We are using the hanging tail command to keep the pod up and running (such as when working with the Redis service in part 1 of the chapter). This allows you to exec into the redis-streams-app container process and run the application.

Create (and schedule) the pod by executing the following command:

```
% kubectl apply -f deployment-manual.yaml
```

A pod/spark-redis-streams-app is created. Now, we can exec into the pod and trigger the application.

Triggering Spark Submit from the Jump Pod

Pop onto the jump pod using the exec command to launch the actual Spark application. For a mental model, consider the jump pod as an authenticated process that resides in your K8s cluster, thus allowing the user to deploy applications (using the spark-controller service account).

```
% kubectl exec --stdin --tty pod/spark-redis-streams-app \
  -n spark-apps \
  -- bash
```

From the exec process, you can submit the application to the Kubernetes cluster using the spark-submit.sh executable that we packaged inside the container. It is located in the following path: /opt/spark/app/conf/spark-submit.sh.

Tip   Before you start the application the first time, I always recommend starting up a second terminal window to observe how the pods are generated by the spark-submit process. You can also use the logs generated by the spark-submit process to observe the actions of the Spark Kubernetes Resource Manager.

```
kubectl get pods -n spark-apps –watch
```

From the exec process on the spark-redis-streams-app pod, run the following command.

```
% /opt/spark/app/conf/spark-submit.sh
```

As the application starts up and switches to a Running state, the spark-submit process you just triggered will make periodic calls to the K8s API to check in on the status of the driver (essentially the same behavior that you'll see from the kubectl process watching the pods in the namespace).

Fun fact: you just used your service account for the first time, which enabled you to create new resources on the fly. This allowed you to delegate control to the driver application process. To view the resources you created, call kubectl get pods -n spark-apps.

NAME	READY	STATUS	RESTARTS	AGE
spark-redis-stream-aggs-app-cf14ec7d5490f37d-exec-1	1/1	Running	0	29m
spark-redis-stream-aggs-app-cf14ec7d5490f37d-exec-2	1/1	Running	0	29m
spark-redis-streams-app	1/1	Running	0	33m
spark-redis-streams-app-driver	1/1	Running	0	29m

You are free to exit the process. You'll return to the exec process since the spark-submit.sh creates a cluster-based deployment. (We covered cluster deploys in Chapter 14, but as a refresher, this means the driver application process runs in the cluster alongside the executors, as opposed to being controlled as an external resource.)

Note   Run ls -l /opt/spark/app/conf on the spark-redis-streams-app pod and you'll notice that a second spark-submit* file exists, called spark-submit-lite.sh. This second config file provides an example of the minimal required configuration to run the aggregation's application. Given the second example isn't an executable (it's missing the -x), you can run the app using bash /opt/spark/app/conf/spark-submit-lite.sh.

Connecting the Deployment Dots

The environment variable-controlled spark-submit.sh executable is shown in Listing 15-20. You'll see the important config settings (just the pod spec-based overrides) that are commonly used to modify the runtime configuration for an application when it is running in different K8s clusters (non-prod versus prod). You run an application using a new checkpointLocation and output path so you don't corrupt the application's state when deploying a replacement application (or canary).

Listing 15-20.   The spark-submit.sh Executable

```
#!/usr/bin/env bash

K8S_MASTER=${K8S_MASTER:-k8s://https://${KUBERNETES_SERVICE_
HOST}:${KUBERNETES_SERVICE_PORT_HTTPS}}
...
$SPARK_HOME/bin/spark-submit \
  --master ${K8S_MASTER} \
  --deploy-mode cluster \
```

```
--class "com.coffeeco.data.SparkStatefulAggregationsApp" \
--conf "spark.kubernetes.driver.pod.name=${SPARK_APP_NAME}-driver" \
--conf "spark.kubernetes.authenticate.driver.serviceAccountName=${K8S_
SERVICE_ACCOUNT_NAME}" \
--conf "spark.redis.host=${REDIS_SERVICE_HOST}" \
--conf "spark.redis.port=${REDIS_SERVICE_PORT}" \
--conf "spark.hadoop.fs.s3a.endpoint=${MINIO_SERVICE_HOST}:${MINIO_
SERVICE_PORT}" \
--conf "spark.hadoop.fs.s3a.access.key=${MINIO_ACCESS_KEY}" \
--conf "spark.hadoop.fs.s3a.secret.key=${MINIO_SECRET_KEY}" \
--conf "spark.kubernetes.container.image=${SPARK_CONTAINER_IMAGE}" \
--conf "spark.kubernetes.container.image.pullPolicy=${SPARK_CONTAINER_
PULL_POLICY}" \
--conf "spark.sql.warehouse.dir=${S3_BASE_PATH}/warehouse" \
--conf "spark.app.sink.options.checkpointLocation=${S3_BASE_PATH}/apps/
spark-redis-streams-app/1.0.0" \
--conf "spark.app.sink.options.path=${S3_BASE_PATH}/warehouse/silver/
coffee_order_aggs" \
local:///opt/spark/app/jars/redis-streams-k8s.jar
```

Let's switch gears and learn to view the runtime information pertaining to the Spark driver application logs and the Spark Driver UI.

Viewing the Driver Logs

The application logs are available on the driver pod generated by calling spark-submit (not on the launch pod itself). This indirection is due to the way we are deploying our application using the cluster deploy mode.

```
% kubectl logs spark-redis-streams-app-driver -f \
  -n spark-apps
```

Now you can follow (-f) the logs on the driver as they are updated in real-time. If you want to dump the logs, skip the -f parameter and you'll see all the available logs that are stored in the kubelet for the driver pod. (The logs for the containers on your pods are memory constrained, so the -f option allows you to read all the logs while working with your streaming applications.)

Viewing the Driver UI

Like the process of viewing the driver logs, you can attach to the remote driver pod and forward the Spark UI using a simple kubectl command called port-forward.

```
% kubectl port-forward {spark-driver-pod} 4040:4040 \
  -n spark-apps
```

The beauty here is that the Spark UI is only available to members of your organization with access to your namespace. In the case here, only you has access to your own minikube cluster.

Sending CoffeeOrder Events

Open a new terminal window and exec over to the Redis service pod from part one of the chapter.

```
% kubectl exec -it redis-deployment-pv-{hash} \
  -n data-services \
  -- redis-cli
```

Then add some records to the CoffeeOrder's stream.

```
% xadd com:coffeeco:coffee:v1:orders  * timestamp 1637548381179 orderId
ord123 storeId st1 customerId ca100 numItems 6 price 48.00
% xadd com:coffeeco:coffee:v1:orders  * timestamp 1637548458800 orderId
ord124 storeId st1 customerId ca101 numItems 3 price 36.00
% xadd com:coffeeco:coffee:v1:orders  * timestamp 1637720977241 orderId
ord125 storeId st2 customerId ca102 numItems 1 price 4.99
% xadd com:coffeeco:coffee:v1:orders  * timestamp 1637783601557 orderId
ord126 storeId st1 customerId ca103 numItems 2 price 10.99
```

You should see some activity in the driver logs (open a window and follow the logs) or head over to the Driver UI to view the Spark Structured Streaming UI or Spark SQL micro-batches. When you are ready and have confirmed that the application is running (and processing data), you can delete the driver and jump pods.

Deleting the Driver and Jump Pods

When you delete your Spark Structured Streaming driver, all resources associated with the driver will also be deleted. These resources are managed through `ownerReferences` automatically for you. Kubernetes will take care of deleting any resource generated by the `spark-submit` process, including auto-generated `configMaps`, driver pod services, and the executor pods, when you are done running your application, or if you need to shut things down during an incident.

```
% kubectl delete pod/redis-streams-app-driver -n spark-apps
pod "redis-streams-app-driver" deleted
```

While the driver pod and all resources are now removed, your jump pod remains. This is because you launched the jump pod, while the `spark-submit` process that you triggered generated the driver and the associated configurations and secrets. When you are ready to shut down your jump pod, you can simply delete it as well.

Automating the Deployment and Automatic Recovery

When you're happy with the stability of your application (things are running as expected), you can apply the simple `deployment-redis-streams.yaml` to essentially deploy everything covered in the past section.

```
% kubectl apply -f ch-15/spark-on-kubernetes/spark-redis-streams/
deployment-redis-streams.yaml

service/redis-service unchanged
configmap/spark-redis-streams-conf unchanged
secret/hivesite-admin configured
secret/minio-access configured
deployment.apps/spark-redis-streams-app created
```

Congrats. You just deployed an idempotent Spark Structured Streaming application on Kubernetes.

Part 2: Summary

The second half of the chapter got you up and running. It put you in the driver seat to run your Spark Structured Streaming applications on Kubernetes. You learned how to create reliable stateful deployments and communicate between namespaces using service aliases (external network service types). This chapter is a beginning, teaching you how to take advantage of fully containerized Apache Spark applications. There are many places to go, now that you have reached the end of the book. I have provided some additional food for thought after the conclusion. So stick around!

Conclusion

First off, thank you for reaching this important milestone and following the book through to the end. It is my hope that the lessons learned throughout the book can help you think differently about the work you do, the engineers you work with, the company you work for (or own), the systems you affect, and how to optimize the results of the time you spend each day as a data engineer. This book is based on real-world experiences while I was employed at a company called Twilio, as well as from numerous workshops (presented internally and externally), countless hours providing guidance and debugging applications during Spark Office Hours (also at Twilio), and from countless other lessons learned over the past six years of working with Apache Spark and interoperating with different data systems and services. I've learned that the road to streaming excellence is not always as smooth as you would like. However, with time, patience, and the ability to test assumptions locally using your local data platform or directly inside your Spark-powered unit tests, you can solve even the most complex problems in less time and with less heartache.

Remember that almost all data problems, big and small, can be deconstructed into smaller, more bite sized problems, and built back up. Take the time to understand the problems at hand, create a plan of attack, and work toward an optimal solution. If the problem warrants a more generic solution enabling widespread reuse, then have at it. Lastly, remember to have fun and seek joy in the work you do.

After Thoughts

Docker, containers, and Kubernetes have changed the way organizations work with data in consistent and isolated ways. Throw Apache Spark into the equation and the doors open to novel new ways of expressing complex data pipelines and transformations through their n-wise relationship to data encompassing the greater data network at scale.

This book was the tip of the modern data engineering iceberg, providing the foundations I've found invaluable to chew on data problems of any size. In terms of where to go next, I have a few thoughts for things you'll want to do next.

- **JVM Tuning**. Apache Spark uses JVM garbage collection. For streaming jobs, the default G1GC settings can create large GC pauses due to the buildup of small objects over time. This is a result of processing many records (per micro-batch) based on IO bound jobs. Use the stage and task-level GC metrics to track GC pauses and the number of GC threads and objects. You can also use -XX:+UseStringDeduplication in your driver and executor and extra Java options to reduce the memory footprint if you are running out of memory and have string heavy objects.

- **Amazon EKS and Managed Node Groups**. Amazon has a feature for their elastic Kubernetes service called *managed node groups* (uses auto-scale groups behind the scenes). This feature enables you to create isolated groups on-demand, or spot instances, that can be explicitly scheduled (using selectors) to run your Spark drivers or application executors. Each managed node group is scaled using the concept of min, desired, and max sizing in terms of the total number of hot nodes. You can have a minimum size of 0, which means nodes will be fetched on-demand when an application is spinning up (this can be beneficial for expensive instance types, such as GPU instances, or gigantic instances that are used for one-off jobs, occasionally). Amazon will attempt to maintain your desired size and cut you off after you've reached your maximum number of allocated nodes. If you need to run a platform based on Amazon EKS, this option can help to carve out resource limits on a namespace basis.

- **Managed Kafka, RDBMS, and Redis**. Managing Kafka at scale
  requires a dedicated team as a company grows. If you are just getting
  started, it is beneficial to run fully managed Kafka (using Confluence
  or Amazon) and spend your time using it rather than running it. The
  same can be said for your database needs (be it OLTP/OLAP/NoSQL/
  etc.). Paying your way into things is the way to move fast without the
  headache of also scaling up and providing 24/7 operational support.
  Remember to run game day scenarios, since uptime SLAs usually
  provide 99.999% style guarantees, which means things can still be
  unavailable or require some manual intervention from time to time.

- **LakeHouse**. We covered the data lakes and the data warehouses in
  this book. There are a few common open-source platforms in the
  market today that mix the scalability of the data lake with strong
  schemas, as well as transactional inserts, merges, and deletes.
  DeltaLake, Apache Hudi, and Apache Iceberg are three competing
  technologies that all support Apache Spark out of the box. I tend
  to be biased toward technologies built by the Spark community, so
  if Apache Spark support is critical to your LakeHouse needs, give
  DeltaLake a try. It requires you to drop in an additional JAR and you
  are good to go.

Index

© Scott Haines 2022
S. Haines, *Modern Data Engineering with Apache Spark*, https://doi.org/10.1007/978-1-4842-7452-1

W, X, Y, Z

Printed in the United States
by Baker & Taylor Publisher Services